NEUROMETHODS □ 17

Neuropsychology

NEUROMETHODS

Program Editors: Alan A. Boulton and Glen B. Baker

1. **General Neurochemical Techniques**
   Edited by *Alan A. Boulton and Glen B. Baker,* 1985
2. **Amines and Their Metabolites**
   Edited by *Alan A. Boulton, Glen B. Baker, and Judith M. Baker,* 1985
3. **Amino Acids**
   Edited by *Alan A. Boulton, Glen B. Baker, and James D. Wood,* 1985
4. **Receptor Binding Techniques**
   Edited by *Alan A. Boulton, Glen B. Baker, and Pavel D. Hrdina,* 1986
5. **Neurotransmitter Enzymes**
   Edited by *Alan A. Boulton, Glen B. Baker, and Peter H. Yu,* 1986
6. **Peptides**
   Edited by *Alan A. Boulton, Glen B. Baker, and Quentin Pittman,* 1987
7. **Lipids and Related Compounds**
   Edited by *Alan A. Boulton, Glen B. Baker, and Lloyd A. Horrocks,* 1988
8. **Imaging and Correlative Physicochemical Techniques**
   Edited by *Alan A. Boulton, Glen B. Baker, and Donald P. Boisvert,* 1988
9. **The Neuronal Microenvironment**
   Edited by *Alan A. Boulton, Glen B. Baker, and Wolfgang Walz,* 1988
10. **Analysis of Psychiatric Drugs**
    Edited by *Alan A. Boulton, Glen B. Baker, and Ronald T. Coutts,* 1988
11. **Carbohydrates and Energy Metabolism**
    Edited by *Alan A. Boulton, Glen B. Baker, and Roger F. Butterworth,* 1989
12. **Drugs as Tools in Neurotransmitter Research**
    Edited by *Alan A. Boulton, Glen B. Baker, and Augusto V. Juorio* 1989
13. **Psychopharmacology**
    Edited by *Alan A. Boulton, Glen B. Baker, and Andrew J. Greenshaw,* 1989
14. **Neurophysiological Techniques: Basic Methods and Concepts**
    Edited by *Alan A. Boulton, Glen B. Baker, and Case H. Vanderwolf,* 1990
15. **Neurophysiological Techniques: Applications to Neural Systems**
    Edited by *Alan A. Boulton, Glen B. Baker, and Case H. Vanderwolf,* 1990
16. **Molecular Neurobiological Techniques**
    Edited by *Alan A. Boulton, Glen B. Baker, and Anthony T. Campagnoni,* 1990
17. **Neuropsychology**
    Edited by *Alan A. Boulton, Glen B. Baker, and Merrill Hiscock,* 1990

NEUROMETHODS

Program Editors: Alan A. Boulton and Glen B. Baker

NEUROMETHODS □ 17

Neuropsychology

Edited by

Alan A. Boulton

University of Saskatchewan, Saskatoon, Canada

Glen B. Baker

University of Alberta, Edmonton, Canada

and

Merrill Hiscock

University of Houston, Houston, Texas

Humana Press • Clifton, New Jersey

Library of Congress Cataloging in Publication Data

Main entry under title:
Neuropsychology / edited by Alan A. Boulton, Glen B. Baker, and Merrill Hiscock.
p. cm. — (Neuromethods : v. 17)
Includes bibliographical references and index.
ISBN 0-89603-133-0
1. Neuropsychological tests. 2. Clinical neuropsychology.
I. Boulton, A. A. (Alan A.) II. Baker, Glen B., 1947-
III. Hiscock, Merrill. IV. Series.
[DNLM: 1. Neuropsychology. W1 NE3378 v. 17 / WL 103 N493353]
RC386.6.N48N49 1990
152—dc20
DNLM/DLC
for Library of Congress 89-26859
rev. CIP

Crescent Manor
PO Box 2148
Clifton, NJ 07015

Printed in the United States of America

Preface to the Series

When the President of Humana Press first suggested that a series on methods in the neurosciences might be useful, one of us (AAB) was quite skeptical; only after discussions with GBB and some searching both of memory and library shelves did it seem that perhaps the publisher was right. Although some excellent methods books have recently appeared, notably in neuroanatomy, it is a fact that there is a dearth in this particular field, a fact attested to by the alacrity and enthusiasm with which most of the contributors to this series accepted our invitations and suggested additional topics and areas. After a somewhat hesitant start, essentially in the neurochemistry section, the series has grown and will encompass neurochemistry, neuropsychiatry, neurology, neuropathology, neurogenetics, neuroethology, molecular neurobiology, animal models of nervous disease, and no doubt many more "neuros." Although we have tried to include adequate methodological detail and in many cases detailed protocols, we have also tried to include wherever possible a short introductory review of the methods and/or related substances, comparisons with other methods, and the relationship of the substances being analyzed to neurological and psychiatric disorders. Recognizing our own limitations, we have invited a guest editor to join with us on most volumes in order to ensure complete coverage of the field. These editors will add their specialized knowledge and competencies. We anticipate that this series will fill a gap; we can only hope that it will be filled appropriately and with the right amount of expertise with respect to each method, substance or group of substances, and area treated.

Alan A. Boulton
Glen B. Baker

Preface

If one envisages neuroscience as a pyramid, with the more molecular disciplines forming the base and the more integrative disciplines positioned above, then neuropsychology clearly would be near the tip. Neuropsychology seeks to find order in the ultimate product of all neural systems, namely behavior, and to relate that product to its neural substrate. Relationships between brain and behavior are sought, but reductionistic explanations are eschewed. Attempting to "explain" complex behaviors in terms of neuronal activity is no more satisfying than attempting to "explain" artificial intelligence in terms of voltages within a computer's central processing unit. If one is to comprehend the functioning of either the brain or the digital computer, one must know something about not only the structure and mechanics of the device, but also the principles according to which components of the device are organized and the context in which the device is operating (e.g., environmental inputs and stored information).

This volume is intended not only for neuropsychologists but also for those scientists whose work involves nonhuman species or whose interests are focused on more molecular aspects of the nervous system. To the extent that these scientists are concerned about the potential relevance of their work to more global aspects of nervous system functioning in humans, they will find something of interest here. Anticipating, therefore, that this volume will reach a broad cross-section of neuroscientists, the editors made two decisions to benefit readers who are not specialists in human neuropsychology. First, we included an introductory section of three chapters to describe how the methods of neuropsychology evolved from disciplines as disparate as physiology and linguistics. These introductory chapters will serve as a bridge between human neuropsychology and other disciplines with which the reader may be more familiar. These three chapters should also broaden the perspective of readers who are specialists in neuropsychology.

Our second decision was to select a modest number of representative topics rather than to attempt encyclopedic coverage of the methods currently being used in neuropsychological research. By choosing authors who are widely known for their work with a major method or an important set of related methods, we are able to depict some of the best research methodology in contemporary neuropsychology.

The introductory chapter provides Bryan Kolb and Ian Whishaw's account of how human neuropsychology evolved from other neuroscientific disciplines and how the parent disciplines have influenced the methods of neuropsychology. Kolb and Whishaw discuss not only the positive contributions of neurology and psychiatry, anatomy, physiology, and comparative and physiological psychology, but they also note the blind alleys and problematic methods that constitute part of neuropsychology's heritage. A somewhat different perspective is provided by John Boeglin, Dan Bub, and Yves Joanette, who trace the development of neuropsychological thinking from its roots in Western philosophy to its current interaction with cognitive psychology. Boeglin et al. emphasize the logic underlying different approaches to the study of normal and brain-damaged humans. In the final chapter of the introductory section, John Ryalls, Renée Béland, and Yves Joanette describe the ways in which the theoretical frameworks of linguistics and the multiple levels of linguistic analysis have influenced neuropsychology in general and aphasiology in particular.

The next two chapters address the application of contemporary brain imaging techniques to neuropsychological research. In the first of these chapters, Terry Jernigan describes the two imaging techniques—computed tomography and magnetic resonance imaging—that provide information about structural characteristics of the human brain in vivo. Jernigan illustrates ways in which these techniques are being used in neuropsychological research and identifies some common pitfalls to be avoided. Frank Wood, in the companion chapter, addresses two other imaging techniques—regional cerebral blood flow measurement and positron emission tomography—that yield information about the physiological state of different brain regions. Wood concludes his chapter with seven specific suggestions for researchers who would use functional imaging techniques to study brain–behavior correspondence in humans.

The neuropsychological methods described in the following three chapters are all associated with the surgical treatment of medically intractable epilepsy. In the first of these three chapters, Rebecca Rausch and Michael Risinger describe the intracarotid sodium amobarbital (ISA) technique, which is used preoperatively to determine which cerebral hemisphere is dominant for language and memory and, thus, to estimate the risk of morbidity following unilateral temporal lobectomy. In the following chapter, Eran Zaidel, Dahlia Zaidel, and Joseph Bogen summarize the myriad of techniques used to assess the mental functioning of patients whose cerebral hemispheres have been surgically disconnected. With its coverage ranging from basic issues of left and right hemisphere competency to the most subtle aspects of methodology, the Zaidel et al. chapter is the most comprehensive work available on methods for examining the split-brain individual. In the next chapter, Catherine Mateer, Richard Rapport, and Don Polly describe the use of intraoperative electrical stimulation to map motor, sensory, and language functions on the cerebral cortex of patients about to undergo epilepsy surgery. Though emphasizing the clinical utility of this technique, Mateer et al. show that it is also an impressive research tool.

Insofar as perceptual asymmetries in the human are thought to reflect the differential specialization of the left and right cerebral hemispheres, researchers have attempted to document and compare perceptual asymmetries obtained with different stimuli and different subject populations. In his chapter, John Bradshaw summarizes this complex and voluminous research literature. Bradshaw examines visual, auditory, and tactile laterality methods in turn, and considers various parameters and procedural variables that may influence the results for each modality.

At the core of neuropsychology is the collection of methods known as the neuropsychological assessment. These methods, although used primarily for clinical evaluation of patients with known or suspected brain dysfunction, also provide the data base for much of the clinical research in neuropsychology. The next two chapters address neuropsychological assessment from two different points of view. Robert Bornstein discusses the neuropsychological test batteries currently available for assessing adults. Bornstein delineates the pros and cons of the "fixed" and "flexible" batteries and contrasts different batteries with respect to theoretical, philosophical, and pragmatic criteria. He then summarizes the empirical evidence pertaining to the most commonly used batteries. In their chapter, Jane Holmes-Bernstein and Deborah Waber illustrate the point made by Boeglin et al. that the term "method" may refer to the rationale of neuropsychological analysis rather than to a specific technique. Holmes-Bernstein and Waber view the neuropsychological evaluation of the child as an attempt to characterize the "child-world system," in which the maturing child and the child's environment exert reciprocal influences on each other throughout development.

The contrasting perspectives of Bornstein and of Holmes-Bernstein and Waber provide insight into the various objectives of neuropsychological assessment, the various criteria against which assessment methods may be judged, and the diverse approaches being used. A similar conclusion applies to the volume as a whole. No set of eleven chapters could cover the vast and rapidly changing landscape of contemporary neuropsychology. Indeed, entire monographs are devoted to relatively narrow topics such as the dichotic listening method and the neuropsychology of motor disorders. The eleven chapters in this volume, through their treatment of some representative methods, reveal the scope and fundamental character of neuropsychological inquiry while, at the same time, showing how neuropsychological methods are derived from and related to methods used in other disciplines.

Merrill Hiscock

Contents

INTRACAROTID SODIUM AMOBARBITAL PROCEDURE
Rebecca Rausch and Michael Risinger

TESTING THE COMMISSUROTOMY PATIENT
Eran Zaidel, Dahlia W. Zaidel, and Joseph E. Bogen

ELECTRICAL STIMULATION OF THE CEREBRAL CORTEX IN HUMANS
Catherine A. Mateer, Richard L. Rapport, II, and Don D. Polly

METHODS FOR STUDYING HUMAN LATERALITY
John L. Bradshaw

Contributors

GLEN B. BAKER • *Neurochemical Research Unit, Department of Psychiatry, University of Alberta, Edmonton, Alberta, Canada*

RENÉE BÉLAND • *Laboratoire Théophile Alajouanine, Centre de Recherche du C. H. C. N., Montréal, Canada*

JOHN BOEGLIN • *Laboratoire Théophile-Alajouanine, C. H. Côte-des-Neiges, Montreal, Canada and Department of Psychology, University of Saskatchewan, Saskatoon, Saskatchewan, Canada*

JOSEPH E. BOGEN • *Department of Psychology, University of California, Los Angeles, CA*

R. A. BORNSTEIN • *Department of Psychiatry, The Ohio State University, Columbus, Ohio*

ALAN A. BOULTON • *Neuropsychiatric Research Unit, University of Saskatchewan, Saskatoon, Saskatchewan, Canada*

JOHN L. BRADSHAW • *Monash University, Clayton, Victoria, Australia*

DAN BUB • *Laboratoire Théophile-Alajouanine, C. H. Côte-des-Neiges and Montréal Neurological Institute, Montréal, Canada*

MERRILL HISCOCK • *Department of Psychology, University of Houston, Houston, TX*

JANE M. HOLMES-BERNSTEIN • *Department of Psychiatry, The Children's Hospital, Harvard Medical School, Boston, MA*

TERRY L. JERNIGAN • *San Diego Veterans Administration Medical Center, Departments of Psychiatry and Radiology, University of California, San Diego, CA*

YVES JOANETTE • *Laboratoire Théphile-Alajouanine, C. H. Côte-des-Neiges and Kaculté de Medecine, Université de Montréal, Montréal, Canada*

BRYAN KOLB • *Department of Psychology, University of Lethbridge, Lethbridge, Canada*

CATHERINE A. MATEER • *Director of Neuropsychological Services, Department of Neurological Surgery and Departments of Speech and Hearing Sciences, University of Washington, Seattle, WA*

DON D. POLLY • *Department of Neurological Surgery and Departments of Speech and Hearing Sciences, University of Washington, Seattle, WA*

RICHARD L. RAPPORT, II • *Department of Neurological Surgery, Group Health Cooperative of Puget Sound and Department of Neurological Surgery and Departments of Speech and Hearing Sciences, University of Washington, Seattle, WA*

REBECCA RAUSCH • *Department of Psychiatry and Biobehavioral Sciences and Department of Neurology, University of California, Los Angeles, CA*

MICHAEL RISINGER • *Department of Neurology, University of California, Los Angeles, CA*

JOHN RYALLS • *Laboratoire Théophile Alajouanine, Centre de Recherche du C. H. C. N., Montréal, Canada*

DEBORAH P. WABER • *Department of Psychiatry, The Children's Hospital, Harvard Medical School, Boston, MA*

IAN Q. WISHAW • *Department of Psychology, University of Lethbridge, Lethbridge, Canada*

FRANK WOOD • *Section of Neuropsychology, Bowman Gray School of Medicine of Wake Forest University, Winston-Salem, NC*

DAHLIA W. ZAIDEL • *Department of Psychology, University of California, Los Angeles, CA*

ERAN ZAIDEL • *Department of Psychology, University of California, Los Angeles, CA*

Methods in Human Neuropsychology

1. Contributions of Physiology, Physiological Psychology, and Neurology

Bryan Kolb and Ian Q. Whishaw

Of all the neurosciences, the task of neuropsychology is by far the most difficult: to account for how the nervous system produces human behavior. As a result, neuropsychologists are found in all branches of neuroscience. It is no accident that a recent survey of the fields of training for basic neuroscientists by the Society for Neuroscience found that more people received their Ph.D. training in neuropsychology (including names such as physiological psychology and psychobiology) than any other single field. Yet, those who actually call themselves neuropsychologists form a small minority of neuroscientists. The multidisciplinary nature of neuropsychology means that the methods of this science must draw heavily on those of other cognate fields. The task of this chapter is to describe how other branches of neuroscience have influenced the current methodology and theory of neuropsychology.

1. Historical Background

1.1. Neuropsychology, the Word

One place to explore the roots of neuropsychology is to begin with the term. According to Bruce (1985), the term was first used by Sir William Osler in 1913. It then received occasional use, without serious definition, by a number of neuroscientists, including D. B. Lindsley, H. Kluver, K. S. Lashley, and D. O. Hebb. Lashley appeared to use the term synonymously with physiological psychology, whereas Hebb used it to refer to an interdisciplinary approach to behavior involving neurology and physiological psychology. The term was given wide publicity when it appeared as a subtitle in Hebb's work, *The Organization of Behavior: A Neuropsycho-*

logical Theory (1949) and in the title of a collection of Lashley's writings published by Beach and others in 1960 *(The Neuropsychology of Lashley)*. Most of the papers by Lashley were studies of rats and monkeys, and none were specifically on neurological patients. The term, then, is of relatively recent origin and was not intended to refer to the exclusive study of brain–behavior relations in humans. Hebb was probably the first psychologist to do interdisciplinary experiments that could be considered exemplary of current human neuropsychology. He studied Penfield's patients, who had unilateral frontal lobe removals for the relief of intractable epilepsy, on standardized intelligence tests. The power of the approach was immediately apparent in his results, because he found that frontal lesions did not impair intelligence as measured by contemporary intelligence tests, a finding that countered the long-held belief, originating with the phrenologists of the early nineteenth century, that intelligence, or as they termed it "the thinking faculties," resided in the frontal cortex. Lashley's students, including Hebb, Pribram, and Teuber, came to embody modern North American neuropsychology as they and their students sought to find a way to account for human behavior within the structure and function of the brain. What typified their approach, especially as it was practiced by Hebb, was its multidisciplinary nature. They exploited physiological psychology, comparative psychology, anatomy, physiology, and cognitive psychology, and they showed no prejudice in their choice of experimental subjects.

1.2. Neuropsychology, the Idea

Modern thinking about the mind began with Descartes. Descartes replaced the Platonic concept of the tripartite soul with that of a unitary mind that is the reasoning or rational soul. Being nonmaterial and having no spatial extent, the mind was therefore distinct from the body. Of equal importance, however, the mind was believed to be an indivisible substance. As such, theories that placed mental activity in the brain could not be correct if they subdivided brain function. This conclusion is significant for neuropsychology for it led to the important question of whether or not functions are localized in the brain, especially in the cortex. Attempts to answer that question have generated much of the data and theory that are the substance of today's neuropsychology. The argument for localization of function began with the phrenological theory of Gall and Spurtzheim. Although phrenology failed because of a number of internal weaknesses, it stimulated a search for evidence on the nature of the neurological organization of behavior, a search that would require the development of new methods of study.

The basic problem of study was simple. If behavior is controlled by the nervous system, then psychological functions ought to be represented in some manner in the nervous system. The solution was necessarily correlational: behavior is correlated with ongoing neurological function or dysfunction. The trick was to measure each.

1.3. Methods

Flourens is usually credited with being the father of experimental brain research. His contribution consisted of using both an experimental approach to evaluate psychological theory and an attempt to evaluate human brain function by the study of brain-damaged animals. Whereas the phrenological methodology of Gall and Spurtzheim claimed that function could be localized in the cortex, Flourens concluded that there was no localization of function in the cerebrum and that all intellectual faculties resided there coextensively. He found that ablation of various parts of the brain of chickens, pigeons, dogs, and other animals led to a general loss in behaviors such as feeding, walking, wing flapping, and the like. Furthermore, Flourens noted that, if the damage was not too severe, the behavioral loss seemed to show almost complete recovery. He concluded that the observed restitution of function was the result of the activity of the remaining brain. Thus, he surmised that any surviving remnant of cortex could fulfill all cortical functions.

It is generally recognized that Flourens' experiments shook the roots of phrenology, for they were clearly not in support of localized functions. This is ironic because his experiments were done on animals with little or no cortex, and his behavioral tests were assessments of such activities as wing flapping and eating, behaviors that bore virtually no relation to the psychological functions described by Gall and Spurtzheim. Although we can be quick to criticize Flourens on these grounds, the methological problems that he faced are still those that pose problems for modern neuropsychology. Consider the following points.

Flourens made virtually exclusive use of preparations with damaged brains, and from this, he generalized to normally functioning human brains. The technique of studying abnormal behavior and brain in human patients became the hallmark of neurology and clearly influenced the development of parallel procedures in neuropsychology. Since then a major portion of the data base of neuropsychology has been collected from animals and patients with brain damage. This leads us to wonder how valid the conclusions and theories generated by this approach are for normal brain function.

Flourens also generalized from animals to humans, and since then, this methodological approach has become the primary source of basic information about brain function. This has raised the question of what generalizations are valid. Whereas many people are ready to accept that the function of a synapse in a crayfish may be very like that of a human or that the actions of a major tranquilizer in a rat may be very like that in a human, they are less ready to accept that purported important human functions, such as consciousness or language, can be modeled in animals. In the extreme, some people even consider that these are the only important questions for psychology, and they therefore conclude that animal research has little to contribute. This influence can be seen in the training given in many programs in neuropsychology where courses in basic brain science are neither required nor offered.

Finally, Flourens did his best with behavior, but he did end up by generalizing from wing flapping in chickens to consciousness in humans. This connection requires a leap of faith that few people today are willing to make. Still, a very important question remains. What are the correct behaviors to study? Much of modern neuropsychology involves administering paper and pencil tests, studying reading and writing, and so on. Little concern is given to the possibility that, although these are intrinsically interesting, they may not lead to an understanding of how the brain works. Consider that the brain was fully evolved before these behaviors developed and that few, if any, humans still live within the biological constraints that evolved that brain. These considerations demonstrate that we still have a problem of choosing behaviors to study and tests to study them. That is, what is the appropriate unit of behavior to correlate with nervous activity?

Although we have no ready prescription for the weakness in the methodology used by neuropsychology, we feel that Flourens' experimental and comparative methodology still provides a potent approach to understanding human brain function. We also feel confident that the problems of species selection, generalization, and choice of behavioral units have their solution in a biological approach.

1.4. The Loss and Recovery of Neuropsychology

By 1900 a field similar to modern neuropsychology had developed. The behavior of various laboratory animals with cortical removals was described in careful detail by several authors, including Goltz, Loeb, and others (cf Luciani, 1915). Similarly, the behavior and behavioral syndromes of human neurological patients was described by Leipmann, Jackson, Wernicke, and Holmes, to name only a few. Consider the following examples.

In the 1890s, Goltz reasoned that, if a portion of the neocortex had a function, then removal of the cortex should lead to a loss of that function. Goltz removed portions of the neocortex of dogs and then studied the behavior of the animals. He discovered they were more active than normal dogs, alternated sleep-waking periods (though these were shorter than normal), and panted when warm and shivered when cold. They walked well on uneven ground and were able to catch their balance when they slipped. If placed in an abnormal posture, a decorticate dog corrected its position. After hurting a hind limb on one occasion, it trotted on three legs, holding up the injured limb. It was able to orient to touches or pinches on its body and snap at the object that touched it, although its orientations were not very accurate. Decorticate dogs also responded to visual and auditory stimuli, although the threshold was elevated.

At about the same time that Goltz was performing his experiments on dogs, Wernicke and his student Leipmann were describing the behavior of human patients with various neurological complaints. One of Wernicke's conclusions was that there were two language zones, which were connected by a large fiber bundle. He reasoned that, if the two areas were disconnected, a speech deficit would occur, even though the language zones themselves were intact. Later Leipmann was able to show that apraxia, an inability to make movements in response to commands, followed disconnection of motor areas from sensory areas. Furthermore, Leipmann posited that the left hemisphere had a special role in movement control that was not shared by the right hemisphere: lesions in the left hemisphere produced apraxia of both limbs; lesions in the right hemisphere had little effect on either limb.

The neuropsychological work of the late nineteenth century is remarkable for, although it was insightful and anticipated modern concepts, it was lost and ignored for more than 50 yr. Thus, in their description of the decorticate rat, Vanderwolf et al. (1978) redescribed 80 yr later the same behaviors that Goltz had originally described in the decorticate dog and, by reinterpreting the research, rekindled an interest in the behavior of the decorticate preparation, an interest that is important in theories of the role of the neocortex in behavioral control. Similarly, the concept of disconnection, which is now central to much neuropsychological theorizing, was ignored until the 1960s when Geschwind reintroduced the concept, with due reference to Wernicke. The special role of the left hemisphere in movement was not reintroduced until the 1970s, by both Geschwind and Kimura.

The original neuropsychological work of the nineteenth century became lost and ignored for at least three reasons. First, much of the work was published in German, and it was English speaking (and reading) scientists who began to dominate neurology and experimental psychology after the turn of the century. Second, as experimental psychology developed in the United States, behaviorism and related environmentally biased schools of thought came to dominate psychological thinking, and there was a strong trend away from the neurology of behavior in human experimental psychology. Third, the theoretical neurology of the late 1800s led to a great debate over the nature of localization of function. There can be little doubt that the "diagram makers," such as the students of Wernicke, overzealously proposed cortical wiring diagrams that went far beyond the data. Head, Marie, Freud, and Jackson wrote pursuasive rebuttals to these theoretical positions, leading to a further loss of interest in the neural mechanisms of cognitive functions.

It was not until the end of the Second World War that interest was renewed in the problems of human neuropsychology, in part because of the study of war veterans with cerebral injuries as well as of patients with frontal lobotomies. Further, it was not until the late 1960s that the neuropsychological work from the turn of the century was rediscovered, largely by neurologists such as Geschwind and physiological psychologists such as Teuber and Kimura, who read German. Physiological and comparative psychologists had been overly influenced by the largely negative results of Lashley, and they too again began to study cortical function, especially as the development of new anatomical tracing techniques in the 1950s began to allow a better appreciation of the nature of cortical organization in the primate. Unfortunately, however, few physiological psychologists displayed any interest in the human brain or in the symptoms of human patients until very recently, possibly because the field became preoccupied with questions of "motivation" and "learning" during the 1940s and 1950s and concentrated upon studies of hypothalamic mechanisms of feeding and drinking, and inferred neural correlates of learning. Given this bias in physiological psychology, North American human experimental psychologists saw little relevance of this work to their own questions regarding human cognitive processes and were very slow to consider the brain as an important variable in their research. Thus, until recently, studies of brain–behavior relationships occupied a secondary role in experimental psychology, and experimental psychology thus drifted away from the various fields of neuroscience. It was the success of a small group of physiological psychologists (e.g., Hebb, Teuber, Milner) who began to study human patients and to borrow methods of human experimental psychology, as well as the demonstration of neural correlates of different forms of memory disruptions, that has brought experimental psychology back to the questions of brain–behavior relationships.

2. Neurology and Psychiatry

The scientific study of human patients with behavioral disorders began in earnest following the observations on aphasia by Dax, Bouillard, Broca and others in the early and middle 1800s. The general idea of correlating behavioral abnormality with subsequent pathology became the fundamental technique of neurology and psychiatry, which were really the same discipline at the turn of the century. Indeed, it is probably the medical model of abnormal behavior, the idea that there is a physical correlate of abnormality, that is the major influence of neurology and psychiatry upon neuropsychology. We note that this approach contrasts with more traditional psychology in which function is usually inferred from studies manipulating variables that affect performance of normal subjects on various tasks. A corollory of the medical model is that it is possible to develop taxonomies of pathology that are functionally meaningful. For example, patients with cerebrovascular accidents of the left hemisphere are seen to have disorders of language function that can be grouped into several categories that are believed to be clinically and theoretically distinct.

The underlying assumptions of the medical model have clearly influenced neuropsychology, both in terms of the development of neuropsychological tests and in the design of basic studies. This influence is probably felt most significantly in the study of aphasia and apraxia, which we shall consider separately.

There has been a second major influence of neurology on neuropsychological methods. The taxonomies of the medical model are based upon clinical impressions of behavioral change. These impressions lead to the grouping of symptoms that become syndromes given names such as Broca's aphasia, ideomotor apraxia, finger agnosia, and so on. There is a second way to describe behavioral deficits, however. This procedure is based upon experimental psychology and requires that a behavioral capacity be quantified so that the behavior can be objectively compared to normal control levels. Indeed, it could be argued that a major difference between neurological and neuropsychological measures of behavior is the "clinical syndrome vs quantification." An important point is that it was the neurological observations that led to the neuropsychological investigation, and in that sense, it is obvious that neurology has had a major role in the development of neuropsychological methodology.

Finally, the medical model has had one very bad influence upon neuropsychology. The medical model emphasizes the abnormality of function. This is in direct contradiction to the mainstream of psychological theory, which is interested in the normal organization of function. Since neuropsychology is more closely allied with the medical model than any other branch of psychology, there has developed over the past two decades an emphasis upon correlating abnormal brain and behavior. In this atmosphere, it has proven rather easy to lose site of the questions regarding how functions are normally organized. Indeed, most contemporary textbooks of neuropsychology emphasize the abnormal, and there is little attempt to discuss the normal. The unfortunate result has been the tendency to neurologize psychology, which is not what the founders of neuropsychology had in mind.

2.1. Aphasia

The standard neurological taxonomy of aphasias derives from the clinical observations and theoretical interpretations of Wernicke (1874) and Lichtheim (1885), and the modern-day revival of them by Geschwind (1965). Although this taxonomy proved useful in organizing the early neurological observations, its influence in modern neuropsychology is not altogether positive. As Marshall (1986) has noted, most of the commonly used neurological aphasia batteries are based on the Wernicke-Lichtheim schema. In basic form, the schema divides aphasias into about seven categories. Unfortunately, careful clinical studies have found it difficult to categorize more than one-third to two-thirds of patients into this schema (e.g., Albert et al., 1981; Benson, 1986). Poeck (1983) suggests that the taxonomic categories may actually be a reflection of vascularization of the cerebral hemispheres rather than of a theoretically significant cerebral organization. This leads to a significant methodological problem for neuropsychologists. How should one study aphasia today?

Marshall (1986) has pointed out that the nineteenth century neurologists who devised the taxonomic categories of aphasia saw their clinical framework as secondary to the functional analysis of normal language processing. Unfortunately, the taxonomy became reified and took on meanings of its own. Meanwhile, the study of normal language became a separate enterprise pursued by linguists. One contemporary upshot of this is that the clinical classification is now taken for granted by neurologists (e.g., Kertesz, 1979) and much of "experimental" study is to "fill in the details" of the linguistic performance of these groups. Several authors have shown that this enterprise *cannot* succeed for several reasons (Badecker and Caramazza, 1985; Marshall, 1986; Schwartz, 1984; Shallice, 1979). First, the process is intrinsically weak. Thus, "the range of overt (symptomatic) impairments and the varied possible underlying deficits that can result in the 'same' symptom within *any* clinically defined group preclude the possibility that such a research programme could be genuinely progressive" (Marshall, 1986, p. 17). Second, the study of aphasia has become divorced from a study of normal language processes. Indeed, it is unlikely that aphasia can be a self-contained unit of inquiry. It shall be necessary to develop a genuine psychobiology of language that is based upon both normal language processing and aphasic deficits (cf Coltheart et al., 1980). Finally, a complete psychobiology of language will have to consider the evolutionary origins of language and language-related neural structures.

2.2. Apraxia

Like aphasia, the study of apraxia dates back to neurology at the end of the last century. Steinthal coined the term apraxia in 1871, but the symptoms had been described by Hughlings-Jackson before that. Two varieties of apraxia, traditionally termed ideomotor and ideational apraxia, were described clinically. Ideomotor apraxia refers to the inability to make voluntary movements of the limbs (limb apraxia) or orofacial musculature (oral or facial apraxia). The most important clinical feature of ideomotor apraxia is a difficulty in selecting elements of movement or in the sequential ordering of movements. Ideomotor apraxia is common and is said to occur in up to 80% of patients with cerebrovascular accidents of the left hemisphere. In contrast, ideational apraxia is quite rare. Clinically, it refers to an impairment in the ability to carry out sequences of actions requiring the use of various objects in the correct way and order, such as in preparing a meal.

Research on apraxia is at an even more primitive stage than that on aphasia. Most textbooks of neuropsychology describe the clinical syndromes and clinical characteristics, and like the taxonomy of aphasia, there is a tendency to use the taxonomy to imply some organization of psychological processes in the brain. There is no compelling evidence for this, however. Indeed, unlike aphasia, there is no standardized battery of tasks available to quantify the deficits in apraxia. As Poeck (1986) points out, the neuropsychological methods in apraxia are still based upon concepts and methods of examination developed at the turn of the century. If research is to go beyond the old concepts, brain-damaged patients and normal subjects should be examined on tasks that carefully measure the actual movements made in different situations. For example, Jeannerod has shown that, when a limb moves to grasp an object, the appropriate final posture of the hand is formed early in the reach (e.g., Jeannerod and Biguer, 1982). Thus, reaching reflects at least two independent processes, one of which recognizes the target object (hand shaping) and one of which recognizes its location (the hand movement). He also reports that apraxic patients fail to assume the correct hand posture early in the movement sequence. Such an observation could only be made by careful slow-motion analysis of videotaped reaching, but more importantly, it requires the recognition that such detailed study is necessary for understanding the neural basis of movement.

2.3. Sensory Systems

Most of the early work (ca. 1910–1930) on alterations of sensory abilities following brain lesions was based on intensive and prolonged studies of single cases by neurologists (e.g., Holmes, 1918). The difficulty with single case studies is that, although they are useful, they have often misled investigators, especially when the autopsy results have been disappointing. Further, single case studies have often led to fanciful theorizing on the basis of truly limited data. In contrast to the studies of aphasia and apraxia, where neuropsychologists have placed undue emphasis upon clinical taxonomies, psychologists were quick to avoid the clinical syndromes (i.e., agnosias) and rather began to study the sensory abilities of subjects with cerebral injuries, a good example being Teuber and his associates (e.g., Milner and Teuber, 1968). Thus, in these studies, the clinical observations, and frequently the clinical tests, were taken and the principles of perceptual psychology used to devise methods of quantifying behavior. An example will illustrate. One common clinical measure of somatosensory function is to move individual fingers and toes upwards or downwards in the blindfolded subject. The subjects' task is to indicate the direction of movement. Clinical examination is usually superficial, the test normally taking less than a minute. This test has been standardized (e.g., Corkin et al., 1970). Each finger is moved according to a prescribed pattern a fixed number of times. This type of assessment has shown deficits relative to normal controls from even rather small lesions, deficits that easily escape informal clinical assessment. Analogous procedures have been devised for the other sensory systems as well. The point here is that many neuropsychological methods for testing sensory capacities have evolved from clinical tests used routinely by neurologists.

2.4. Affective Behavior

In contrast to disruptions of language, movement, and perception, which were studied by neurologists 100 yr ago, proposed correlations between changes in affective behavior and the brain have been a relatively modern development. In 1937, Papez proposed that the structures of the limbic lobe form the anatomical basis of emotions. The idea of an emotional brain gained instant approval because of the predominance of Freudian thinking, but it proved difficult to find clinical evidence in support of this model. In the 1930s, clinicians were reporting detailed observations of patients with large unilateral lesions, noting an apparent asymmetry, in the effects of left and right hemisphere lesions. The best-known descriptions are those of Goldstein, who suggested that left hemisphere lesions produce "catastrophic" reactions characterized by fearfulness and depression, whereas right hemisphere lesions produce "indifference" (Goldstein, 1939). This distinction was really based upon clinical impression, and neither Goldstein nor his contemporaries made any attempt to devise clinical measures of this impression. In fact, there was not even an attempt to produce a taxonomy of affective disorders until the 1980s (cf Ross, 1981). It is only very recently that psychologists have begun to follow up the clinical impressions of Goldstein and others. As in other functions, there has been an attempt to quantify some of the aspects of affective behavior that contribute to the clinical syndrome (*see* Kolb and Whishaw, 1985b, for a review), but like the studies of aphasia and apraxia, very little is known about the normal neural basis of affective behavior.

Perhaps the greatest contemporary effect of psychiatry on neuropsychological methods has been the medical model of schizophrenia. The hypothesis that schizophrenia is somehow related to changes in some part of the brain (e.g., dopamine hypothesis; left hemisphere hypothesis) has led to the publication of dozens of studies on schizophrenics tested on a wide array of psychological measures. Unfortunately, to date very little has come of this. Like the study of aphasics, the study of schizophrenics by neuropsychologists may be doomed to failure. Like aphasics, schizophrenics are grouped clinically and the studies are dependent upon the classification, which is at least as difficult as the classification of aphasia. In order for real progress to be made, it will be necessary to produce better descriptions of behavior against which to compare schizophrenic behavior. At this writing, little has been learned about how the normal brain works from the neuropsychological study of schizophrenics. The medical model of schizophrenia may have proven overly seductive to many neuropsychologists.

2.5. Summary: Numbers Are the Currency of Science

We have seen that the medical model has had a major effect upon the development of neuropsychological methods and theory. First, there has been an emphasis upon the idea that abnormalities in behavior are somehow correlated with physical pathology. Second, there has been an emphasis upon clinical syndromes, which have inadvertently become reiffied. Finally, there has been a tendency to be concerned with the abnormal, rather than upon the normal, organization of brain and behavior. One of the strengths of experimental psychology is that it has devised sophisticated ways of quantifying and analyzing behavior. Indeed, this is what makes it different from neurology and psychiatry. It is the job of neuropsychologists to take the good from the medical model and to devise ways of quantification. This will nicely complement the medical behavioral sciences, and will lead to a behavioral technology that will be theoretically and practically useful.

3. Anatomy

Of the two current major neuropsychological theories of cortical function, one has been closely linked to the study of anatomy from the outset, whereas the second had its origins in biological philosophy. The originators of the first concept, Gall and Spurtzheim, were leading anatomists of their day, and they were quite familiar with individual differences in morphology, a knowledge that led them to develop phrenology. Although phrenology was based on a silly anatomical proposition, that bumps on the outside of the skull were correlated with underlying well or poorly developed areas of the brain, the idea that individual differences in behavior might be related to morphological differences had a major impact upon the development of neuropsychology. Much of the study of cortical function is still centered around the functions of cortical lobes, i.e., frontal, temporal, parietal, and occipital, but it is often not recognized that the lobes receive their major definition and names from the overlying bones of the skull. Many people naively accept that the lobes are functional entities and then assume that their relative sizes give some insight into individual differences. Much of the history of this sort of misguided approach has been reviewed by Gould (1981).

There are, however, interesting anatomical differences in the lobes, the best known of which include differences between the left and right hemispheres in the slope of the Sylvian fissure, which is steeper in the right hemisphere than in the left hemisphere, in the size of the parietal language-related cortex, the Planum temporale, which is larger on the left hemisphere than on the right hemisphere, and in Heschl's gyrus, or auditory cortex, which consists of one gyrus on the left hemisphere and two on the right hemisphere (Geschwind and Levitsky, 1968). Interestingly, this arrangement is present in only 65% of right-handed people. Therefore, just as the anatomical differences between hemispheres have been thought to underly functional differences between the hemispheres (i.e., language on the left), the anatomical differences between people have been thought to signify difference in behavioral abilities in the use of language. The same form of analysis has been applied to many other anatomical differences between individuals and even sexes (*see* Kolb and Whishaw, 1985a). Anatomical findings from animal research seemingly give support to this kind of approach. Nottebohm and his coworkers have found a good correlation between the number of neurons in the hyperstriatum of the left hemisphere of song birds and the complexity of the song of individual birds (Nottebohm et al., 1981). If this relation exists in birds, one cannot help but think that the number of neurons in the language cortex of individual humans may be related to the individual's language ability.

The study of cytoarchitectonics, or the size and shape of cells, of the cortex has had a notable influence on this theoretical approach. Brodmann (1909), for example, has described over 50 different areas in the cortex that have distinctive types of cells and arrangements of cells. Brodmann's numbered areas have been found to correspond quite closely to the functional areas of the cortex. For example, area 17 is primary a visual cortex, area 4 is a motor cortex, area 40 is a posterior association cortex, and so on. Brodmann's anatomy has lent such strength to theories of localization of function that it is common to find gyri, numerical zones, and functional areas equated.

The second major theory of cortical function was first suggested by Herbert Spencer. Influenced by Darwin's theory of evolution, Spencer proposed that the brain evolved in a series of steps with the result that the more recently evolved structures were involved in the highest functions. This idea was further developed by the neurologist John Hughlings Jackson, who related the idea to human brain anatomy. Jackson suggested that the three major evolutionary steps were the development of the spinal cord, the development of the basal ganglia and motor cortex, and finally the development of the frontal cortex. Jackson also exploited this concept to explain neurological disorders. He reasoned, for example, that brain disease could reverse the evolutionary process, resulting in what he called dissolution of behavior.

The Jacksonian concept of hierarchical function was adopted and integrated with modern anatomy by the pioneering Russian neuropsychologist, A. R. Luria. Luria recognized the significance of the anatomical zones proposed by the cytoarchitectonic studies of Brodmann and others, but rather than thinking that they represented houses of different functions, he suggested the zones were related hierarchically such that incoming sensory information was progressively elaborated as it passed from zone to zone until it was ultimately expressed as motor output. For example, a visual input into area 17 was passed along to other zones in occipital cortex, passed from there to the association cortex of the parietal area, from there to the association zones in frontal cortex, and from there it was finally fed into motor cortex, which was able to execute some response to the visual stimulus. A little thought suggests that the implications of the hierarchical model are quite different from those of strict localizationalist models. For example, a hierarchical model suggests that a given function can be impaired by damage to a number of brain sites in different lobes. Recent anatomical findings support the idea of transcortical anatomical subsystems (e.g., Pandya and Yeterian, 1985), and parallel behavioral studies using monkeys demonstrate that these anatomical subsystems do support individual functions (e.g., Ungerleider and Mishkin, 1982). Mishkin's research suggests, for example, that one visual subsystem feeds into the hippocampus where it is elaborated for use in spatial navigation, while another subsystem feeds into the amygdala where it is elaborated for object identification.

Anatomical investigations have had an influence on neuropsychological thinking in another way, through the study of the structure and interrelations of individual neurons. Had Golgi's nerve net idea, which postulated that brain cells were all physically interconnected, been supported, it would have led in turn to support of holistic or Gestalt approaches to cortical function. Neuropsychology today would be heavily biased toward theories of mass action and equipotentiality that hypothetized that all parts of the cortex worked together on every function. Cajal's neuron theory, that each cell was a functional entity and separate from other cells, however, eventually triumphed, and it seemingly supported localization of function notions. It seemed reasonable, for example, that, if neurons were individual and relatively autonomous, then they could have an individual and autonomous role in supporting individual functions or even individual memories.

The idea of the individualization of memories within assemblies of a few cells was in fact elaborated by D. O. Hebb (1949). In Hebb's model, individual cells were hypothetized to form connections with each other if they were activated together. These connections in turn formed the substrate of enduring memories. This model is extremely influential in modern neuropsychology, and seems to receive additional support from studies of the chemistry and structure of the junctions or synapses between neurons. In a number of laboratories a major thrust is now being made to clarify how neurons make and reinforce connections between each other. We must note, however, that we and many others feel that these studies are unlikely to uncover anything more than what the nervous system does when a memory is formed. The neural substrate of individual memories is unlikely to be found, since it likely involves many hundreds of neurons in a number of brain areas. Still, an understanding of the chemical process involved in memory formation will be significant, for it will likely help us understand and remediate disorders of memory produced by accident or by aging.

The lesson that neuropsychology must learn from anatomy is that it provides the boundaries on possible neuropsychological methods and theories. It is unfortunate, therefore, that some training programs in neuropsychology do not emphasize the study of anatomy, and many neuropsychologists have never had a course in anatomy. As a result, many clinically oriented neuropsychologists are poorly equipped to critically evaluate the almost overwhelming number of theories that are advanced to explain various behavioral phenomena.

4. Physiology

Because the activity of nerve cells has an electrochemical basis, the activity can be recorded with instruments sensitive to small changes in electrical or chemical activity, and can be altered with the application of electrical or chemical stimuli. This feature of the nervous system has proven attractive to neuropsychologists, because it provides a means of manipulating neural activity with which behavioral change can be correlated. Since many techniques are applied by neurosurgeons in awake patients undergoing elective surgery, or in preparation for elective surgery, it has even proven possible to study brain–behavior relationships in humans directly. These techniques have subsequently led to the parallel study of normal brains. Such studies are necessarily correlational, however, and like the techniques borrowed from neurology, those from physiology have significant traps for the unwary.

The major distinction between neurophysiology and neuropsychology can be seen in the basic question that each field asks. For the physiologist, the question is "how does the nervous system work?" For the psychologist, the question is "how does the working of the nervous system produce behavior, including inferred cognitive processes?" The answers to the second question are clearly constrained by the answers to the first. Physiologists have devised six principal methods to study the brain that have had an impact upon psychological thinking and research:

1. Brain stimulation
2. Electroencephalography (EEG)
3. The evoked potential (EP)
4. Single cell recording
5. Neuropharmacological manipulations and
6. Techniques for measuring cerebral blood flow and metabolic activity.

The direct application that neuropsychologists can make of these techniques is limited necessarily by the complexity of the methodology as well as the constraints against using invasive procedures in human subjects. As a result, the most basic neuropsychological application of neurophysiological techniques is done using laboratory animals or is limited to EEG and EP studies with scalp electrodes. The major impact of physiology on neuropsychology is therefore largely theoretical, as we shall see in the following sections.

4.1. Brain Stimulation

Brain stimulation produces different behavioral effects depending upon where the stimulation is applied. Thus, stimulation of the cortex produces qualitatively different effects from stimulation of the brainstem, and for this reason, brain stimulation research is seen to be divided into two different fields in most textbooks. In keeping with this unwritten tradition, we will follow the same practice here.

4.1.1. Cortical Stimulation

Attempts to evoke behavior by stimulating the cortex can be traced back to the early 1800s. Notably, Flourens stimulated the cortex of animals, but failed to find any response, and so the cortex was thought to be silent with respect to evoked behavior. In 1870 Fritsch and Hitzig, reported that they could evoke movements in contralateral body parts by cortical stimulation. They also reported that the body appeared to be topographically represented in the cortex. Thus, they produced one of the first functional maps of the neocortex and stimulated the more detailed studies of cortical organization in many species by Ferrier, Sherrington, and others. The first formal report on effects of brain stimulation on humans was made by Bartholow in 1874. He reported that, although brain stimulation could elicit both behavior and sensation, it was not painful. Subsequently, the technique was used to identify cortical areas in humans by Penfield and his colleagues (e.g., Penfield and Roberts, 1959), who found it to be an invaluable method for identifying speech areas and primary motor cortex, so that they could be avoided, if possible, during elective surgery.

The cortical stimulation studies of the first half of this century had a significant impact on neuropsychological thinking during a time that behaviorism was emphasizing an S-R connectionist view of behavior. First, stimulation data implied that there was a topographical representation of the body muscles and movements in the frontal cortex and discrete sensory areas in the posterior cortex.

Since large areas in posterior and frontal cortex appeared to produce no movements or sensations, they were called "silent" and then eventually received the name "association" cortex. It was logical to conclude that the silent association cortex collected information from different sensory systems to form ideas or concepts. In theory, these could be found in the hyphen of the S-R or S-S theories, and thus held theoretical appeal to psychologists. A second impact came from Penfield's observation that stimulation of some cortical sites, especially temporal cortex, appeared to trigger thoughts, memories, or ideas. This led to the view that not only did memories have a cortical representation, but that a great deal more was stored in memory than could be recalled readily. Psychological theories of memory were clearly influenced by Penfield's observations, but as we shall see, the data were accepted too readily and uncritically: recent multidisciplinary work has challenged the earlier views and must lead to a revision of psychological theories of sensation, cognition, and memory. Consider the following facts.

First, rather than there being a single representation of the motor or sensory systems in the cortex, each system appears to have multiple representations, some of which extend into zones that previously appeared to be silent. These representations appear to code different aspects of sensory experience (e.g., color, form, size, movement, and so on, in the visual areas) and are activated simultaneously. Clearly, conscious experience must result from an integration of the nearly simultaneous activity of these areas and not a simple S-R or S-S arrangement. Furthermore, removal of the association cortex does not abolish sensory experiences or memories, results that again lead to a need to revise psychological thinking. Second, there are many connections between anterior and posterior cortex that clearly interrelate their activity. These extensive connections seriously compromise the view that either can be viewed as either mainly sensory or motor (Pandya and Yeterian, 1985). Third, the thoughts and memories that are elicited by electrical stimulation appear more parsimoniously attributed to electrographic abnormalities produced by the stimulation, rather than to the activation of memory banks located in the areas around the electrode tip (Loftus and Loftus, 1980).

There have been two main effects of these reevaluations on neuropsychological thought. First, it appears that the organization and function of the cortex is considerably more complex than originally thought. Current neuropsychological accounts of complex cognitive processes will require considerable modification as physiological methods discover new information. Second, the idea that electrical stimulation can activate normal nerve cells to approximate normal behavior must be viewed more critically.

4.1.2. Subcortical Stimulation

The most influential studies on subcortical stimulation were those of Hess (e.g., Hess, 1957) and Olds and Milner (1954). Hess demonstrated that stimulation of the brainstem of cats elicits a large number of well-integrated behaviors and that there was good localization of sites producing particular behaviors. For example, certain hypothalamic sites could elicit eating, others could elicit predatory attack, whereas still others could produce avoidance reactions. These behaviors were thought to result from the activation of brainstem "centers" that were responsible for the normal production of the behavior. The demonstration of evoked behaviors in cats quickly led to the view that brain stimulation could be used to control behavior, an idea that still can be seen in science fiction.

The research of Olds and Milner was rather different: they demonstrated that brainstem stimulation could be reinforcing. Thus, rats with septal or hypothalamic electrodes would energetically press bars in a Skinner box to obtain short bursts of electrical brain stimulation. This phenomena, which has come to be called "self-stimulation" has been demonstrated in virtually every species of animal tested, including humans. The discovery of self-stimulation quickly led to the view that the brainstem contained pleasure centers that normally reinforced behavior. It followed that certain behavioral disorders such as depression or schizophrenia, in which there appears to be a loss or change of the affective properties of stimuli, might result from abnormalities in the pleasure centers.

The view that the brainstem contained centers that were the substrate for both complex behavior and for reinforcement of behavior stimulated a large number of theories that are beyond the scope of the present discussion. Nevertheless, most textbooks of physiological psychology subscribed to the view that the brainstem contained "centers" for organized behavior as well as centers that served as substrates for reinforcement. This view has been weakened, however, by a number of recent lines of research. For example, Valenstein (1975) has demonstrated that the same brain site can produce different behaviors depending upon the context in which an animal is tested. Szechtman and Hall (1980) demonstrated that slight pressure on the tail of a rat can elicit behavior almost as effectively as brain stimulation. Although there is some dispute concerning the interpretation of these and similar studies, it is agreed that the classical concept of localizaed centers requires reevaluation, a reevaluation that will certainly affect neuropsychological theorizing and experimentation.

4.2. Electroencephalography and Evoked Potentials

In 1875, Robert Caton published the first documentation that electrical activity could be recorded from outside the brain of animals. Similar electrical activity of the human brain was successfully recorded from the scalp surface by an Austrian psychiatrist, Anton Berger, in 1924, and was reported by him in 1929. The record of these fluctuating electrical signals emanating from the brains of humans and other animals is called the electroencephalogram. The considerable technical advances made since Caton's and Berger's studies have provided a technique for unobtrusively measuring brain activity for experiments on the relationship between brain and behavior, and for assessing various clinical conditions such as epilepsy and brain damage.

The recorded electroencephalogram is actually a reflection of the activity of many millions of neurons located in a large volume of tissue in the brain. It has a wave-like character, and can be characterized in terms of amplitude and frequency. The electroencephalogram observed in scalp recording from awake, alert subjects is typically composed of desynchronous fast waves of low amplitude. Different areas of the human cortex, however, do have distinctive patterns, as documented by Penfield and Jasper (1954).

Deviations from the normal waking pattern have been used to diagnose epilepsy, which frequently has a distinctive spike-and-wave character; tumors, which produce no electrical signal; and brain death, which is also characterized by an absence of an electrical signal. Deviations from the desynchronized pattern toward a pattern of slow large amplitude waves can also be used to judge the depth of anesthesia and sleep, and so EEG recording is routinely used for surgery and for sleep studies.

Since the application of EEG to the solutions of practical problems has been so successful, there has been a concerted attempt to use EEG to evaluate mental states, levels of arousal, the efficiency of cortical function, differences in function of the two hemispheres or of different lobes, and so on. There has been far less success in these endeavors. There are probably three reasons for this. First, EEG does not provide a good index either of the area of tissue generating a signal or the number of cells involved in generating a signal. For example, Whishaw et al. (1978) removed over 90% of the granule cells from the hippocampus of rats without producing a change in the distinctive electrical signal that they generate. Second, changes in EEG reflect changes in overt behavior, and for laboratory animals, in which the most comprehensive studies have been carried out, virtually all changes in the electroencephalogram can be accounted for in terms of motor activity. That is, certain EEG patterns seem to occur only when animals are immobile, and these patterns change if the animals make a movement, even though the movement may be as small as a slight head tilt. Third, the cortex and some subcortical structures seem to produce more than one seemingly identical EEG pattern that have completely different neural bases. In the rat, for example, the activated EEG pattern that occurs when the animal is sitting still is pharmacologically different from the identically appearing activated pattern that occurs when the animal walks. Humans may also have seemingly identical EEG patterns that are state-related. This may explain the puzzling occurrence of activated EEG patterns in some coma patients. The pattern may be pharmacologically quite different from the activated electroencephalogram that occurs in normal alert people (*see* Vanderwolf and Robinson, 1981, for a review of some of these issues).

Evoked potentials, or EPs, consist of a short trace of the electroencephalogram recorded immediately before, during, and immediately after the presentation of a sensory stimulus. If a bright light or noise is presented for about 250 ms, electrodes placed on the scalp over the visual cortex will record a large and rather complex slow wave. Typically, many segments of electrical activity recorded during repeated presentations of the stimulus are averaged to obtain an adequate evaluation of neural change to the stimulus. This average is often called an averaged evoked potential (AEP). More recently, the terms EP and AEP have given way to the term "event-related potential" (ERP), as attempts have been made to relate neural activity to cognitive events as well as to sensory events.

Ideally, ERPs could be used to study the activity of discrete areas of the brain, but this is unfortunately not so. The physiologist has the advantage of placing an electrode in the brain and recording a signal of relatively high amplitude, the source of which can often be localized in a particular cell layer. Recording from scalp electrodes is far more complex and subject to considerable methodological artifact. We concur with a recent review by Gevins (1986), who concludes that, although they continue to report interesting and useful results, both currently popular paradigms, EEG and ERP, are based on very simplified models of the brain and cognition.

4.3. Single-Unit Recording

If a small wire that is insulated except for a very small portion of its tip is inserted into the brain so that the tip is placed near or in a nerve cell body or axon, the change in the cell's electrical potential, i.e., unit activity, can be recorded. Intracellular recordings are made from electrodes with very tiny tips, less than one-thousandth of a millimeter in diameter, which are placed in the cell, whereas extracellular recordings are made when an electrode tip is placed adjacent to one cell or a number of cells. The technique requires amplification of the signal and some kind of display. The cell's activity is either displayed on an oscilloscope for photographing or recorded on a tape recorder for computer analysis. In many experiments, the signal is played through a loudspeaker, so that cell firing is heard as a "beep" or "pop." Both recording techniques require considerable skill to perform because it is difficult to place the electrode in or sufficiently close to the cell without killing it, and when a cell is "captured," it is often difficult to hold it for more than a few minutes or hours before the signal is lost.

Unit recording techniques provide a particularly interesting insight into the brain's function. For example, cell records obtained from the visual cortex of cats, monkeys, and other animals reveal that cells have a preferred visual stimulus and a preferred response pattern, i.e., some cells fire to horizontal lines, some to diagonal lines, others show preferences for colors, and so on. In the hippocampus of rats, cells have been found that respond only when the animal is in a given location. Cells appear to code information by alterations in speed of firing, frequency of firing, the pattern of bursts of firing, and so on. In reality, single cell recording techniques provide the most effective insight into how the nervous system codes information, but the methodology is currently limited by the difficulty in recoding from many cells at once. This is important, because it is recognized that much of the processing of the nervous system must reside in the relational activity of many cells located in many brain areas rather than in the activity of a single cell. At the present time, methodology for recording from many cells is limited, it is difficult to "hold" cells for long periods of time, and it is also very hard to identify precisely what cell is being recorded from. Nonetheless, the results of recording studies are significant for neuropsychological theories and experiments. For example, it has been shown that there are cells in the temporal cortex of the monkey that are maximally excited by very complex visual stimuli, such as faces or hands. This observation is important, because it has long been known that right temporal lobe patients are impaired at recognition of complex visual information, including faces (e.g., Milner, 1980). Thus, although unit recording studies are unlikely to be performed on human subjects, such studies have an important influence upon neuropsychological theorizing about cortical activity.

4.4. Neurotransmitters and Neuromodulators

Following Otto Lowi's early experiments in the 1920s, which demonstrated that the vagus nerve changes heart rate by secreting small amounts of a chemical substance onto it, there has been a truly amazing growth in our knowledge of how neurons communicate. Some excite or inhibit the activity of other neurons by secreting a transmitter chemical directly onto special receptors on the surface of the neuron. Chemicals used in this way are called neurotransmitters, and their action is thought to be relatively discrete. Some neurons secrete their chemicals rather diffusely into extracellular space, and these chemicals are called neuromodulators because their action is thought to be rather general.

A general understanding of chemical neural communication is essential for understanding many facets of contemporary neuroscience. The reasons for this are so numerous that they are difficult to list, but the following three are perhaps the most important. First, the brain appears to be organized into chemical systems and malfunctions, or diseases in these systems can now be related to many kinds of illness. For example, one group of cells in the midbrain project to a number of forebrain areas where they secrete the chemical dopamine. If these cells are destroyed, as they can be by certain viruses or environmental toxins, a condition called Parkinson's disease ensues. A major symptom of the disease is difficulty in making movements. Other diseases such as Huntington's chorea, Alzheimer's disease, and schizophrenia are postulated to be the result of abnormal function or damage in other brain chemical systems. Second, many food substances or drugs have a selective action on specific neural transmitter systems. This selectivity now provides much of the theoretical basis of pharmacology and therapeutics. Since about 30 chemicals are thought to be neurotransmitters, explanations of action, in terms of neurotransmitter effect, are now being developed for the thousands of pharmacological agents that have been found to affect physiological functions and behavior. Third, changes in the effectiveness of neural transmission is thought to underly learning. Since learning is a central focus of many approaches in neuropsychology, its neural bases is of special interest. Perhaps the whole matter of neurotransmission is also of interest to neuropsychologists in a way in which many other physiological methodologies are not. Pharmacological agents are widely used by humans and this provides neuropsychologists special opportunities for evaluating their functions. Pharmacological studies are also relatively easily performed in laboratory settings, with both humans and animals, in a way that other methodological approaches are not. Thus, this methodology can be more easily accessed and more widely used by neuropsychologists than can other physiological methodologies.

4.5. Cerebral Blood Flow and Metabolic Activity

The first evidence that mental activity might change cerebral blood flow came from an incidental observation in 1928 by Fulton who observed an increased bruit (i.e., sound or murmur) over an arteriovenous malformation in the occipital pole of a patient who was reading. This report was ignored as it was generally assumed that changes induced by mental effort would be too small to measure. The first evidence indicating that mental activity was correlated with changes in blood flow was published in a series of papers in the mid-1960s by Ingvar, Risberg, and their colleagues (e.g., Ingvar and Risberg, 1967). By the mid-1970s, procedures had been devised that were noninvasive and nontraumatic, and it became possible to construct functional maps of the cortex. Similarly, parallel procedures were devised to measure metabolic activity by using positron emission tomography. These procedures constituted a major breakthrough for neuropsychology, because they provided a method, albeit expensive, of measuring changes in cerebral activity during the performance of neuropsychological tests.

Measures of blood flow and metabolic activity have proven more valuable in understanding the activity of the abnormal than the normal brain. For example, it has proven difficult to demonstrate clear cerebral asymmetries in blood flow, although intrahemispheric specialization has proven fairly reliable (e.g., Risberg, 1986). Furthermore, it is virtually impossible to record rapid changes in activity that would correlate with much of our normal mental activity.

4.6. Conclusions

The major effect of physiology on neuropsychology has been the development of a window on the activity of the normal brain. This research still must overcome serious technical problems in the study of the normal brain and behavior. Indeed, to date, little new knowledge has been gained, since virtually all of the results could be predicted from previous studies of lesion patients. Significantly, however, the complementary results from lesion studies and physiological studies have shown that many of the inferences about normal brain function that were made from the study of the diseased brain were in fact valid. Furthermore, the relatively symmetrical activity of the two hemispheres in tasks that *a priori* would be expected to heavily load one hemisphere is likely to have a significant impact upon neuropsychological theorizing that has almost certainly *overemphasized* the unique contributions of the two hemispheres to behavior.

5. Comparative and Physiological Psychology

As we saw at the beginning, neuropsychology as a term and as a field can really be traced back to Lashley and colleagues like

Kluver in the 1920s. It was the ideas of Lashley and Kluver, and later Hebb, that shaped the ideas and methods of modern-day neuropsychology. This influence can be seen in a broad range of ways that are still having a significant impact upon the way human neuropsychological research is conducted. It is of interest that, although today most psychologists who call themselves neuropsychologists are clinical psychologists by training, clinical psychology has had negligible impact upon theory or methodology in neuropsychology. The primary influence has come from the basic neurosciences, especially physiological and comparative psychology.

5.1. Lesion Technique

The study of patients with lesions clearly comes from nineteenth century neurology. The study of groups of patients with similar etiology is a uniquely psychological approach. As Lashley and others began to experimentally manipulate lesion locus and behavior, they recognized that there were individual differences in behavior and in the brain. As a result, they used multiple subjects, which formed groups, and used statistical procedures to reduce the intersubject variation. This approach to behavioral neuroscience was subsequently used by Hebb and others as they began the first truly neuropsychological investigation of human patients in the 1930s (e.g., Hebb, 1939). A methodological principle that evolved in this work was that of *double dissociation.* This refers to an inferential technique whereby two lesion groups are functionally dissociated by two behavioral tests, each lesion group being uniquely impaired at the performance of one test but not the other. The technique of double dissociation is important to neurological experimentation, for it ensures that an observed behavioral deficit is a result of a unique effect of damage to a particular region of the cortex, and not because of nonspecific factors associated with brain injury. This procedures differs from the classical neurological approach to behavior in which individual case histories are given preeminance (e.g., Geschwind, 1965). There is still a place for the intensive study of individual cases, especially where the patient's syndrome appears itself to be unique, or in those cases of brain injury where the lesion is known to be unique (*see* Milner and Teuber, 1968, for examples). Nonetheless, the data of neuropsychology come from lesion studies employing the procedure of double dissociation.

5.2. Neuropsychological Testing

The development of the lesion technique was paralleled by the development of psychometric test procedures (*see* Chapter 2). During the 1920s, 1930s, and 1940s, experimental psychologists placed particular emphasis upon the study of learning by laboratory animals, principally rats. In doing so, they devised various batteries of maze tests to assess learning in rats. These tests were subsequently used by physiological psychologists in their lesion studies. Although most tests suitable for laboratory animals cannot be directly transferred to human subjects, the general principle of using learning tests to assess cortical function was generalized to the study of people. Nonetheless, some testing methods have transferred nicely from the animal laboratory, just as clinical observations have often been the source of new animal experiments *(see below)*.

5.3. Comparative Method

Flourens introduced the use of nonhuman species to the study of brain–behavior relationships. The appropriateness of nonhuman species as models of human brain function remains a legitimate issue, which we have discussed elsewhere (Kolb and Whishaw, 1983, 1985a). Historically, there is little doubt that research with nonhuman species has strongly influenced the methods of human work and vice versa. For example, in the 1880s Loeb tested dogs with unilateral cortical lesions by presenting two lures, one to each side of the animal, to show that when confronted with two simultaneous stimuli the dogs showed neglect of the contralateral stimulus, even though they responded normally when only one lure was presented. This experiment soon led to the development of the procedure of double simultaneous stimulation in testing human patients. Similarly, observations on visual-field defects in humans guided the early animal studies involving ablation of the visual pathways. These experiments led in turn to the development of new methods of assessment of human patients. We note, however, that direct extrapolations of methods derived from nonhumans have also been far from satisfactory. A good example is the delayed response task that has become a classic test of frontal lobe damage in nonhuman species. In this task, the subject must recall, after a short delay, which of two containers conceals a reward. Whereas nonhuman species with frontal lobe lesions are impaired at this task if the delay exceeds a few seconds, analogs of this test for humans have proven unrevealing: humans with frontal lobe lesions are not normally impaired at such a test. Thus, we see that the direct transfer of tests across species is difficult. Rather, the main influence of comparative work has been more general in terms of basic techniques of analysis, rather than in the transfer of specific behavioral tests.

5.4. Memory

It is in the study of learning and memory that physiological and comparative psychologists probably have had their greatest effect upon contemporary human neuropsychology. The neuropsychological study of memory dates back to about 1915, when Karl Lashley embarked on a lifetime project to identify the neural locations of learned habits. In most of his experiments, he either removed portions of the neocortex or made cuts of fiber pathways in hopes of preventing transcortical communication between sensory and motor regions of the cortex. After hundreds of experiments, Lashley was still unable to interfere with specific memories. In 1950 he concluded that memories must be distributed throughout large regions of the cortex and not localizable to any specific place. Further, Lashley concluded much earlier that memory loss is directly related to the size of cerebral injury: the larger the damage, the greater the memory loss. Lashley's experiments had a major impact upon neuropsychology both because they shaped the thinking of a generation of psychologists and because they legitimized the idea that some area of the brain would be responsible for controlling some inferred process, such as memory. This set the stage for the interpretation of an amazing discovery in 1953 by William Scoville, when he operated on the now famous patient H. M. to relieve intractable and debilitating temporal lobe epilepsy (Scoville and Milner, 1954). Scoville bilaterally removed the medial temporal lobes, including the hippocampus, leading to a dense amnesia for all events after the surgery. In view of Lashley's experiments, it was logical to conclude that the hippocampus played a major role in memory, although the cortex did not. (Indeed, it was often incorrectly believed that the H. M. case showed that memories were stored in the hippocampus.) The data from H. M. and the subsequent studies on a limited number of similar patients led to intense study of the hippocampus and memory in nonhuman species. The difficulty that soon arose was that hippocampal lesions in rats and monkeys failed to produce the dense anterograde amnesia observed in H. M. This soon led to a question of whether or not there was a significant species difference in the function of the hippocampus. In the last decade, the animal studies have finally come to fruition and led to a rethinking of the role of the hippocampus in human memory. Two results are particularly influential.

First, H. M. and other bitemporal patients do not have selective hippocampal lesions. The amygdala and other medial temporal regions are also removed. It is now clear that it is the combined removal of these structures that is crucial. Thus, combined amygdala and hippocampal lesions in monkeys produce clear performance deficits (e.g., Mishkin, 1978) similar to those in human patients with similar damage. Second, O'Keefe and Nadel (1978) proposed that the hippocampus had a special role in spatially guided behavior. This novel proposal was based initially upon the observation that there were cells in the hippocampus of rats that fired selectively in certain spatial locations and not in others, regardless of the rats' behavior. O'Keefe and Nadel went on to write an extensive monograph in which they interpreted behavioral change after hippocampal lesions in terms of a defect in spatial processing. This led to a reevaluation of space and memory in temporal lobe patients, which is still continuing.

Two important methodological messages come out of the memory studies. First, it is very difficult to study inferred processes like learning and memory. These are not observable, and it is likely that they do not exist as entities in the brain. They are, however, very tempting constructs to try to study. Second, damage to any part of the brain will change behavior and may produce poor performance on a test that involves what people normally call memory. This is not proof that memory processes can be localized. The clear message from Lashley is that they cannot. Rather, whenever a lesion produces a phenomenon as dramatic as that in H. M., it is safe to assume that there are multiple behavioral changes, and psychologists must not be taken in by the appeal of flashy inferred processes. It is behavior that is observable, and it is behavior that must be carefully studied. Many physiological psychologists still fail to appreciate this even today and continue to seek the locale of inferred functions. This is doomed to fail.

6. Future Directions: Neuroethology and Neuropsychology

The study of behavior evolved in parallel in two distinctly different traditions, ethology and psychology. Ethologists study behavior with the primary aim being the study of the functional importance of behavior to the organism. This is achieved by considering the evolutionary significance of behaviors as well as the immediate causation of behaviors. In contrast, psychologists study behavior with the primary interests being the development of behavior (including environmental contingencies in shaping behavior) and the physiological mechanisms underlying behavior (cf Lehrman, 1970). The psychological bias easily can be seen in the lesion study, which has traditionally been the principal method of neuropsychology. Thus, patients are chosen to study after they acquire discrete lesions. The behavioral analysis normally involves studying either symptoms that are obviously abnormal (e.g., aphasia, apraxia) or studying performance on behavioral tests that have been chosen because they are of theoretical interest (e.g., maze tests, memory tests). Performance then is compared to that of a matched control group. Consider what might be done, however, if a neuroethological approach were taken. First, a behavioral taxonomy of the normal person is prepared. This would include especially those behaviors that are typical of our species, beginning with behaviors that function for personal survival (eating, grooming, moving) and then including behaviors that function primarily in group survival (social and sexual behavior, maternal behavior, communication, and so on). The behavior of patients would then be examined according to these behaviors.

There are several differences between the neuropsychological and neuroethological approach that are significant. First, it is apparent that the number of behaviors to study in either case is overwhelming. Note, however, that the behaviors studied in each case are nearly nonoverlapping! Second, it is obvious that the rationale for choosing behaviors to study is very different in the two approaches. In the former, behaviors are chosen because of observed symptoms or an interest in inferred mental processes. In the latter, behaviors are chosen because normal humans have them, and thus, they must have a function. Virtually no neuroethological studies have been done on people, in part because so little is known about the neuroethology of people. Many neuropsychologists who work with nonhuman species have begun to look at brain–behavior relations in a more neuroethological way, however, and this is sure to have an impact in the near future. Of particular importance is that there are likely to be very different theoretical conclusions when one considers the data from the two approaches. Consider an example. We examined the behavior of rats with damage to the frontal cortex at different ages by using both a neuropsychological and a neuroethological approach (Kolb and Whishaw, 1981). Our results showed that when compared to rats with adult removals, animals with damage early in life showed dramatic recovery of function when tested on various maze tests. In contrast, the same rats showed no recovery at all on tests of species' typical behavior. The conclusions regarding brain plasticity following early cortical injury are very different depending upon which behaviors are studied. Perhaps most important, when both procedures are used together, the conclusions are different again and probably closer to being correct.

References

Albert M. L., Goodglass H., Helm N. A., Rubens A. B., and Alexander M. P. (1981) *Clinical Aspects of Dysphasia* (Springer, New York).

Badecker W. and Caramazza A. (1985) On considerations of method and theory governing the use of clinical categories in neurolinguistics and cognitive neuropsychology: the case against agrammatism. *Cognition* **20,** 97–125.

Beach F. A., Hebb D. D., Morgan C. T., and Nissen N. W., eds. (1960) *The Neuropsychology of Lashley* (McGraw-Hill, New York).

Benson O. F. (1986) Aphasia and the lateralization of language. *Cortex,* **22,** 71–86.

Brodmann K. (1909) *Vergleichende Lokalisationlehre der Grosshirnrinde in thren Prinzipien dargenstellt auf Grund des Zellenbaues* (J. A. Barth, Leipzig).

Bruce D. (1985). On the origin of the term "neuropsychology." *Neuropsychologia,* **23,** 813–814.

Coltheart M., Patterson K., and Marshall J. C., eds. (1980) *Deep Dyslexia.* (Routledge & Kegan Paul, London).

Corkin S., Milner B., and Rasmussen T. (1970) Somatosensory thresholds. *Arch. Neurol.* **23,** 41–58.

Geschwind N. (1965) Disconnexion syndromes in animals and man. *Brain,* **88,** 237–294, 585–644.

Geschwind N. and Levitsky W. (1968) Left-right asymmetries in temporal speech region. *Science,* **161,** 186–187.

Gevins A. S. (1986) Quantitative human neurophysiology, in *Experimental Techniques in Human Neuropsychology* (Hannay, H. J., ed.), Oxford University Press, New York, pp. 125–162.

Goldstein K. (1939) *The Organism: A Holistic Approach to Biology, Derived from Pathological Data in Man* (American Book, New York).

Goltz F. (1960) On the functions of the hemispheres, in *The Cerebral Cortex.* (von Bonin, J. G., ed.) Charles C. Thomas, Springfield, Ill., pp. 118–158.

Gould S. J. (1981) *The Mismeasurement of Man* (Norton, New York).

Hebb D. O. (1949) *The Organization of Behavior* (McGraw-Hill, New York).

Hebb D. O. (1939) Intelligence in man after large removals of cerebral tissue: Report of four left frontal lobe cases. *J. Gen. Psychol.* **21,** 73–87.

Hess W. R. (1957) *The Functional Organization of the Diencephalon* (Grune & Stratton, New York).

Holmes G. (1918) Disturbances of vision by cerebral lesions. *Br. J. Ophthalmol.* **2,** 353–384.

Ingvar D. H. and Risberg J. (1967) Increase of regional blood flow during mental effort in normals and in patients with focal brain disorders. *Exper. Brain Res.* **3,** 195–211.

Jeannerod M. and Biguer B. (1982) Visuomotor mechanisms in reaching within extrapersonal space, in *Analysis of Visual Behavior* (Ingle D. J., Goodale M. A., and Mansfield R. J. W., eds.), MIT Press, Cambridge, MA.

Kertesz J. A. (1979) *Aphasia and Associated Disorders* (Grune & Stratton, New York).

Kolb B. and Whishaw I. Q. (1981) Neonatal frontal lesions in the rat: sparing of learned but not species-typical behavior in the presence of reduced brain weight and cortical thickness. *J. Compar. Physiol. Psychology,* **95,** 468–483.

Kolb B. and Whishaw I. Q. (1983) Generalizing in neuropsychology: problems and principles underlying cross-species comparisons, in *Behavioral Approaches to Brain Research* (Robinson T. E., ed.) (Oxford University Press, New York), pp. 237–263.

Kolb B. and Whishaw I. Q. (1985a) Can the study of praxis in animals aid in the study of apraxia in humans? in *Advances in Psychology: Neuropsychological Studies of Apraxia and Related Disorders,* (Roy E. A., ed.) North Holland, Amsterdam, pp. 203–224.

Kolb B. and Whishaw I. Q. (1985b) *Fundamentals of Human Neuropsychology* (2nd Ed.) (W. H. Freeman & Co., New York).

Lehrman D. S. (1970) Semantic and conceptual issues in the nature-nuture problem, in *Development and Evolution of Behavior* (Aronson L. R., Tobach E., Lehrman D. S., and Rosenblatt J. S., eds.) W. H. Freeman & Co., San Francisco, pp. 17–52.

Lichtheim L. (1885) On aphasia. *Brain,* **7,** 433–484.

Loftus E. F. and Loftus G. R. (1980) On the permanence of stored information in the human brain. *Amer. Psychologist,* **35,** 409–420.

Luciani L. (1915) *Human Physiology* (Macmillan, London).

Marshall J. (1986) The description and investigation of aphasia language disorder. *Neuropsychologia,* **24,** 5–24.

Milner B. (1980) Complementary functional specializations of the human cerebral hemispheres. *Pontificiae Academiae Scientiarum Scripta Varia,* **45,** 601–625.

Milner B. and Teuber H.-L. (1968) Alteration of perception and memory in man: Reflections on methods, in *Analysis of Behavioral Change* (Weiskrantz L., ed.) Harper & Row, New York, pp. 268–375.

Mishkin M. (1978) Memory in monkeys severely impaired by combined but not by separate removal of amygdala and hippocampus. *Nature,* **273,** 297–298.

Nottebohm F., Kasparian S., and Pansazis C. (1981) Brain space for a learned task. *Brain Res.* **213,** 99–109.

O'Keefe J. and Nagel L. (1978) *The Hippocampus as a Cognitive Map* Oxford: (Oxford University Press, Oxford).

Olds J. and Milner P. (1954) Positive reinforcement produced by electrical stimulation of the septal area and other regions of the rat brain. *J. Compar. Physiol. Psychol.* **47,** 419–427.

Pandya D. N. and Yeterian E. H. (1985) Architecture and connections of cortical association areas, in *Cerebral cortex,* Vol. 4. (Peters A. and Jones E. G., eds.), Plenum Press, New York, pp. 3–62.

Papez J. W. (1937) A proposed mechanism of emotion. *Arch. Neurol. Psychiatr.,* **38,** 725–744.

Penfield W. and Jasper H. H. (1954) *Epilepsy and the Functional Anatomy of the Human Brain* (Little, Brown & Co, Boston).

Penfield W. and Roberts L. (1959) *Speech and Brain Mechanisms* Princeton: (Princeton University Press, Princeton).

Poeck K. (1983) What do we mean by aphasic syndromes? A neurologist's view. *Brain and Language* **20,** 79–89.

Poeck K. (1986) The clinical examination for motor apraxia. *Neuropsychologia* **24,** 129–134.

Risberg J. (1986) Regional blood flow in neuropsychology. *Neuropsychologia,* **24,** 135–140.

Ross E. D. (1981) The aprosodias: Functional-anatomical organization of the affective components of language in the right hemisphere. *Arch. Neurol.* **38,** 1344–1354.

Schwartz M. F. (1984) What the classical aphasia categories can't do for us, and why. *Brain and Language* **21,** 3–8.

Scoville W. B. and Milner B. (1954) Loss of recent memory after bilateral hippocampal lesions. *J. Neurol., Neurosurg. Psychiatr.* **20,** 11–21.

Shallice T. (1979) Case study approach in neuropsychological research. *J. Clin. Neuropsychol.* **1,** 183–211.

Szechtman H. and Hall W. G. (1980) Ontogeny of oral behavior induced by tail pinch and electrical stimulation of the tail in rats. *J. Comp. Physiol. Psychol.* **94,** 436–445.

Ungerleider L. and Mishkin M. (1982) Two cortical visual systems, in *Analysis of Visual Behavior* (Ingle D. J., Goodale M. A., and Mansfield R. J. W. eds.), MIT Press, Cambridge, MA.

Valenstein E. (1975) Persistent problems in the physical control of the brain. *American Museum of Natural History.*

Vanderwolf C. H. and Robinson T. E. (1981) Reticulo-cortical activity and behavior: A critique of the arousal theory and a new synthesis. *Behav. and Brain Sci.* **4,** 459–514.

Vanderwolf C. H., Kolb B., and Cooley R. (1978) Behavior of the rat after removal of the neocortex and hippocampal formation. *J. Comp. Physiol. Psychol.* **92,** 156–175.

Wernicke C. (1874) *Der aphasische Symptomenkomplex* (Cohn & Weigart, Breslau).

Whishaw I. Q., Bland B., and Bayer S. (1978) Postnatal hippocampal granule cell agenesis in the rat: effects on two types of rhythmical slow activity (RSA) in two hippocampal generators. *Brain Res.* **146,** 249–268.

Methods in Human Neuropsychology

2. Contributions of Human Experimental Psychology and Psychometrics

John Boeglin, Dan Bub, and Yves Joanette

1. Introduction

Neuropsychology can be broadly defined as the study of brain–behavior relationships. The methods on which this discipline is founded are equally as broad, a fact to which the present volume attests. Indeed, neuropsychology is at the crossroads of the neurosciences, which include neurology, neuroanatomy, neurophysiology, neurochemistry, and the behavioral sciences, which include psychology and linguistics (Hécaen and Albert, 1978). The purpose of this chapter is to expose the nature and origin of methods issued from human experimental psychology and, to a minor degree, from psychometrics to the systematic study of brain–behavior relationships. In doing so, this chapter will not focus on fundamental issues, like the still problematic question of the relation between function and structure, but rather on the methods used to examine these issues.

Before going any further, it is essential that one understands what is meant exactly by methods. Within the present context, a suitable definition would be that methods form the logic or rationale, as seen from the viewpoint of psychological theory, underlying the scientific study of brain–behavior relationships. As we will attempt to show in this chapter, these methods may fall into one of three categories. First of all, such methods may be of a purely clinical value. In this case, the researcher or clinician is basically interested in determining how human behavior *per se* can be used to localize brain damage. Secondly, these methods may entail the use of statistical models (i.e., models derived from the factor analysis of a patient's performance on various cognitive tasks) to fractionate performance into more basic elements. Finally, statistical models may be replaced by processing models (i.e., models put forward by cognitive neuropsychology). Here the focus is on the extreme cases, thereby making this approach more suitable for studies of the single-case type.

2. The Mind–Body Problem

In examining the nature and the origins of the contributions of experimental psychology and psychometrics to the foundations of human neuropsychology, one must bear in mind the fact that psychology *per se* is a relatively young science. Although scientific methods have been employed for centuries within the realm of the natural sciences, it is only since the latter half of the nineteenth century that such methods have been systematically applied to the study of human behavior. This is not to say, however, that the prescientific antecedents of psychology, in particular the writings of the British and French schools of philosophy, did not have any significant impact on the future of psychology and, subsequently, neuropsychology.

One of the most important and controversial issues in the history of philosophy has been that of determining whether the mind and body are essentially the same or different in nature. Among the first to have tackled this issue was the seventeenth-century French philosopher René Descartes. In his 1637 book entitled *Discours de la Methode,* Descartes expressed the belief that the mind was different from the body. Indeed, Descartes regarded the mind as being unextended, free, and lacking in substance, and the body as being extended, governed by physical laws, and having substance (Hothersall, 1984). From this point of view, often referred to as the dualist solution to the mind–body problem, there was little sense in applying scientific methods in an attempt to uncover the basic elements of the human mind. Conversely, the human body was viewed as a highly complex machine and, as such, could be studied by rational, scientific methods similar to those applied to the inanimate objects of the natural sciences.

However, as time passed, it became more and more apparent that even the human mind could be studied by these same scientific methods. The writings of the British empiricist philosophers played an important role in this respect. For example, three years after the publication of Newton's 1687 treatise *Principia,* in which he describes a universe that follows a single set of rules, John Locke sought to establish a similar set of rules for the human mind in a book entitled *Essay Concerning Human Understanding* (Hothersall, 1984). This was to be achieved by fractioning the human mind, so to speak, into its basic elements or "ideas." According to Locke, such ideas could be either of external origin (i.e., by way of sensation through contact with physical objects in the environment) or of internal origin (i.e., by way of association with existing ideas). From this point on, it became more and more obvious that the human mind was also a highly complex machine, and being so, it too could be scrutinized by scientific methods. However, although the philosophers of the seventeenth and eighteenth centuries are to be credited for opening the way for a scientific study of the human mind, it should be emphasized that their approach, being more based on anecdote and introspection than on demonstrable facts *per se,* was no more than rudimentary. This does not mean that contemporary approaches are problem-free. In fact, one of the biggest problems of all time is the implicit metaphor used when trying to describe the human mind as a complex machine. Whereas at one point the metaphor was hydraulic (e.g., ventricular theories), the analogy is nowadays modeled on computer architecture given that it represents the most complex machine available. The problem with such metaphors is that most are implicit rather than explicit and that they impose a limitation of the possible conceptualization of a given function. For example, the computer metaphor has imposed the notion of sequential processing in most models currently debated in cognitive neuropsychology. However, this metaphor might not be the most appropriate for the interpretation of higher-level functioning.

3. Human Neuropsychology: Classical Views

During the course of the nineteenth century, major advances within the fields of anatomy, clinical neurology, and sensory physiology, in addition to the founding of a new science—psychology—set the tone for a more scientific approach to the understanding of brain–behavior relationships. The work of the early anatomists (e.g., Gall and Flourens) provided insight as to the anatomy and function of the central nervous system. Furthermore, they, like many others who were to follow from the mid-1800s to the early 1900s, were primarily concerned and influenced by general assumptions about the ability to localize specific functions within the cortex of the human brain. Those who became involved in such a practice were eventually dubbed the "localizers." Their work was inspired to a large extent by that of the "phrenologists," led by the Austrian anatomist Franz Joseph Gall. In short, Gall's phrenology was based on the notion that each of the various intellectual and affective functions that make up human behavior was localized in a specific surface area of the brain. If, in a given individual, one or several of these functions were particularly well-developed, then the corresponding brain area was overdeveloped and this, in turn, was reflected in bumps on the skull overlying the relevant area. Conversely, indentations on the skull were thought to reflect less developed functions (Boring, 1950). Although "reading the bumps" remained in vogue throughout most of the nineteenth century and well into the twentieth century, the scientific basis of phrenology was too unsound to assure its longevity.

Nevertheless, Gall should at least be credited for having been the first to direct attention to the cerebral cortex as well as for having promulgated the idea that human behavior can be broken down into a number of components, and each component associated with a specific area of the cortex. Although specific localization of function is not emphasized by contemporary neuropsychology, the fractioning of human behavior remains a key issue. Even so, it is considered by some (e.g., Lecours et al., 1984) that it was Gall's concept of phrenology that instigated the systematic study of brain–behavior relationships.

As for the early clinical neurologists (e.g., Bouillaud and Broca), they too were interested in issues concerning cerebral localization of function. However, in marked contrast to the phrenologists, their overall approach to the understanding of brain–behavior relationships was basically one of establishing clinicopathological correlations. In addition, their focal point of interest shifted from the so-called bumps of the skull to the convolutions of the cerebral hemispheres. Finally, whereas Gall focused his attention on individual traits (e.g., personality), the clinical neurologists were more concerned with higher cognitive functions (e.g., speech). As a result, patients displaying impaired performance were examined in detail and inferences were then drawn as to the possible locus of the brain insult underlying their behavior dysfunctions. Subsequent post-mortem examinations of these patients provided the necessary evidence to either support or refute these inferences. This approach is best exemplified by Paul Broca's 1861 descriptions of two of his most famous patients, Leborgne and Lelong (*see* Schiller, 1979). Although the contribution of the clinical neurologists was undoubtedly an improvement over phrenology, its scientific value was questionable. Indeed, a great deal of emphasis was placed on the individual clinician's personal intuition concerning the functional organization of the brain. In addition, the notion of subject heterogeneity, which only recently emerged as a major factor in neuropsychological research, has cast a serious doubt on making any generalizations from such intuitions. Nevertheless, the views of Broca and others concerning the cerebral localization of function did serve as the forerunner of a new enterprise whose proponents were the diagram makers.

Like the localizationists, the "diagram makers" or "associationists" (e.g., Bastian, Lichteim, and Wernicke) also postulated (or presumed) a specific function for each anatomically defined area of the brain. Furthermore, so as to account for the possible links between these so-called "brain centers," some intracerebral connections or pathways were also postulated (or presumed), thereby leading to the elaboration of diagrams accounting for brain function (*see* Morton, 1984). As a result of diagram making, behavior dysfunctions, such as those of speech and language, came to be viewed as deficits in the centers subserving speech and language functions and/or the pathways connecting them. Though the diagram makers were undoubtedly influenced by Gall's ideas, it is likely that they were even more inspired by the work of the physiologists (e.g., Bell, Ferrier, Helmholtz, and Mueller) who focused their attention on the transmission of sensory and motor information to the brain as well as on the localization of sensory and motor functions within the brain itself (*see* Boring, 1950). Judging by the ". . . proliferation of maps and diagrams showing the supposed location of all types of functions . . ." (Kolb and Wishaw, 1985, p. 315), localizing and diagram making remained popular well into the early 1900s.

At the time, the only strong opposition to the concept of cerebral localization of function came from the prominent English neurologist, John Hughlings Jackson (*see* Taylor, 1932). In Jackson's theory, localizing *damage* to a part of the brain associated with a disturbance of function was one enterprise; localizing the *function* itself was quite another matter. Furthermore, Jackson contended that brain damage did not result in the total loss of a given function. A follower of Herbert Spencer, a British evolutionary associationist, Jackson postulated an evolutionary process within the nervous system: complex functions were thought to be built up in stages from a number of more elementary functions, each higher level expanding on the functions present at the lower levels. The higher the level, that is, the more complex a function was, the more widespread the involvement of the cortex was. As for diseases of the nervous system, they were viewed by Jackson as reversals of evolution, that is, as dissolutions. In other words, brain damage could result in the disturbance of a function at its highest level of evolution, thereby giving way to the expression of its more elementary components present at the lower levels of evolution. Unfortunately, Jackson's views were to be largely ignored by his contemporaries (*see* Head, 1926).

Staunch opposition to the notions of brain centers and of their connections postulated by the classical neurologists did not surface again until the mid-1920s. Spurred on by the antilocalizationist views of John Hughlings Jackson, Henry Head (1926) advocated a "holistic" approach to the study of brain–behavior relationships. Although he did not entirely dismiss the notion of anatomical localization, admitting that topographical relationships within the brain did exist between the different parts of the body, Head argued that, as far as functions were concerned, be they low-level (e.g., sensory or motor) or high-level (e.g., language), these could not be localized. As for the diagram makers, they too were subject to Head's criticism. For example, Head relates the now-familiar story of Bastian who, for a number of years, had apparently presented an aphasic patient to his students, explaining the patient's deficit by way of a diagram revealing which cortical areas and which pathways believed to subserve speech were affected or not. Unfortunately for Bastian, but obviously to the delight of Head, post-mortem examination of this patient revealed much more diffuse and extensive damage than had been postulated in the diagram (*see* Head, 1926, pp. 56–57).

Another well-known proponent of the holistic viewpoint was Kurt Goldstein (1948). Personal observations of numerous physiological and psychological phenomena, both normal and pathological, led Goldstein to formulate his so-called "organismic approach" to the function of the human organism. The focal point of Goldstein's theory was that pathological behavior could be understood ". . . only from the aspect of its relation to the function of the total organism . . ." (Goldstein, 1948, p. 21). In other words, attention should be focused not only on the nature of the deficit itself, but also on the manner in which the individual reacts to the deficit.

This new approach to the understanding of brain–behavior relationships was undoubtedly influenced to some extent by certain developments within the realm of psychology, in particular that of Gestalt psychology. In much the same way as Head and Goldstein had rejected the localizationist and associationist theories of brain–behavior relationships, the proponents of Gestalt psychology rejected the associationist explanations of human behavior that had dominated the scene since the time of the eighteenth-century philosophers of the British school. The Gestaltists argued that the classical approaches to the understanding of human behavior were too atomistic (Hothersall, 1984). Their main tenet was that "wholes" were more than simple aggregates of their individual "parts." As a result, the properties and qualities of each part could be defined only with respect to the relation of the parts to the whole. In much the same way, the activity of the brain itself could be viewed as the involvement of more than a simple summation of the activities of highly specialized centers and their connections. From its onset, Gestalt theory had profound implications with respect to existing views on visual perception, though it was eventually extended to other cognitive functions as well, such as learning and memory (*see* Kohler, 1947).

4. The Birth of Experimental Psychology

In addition to the various issues concerning the neurobiological basis of human behavior, the latter part of the nineteenth century was marked by yet another important development, namely, the founding of experimental psychology. Indeed, it was in 1879 that psychology came to be officially recognized as a science with the establishment of the first laboratory of experimental psychology by Wilhelm Wundt at the University of Leipzig. Although Wundt was instrumental in establishing experimental psychology as a separate scientific discipline, he did not believe that the scientific methods in use at the time (i.e., those of the natural sciences) could be applied in a psychological investigation (Hothersall, 1984). According to Wundt, the problem of psychology was the analysis of "conscious processes"; *Selbstbeobachtung,* that is, "self-observation" or "introspection," was the strictly controlled procedure put forward by the Wundtians to study such processes. However, a good part of the experimental work that was actually carried out in the Leipzig laboratory had less to do with introspection than it did with various topics within the field of sensation and perception (Boring, 1950).

5. Human Neuropsychology: The Modern Era

The major advances that have occurred within the neurosciences and the behavioral sciences during the course of the twentieth century have had two, albeit opposite, effects on the study of brain–behavior relationships. On the one hand, these two fields of study have merged, so to speak, and in doing so, have created a new field of study, namely, neuropsychology. This so-called merger is implicit in the definition of neuropsychology that was given at the beginning of this chapter. It is also evident in comments made by Sir William Osler, who supposedly coined the term "neuropsychology" during the course of an address to the Johns Hopkins University Hospital in 1913 (Bruce, 1985). According to Bruce, this term was employed by Osler to suggest that all mental disorders were a disease of the central nervous system. Whether Osler was casting a personal vote in favor of uniting the neurosciences and the behavioral sciences as a single discipline, or whether he was expressing an opinion widely shared by fellow clinicians and researchers is unclear. However, if one considers the fact that neuropsychology has only recently acquired the status as a field of study in its own right, then it is obvious that Osler's views were not shared by many of his contemporaries.

On the other hand, the issues of human neuropsychology have become so complex that two distinct branches, which are not entirely independent of each other, have emerged: clinical neuropsychology and experimental neuropsychology. Clinical neuropsychology deals almost exclusively with individuals who display deviant behavior patterns subsequent to injury of the brain (e.g., from disease or from physical damage). According to Luria, clinical neuropsychology has two principal objectives:

> *First, by pinpointing the brain lesions responsible for specific behavioral disorders we hope to develop a means of early diagnosis and precise location of brain injuries . . . Second, neuropsychological investigations should provide us with a factor analysis that will lead to better understanding of the components of complex, psychological functions for which the operations of the different parts of the brain are responsible* (Luria, 1970, p. 66).

5.1. Psychometrics

Since its earliest phases, the main concern of clinical neuropsychology has been with the behavioral expression of brain dysfunction (Lezak, 1983). Although it is difficult to pinpoint exactly at what point in time clinical neuropsychology emerged as a separate discipline, it is apparent that its evolution has been spurred on, at least in part, by the historical events of the twentieth century. For example, various international conflicts have yielded, so to speak, vast numbers of individuals with war-inflicted wounds. The great demands made on clinical neurologists, on the one hand, and psychologists, on the other hand, to provide assessment and diagnosis of these individuals exposed the necessity of a new breed of clinicians to assist in handling this large influx of patients.

The task of the early clinical neuropsychologist, in many ways similar to that of the clinical neurologist of the nineteenth century, consisted essentially of providing detailed and accurate descriptions of behavioral change in individuals who had supposedly sustained damage to one part of the brain or another. On the basis of these descriptions, the clinician was then able to identify the nature of the disturbed function and then deduce the possible locus of the neurological insult responsible for the dysfunction. The task of the contemporary clinical neuropsychologist has changed somewhat with respect to that of his predecessors: the assessment of behavior dysfunction subsequent to brain damage is still the focal point of clinical neuropsychology, but the localization of the damage itself is now established through highly precise electronic scanning methods. In addition, the contemporary clinical neuropsychologist has been given yet another task, that is, the development and application of rehabilitation procedures for brain-damaged individuals whose functional capacity has not been improved by interventions within the traditional health care system (Diller and Gordon, 1981).

By today's standards, the early clinical neuropsychology examination of brain-damaged patients was rather rudimentary: in most cases, it involved nothing more than establishing a repertoire of the patients' abilities and/or disabilities by way of verbal descriptions. In the best of cases, informal testing procedures may have been used (e.g., Head, 1926). However, as the knowledge of behavior dysfunction following brain damage became increasingly more complex, the need for more appropriate assessment methods began to surface. The methods in question were not developed by clinical neuropsychologists themselves, but instead were borrowed from yet another specialization within the field of psychology, namely, psychometrics.

The development of psychological testing in the late nineteenth century and early twentieth century, by such prominent psychologists as James Cattell and Alfred Binet (*see* Anastasi, 1976), opened a new avenue within the field of clinical neuropsychology. The use of psychological tests, that is to say, assessment procedures with demonstrated validity and reliability, to evaluate specific aspects of human behavior (e.g., intelligence, various abilities, and personality), provided clinical neuropsychologists with the necessary tools not only to describe but also to quantify such behavior, thereby offering a new approach in the study of brain–behavior relationships.

A particular impetus to neuropsychological assessment or neuropsychometrics was provided when Ward Halstead founded the first laboratory of clinical neuropsychology at the University of Chicago in 1935. Whereas the early attempts at providing neuropsychological assessments of behavior dysfunction led to a proliferation of single-function tests (Lezak, 1983), Halstead's work resulted in the elaboration of the first full-scale neuropsychometric battery (Halstead, 1949), now known as the Halstead-Reitan Neuropsychological Test Battery.

It would appear that Halstead elaborated his battery with two specific goals in mind. The first was the ability of the battery to discriminate between organic vs nonorganic modifications of behavior. The second was the ability of the battery to determine in which hemisphere and, more precisely, in which lobe the neurological insult had occurred. In short, an individual's performance on the Halstead battery informed neurologists and neurosurgeons as to the possible existence and location of brain damage, and in doing so, constituted the functional equivalent of modern brain imagery techniques. Other neuropsychological test batteries were to follow, although their objectives were not always the same as those of Halstead's original test battery. For example, the Luria-Nebraska Neuropsychological Battery (Golden, 1981), which is based on Luria's functional conception of the brain, provides a detailed qualitative analysis of various neuropsychological functions (e.g., motor functions, higher visual functions, speech, and mnestic processes) subdivided into their most basic components.

As mentioned earlier, the development of neuropsychometrics was also influenced, albeit indirectly, by various international conflicts. Indeed, the vast numbers of individuals with war-inflicted brain injuries rendered simple verbal descriptions of their behavior time-consuming. In addition, this approach did not allow for any between-subject comparison. The systematic analysis of their abnormal behavior could only be achieved through the adoption of quantitative methods. Batteries, such as those mentioned above, as well as numerous other tests (*see* Lezak, 1983) have come to form the basis of contemporary clinical neuropsychology. Today, these batteries and tests are employed by clinicians in the practical clinical work of diagnosing brain damage and, more and more, in the rehabilitation of brain-injured patients as well as by some researchers in the scientific study of brain–behavior relationships.

5.2. Cognitive Neuropsychology

Following the harsh criticisms of Jackson, Head, and Goldstein, the debate surrounding the concept of cerebral localization of function subsided. It was not until the mid-1960s that a new controversy emerged, mainly because of the efforts of Norman Geschwind (1965) to revive the older localizationist theory. Geschwind's extensive investigations of the so-called "disconnection syndrome" (i.e., the effects of lesions of the inter- or intrahemispheric associative pathways) led him to suggest disturbances of the higher functions of the nervous system were the result of the ". . . anatomical disconnection of of primary receptive and motor areas from one another" (Geschwind, 1965, p. 640). In other words, complex behavior, according to Geschwind, results from the connections that exist between the different regions of the brain.

With the advances that have occurred since the Second World War in the fields of cybernetics and psychology, particularly within that branch of psychology now known as cognitive psychology, there has been a recent trend to resort once again to diagram making. However, in contrast to many of the classical diagram makers, these theoreticians have been more concerned with a functional analysis of higher cognitive skills than with issues pertaining to cerebral localization.

5.2.1. Modularity of the Mind: Functional Modules

Complex skills like reading, writing, face recognition, and so on, are mediated by the combined interaction of many different processing components that together make up the collective function. This claim, known as the modularity principle, has been elegantly summarized by Marr (1976):

> Any large computation should be split up and implemented as a collection of small sub-parts that are as nearly independent of one another as the overall task allows. If a process is not designed in this way, a small change in one place will have consequences in many other places. This means that the process as a whole becomes extremely difficult to debug or improve, whether by a human designer or in the course of natural evolution, because a small change to improve one part has to be accompanied by many simultaneous compensating changes elsewhere.

Shallice (1981) points out that, if one also makes the reasonable assumption that functional independence is correlated with a physical separation of processing modules in the brain, then we would expect that a single part of a complex system can be damaged without any disturbance to the remaining components. The major goal of cognitive neuropsychology, then, is to identify the modules that together carry out a global function, and the flow of information between them by studying the performance of theoretically appropriate brain-damaged cases. We should reemphasize that the quest for a detailed functional architecture—the organization of the different components and their algorithmic content—is completely distinct from the question of how these modules are physically represented in the brain. Indeed, many researchers are of the opinion that neuroanatomical considerations could not, in principle, be used to adjudicate between rival claims about functional mechanisms (Morton, 1984).

5.2.2. Functional Diagram Makers

The relevant evidence from neurological populations supporting one or the other processing model demands a proof of the existence of a particular *module* at a particular *location* in the functional architecture, or a proof of a *bifurcation* within a functional subsystem such that two different routines are independently and concomitantly involved in a given task. We shall presently find that these two kinds of evidence entail slightly different methodological approaches. For now, we illustrate the concept of a functional architecture by turning to a popular, though rudimentary, model of the processing components that determine the translation of a written word into sound. (*see* Fig. 1).

The main assumptions underlying the diagram are as follows: Letters are perceptually extracted from basic visual features and then make contact with the visual word-form system, a component that stores the permanent orthographic description of whole words. The visual word-form system gains direct access to the semantic representation of the word, and the pronunciation of the word is then retrieved from its meaning. A second routine involves the mapping of whole-word orthography onto pronunciation *without* first making contact with the semantic description. Finally, *subword* units are extracted from the letter string and translated into sound via a knowledge of the correspondences between orthographic patterns (e.g., INT) and their phonemic values (*see* Henderson, 1985; Coltheart, 1985 for reviews). Thus, the model is based on the claim that reading aloud takes place:

1. By mapping the letters onto an orthographic description of an entire word, which then activates the meaning and then the pronunciation
2. By translating the orthography of the whole word into pronunciation without first recovering the meaning and
3. Extracting subword units and assembling them into a response.

The performance of a normal reader is assumed to be mediated by all three procedures acting simultaneously and in parallel.

5.2.3. Testing a Model

Brain-damaged patients may experience selective damage to a particular processing component, and the resulting performance

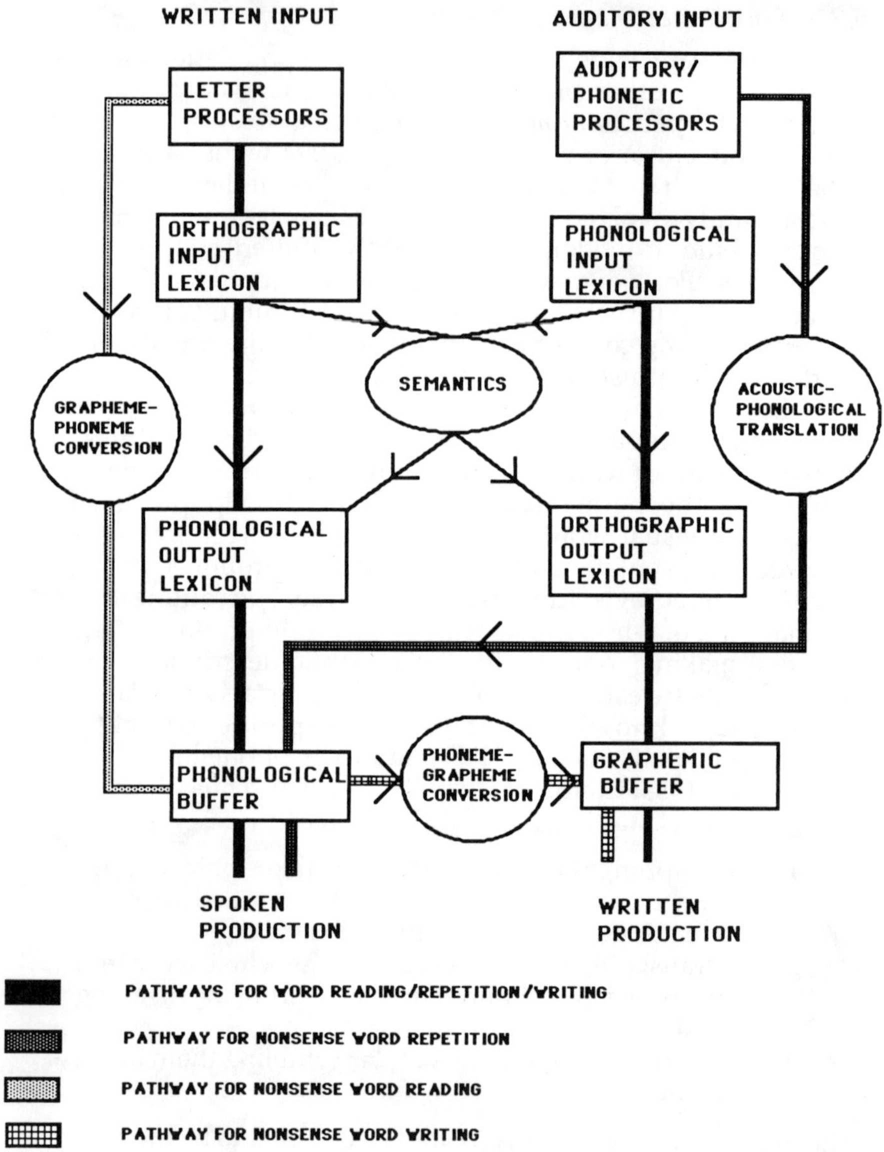

Fig. 1. Some basic functional components of single-word production and recognition.

can be used to test hypotheses derived from the model. The existence of a separate routine dealing with subword units entails that certain patients should be observed with impairment to this branch of the reading mechanism. The subword routine is inevitably required for translating any spelling pattern into sound that lacks a permanent description in the visual word-form system. The damage should therefore prevent the accurate reading aloud of nonsense words, even though legitimate words are pronounced without difficulty. The reading performance of phonological dyslexics (Beauvois and Derouesné, 1979; Funnell, 1983) confirms a theoretical distinction between the processes mediating the pronunciation of written words and nonsense words; these patients, in the purest cases, are unable to translate even the simplest nonsense word into sound, but can easily read aloud a full range of orthographically complex, low frequency words.

Other patients demonstrate the reverse of the dissociation observed in phonological dyslexics; they can read nonsense words accurately, but their performance reveals that they are impaired in their ability to translate *whole-word* orthographic units into sound. The subword routine operates by using the most general correspondence associated with a spelling pattern. In English and many other languages, a large number of words do not obey these regular principles of translation (e.g., PINT as opposed to HINT, MINT, STINT). Surface dyslexics (e.g., Patterson et al., 1985) make numerous errors when asked to read orthographic exception words aloud, inappropriately applying the regular phonemic value to the spelling. Comprehension is based on the pronunciation of the target rather than its visual form (e.g., RODE could be defined as "a pathway"), consistent with the failure to obtain the orthographic description of the entire word for access to meaning.

We cannot provide a complete review of the literature on acquired dyslexia in this chapter (*see* Coltheart, 1985 for a review). However, we do wish to note that the kind of evidence we have described is characterized by dissociations in performance: A patient is very good at reading words aloud, for example, but cannot read nonsense words, or is very good at reading nonsense words but cannot read words of a certain type. Providing we can argue that the results are not caused by some irrelevant property of the stimuli (e.g., their ease of pronunciation rather than their orthographic characteristics), the dissociation itself leads directly to the inference that words and nonsense words recruit functionally separate reading procedures. An interesting methodological point is that we do not require any detailed background statements about the internal structure of the hypothetical processing routines to arrive at this conclusion. We did not stipulate, for example, what are the actual units of translation used by the subword routine (graphemes, syllables, word endings?), other than that they must be, by definition, smaller than whole words. The dissociation in performance allows us, without further analysis, to claim that a bifurcation must occur at a particular location in the functional architecture.

The situation is rather different when we seek neuropsychological evidence for a discrete component at a given location (Bub and Bub, in press). Here, a two-step procedure is required: First, we obtain evidence that locates the damaged component by demonstrating impairment on a variety of related tasks. Second, we derive some prediction from our knowledge of the damaged component regarding the expected pattern of performance. Unlike the method of arguing from dissociations in performance to reveal a functional separation between related procedures, the use of associated deficits to isolate a discrete processing component demands a more complex methodology. To illustrate this point with a concrete example: Fluent speech includes the activity of a processing component that allows several words to be maintained in prearticulatory form prior to overt production. The phonological buffer, which serves this function, is needed whenever speech segments are assembled for output—pronouncing written words or nonsense words, repeating them to dictation, and determining the spelled form of nonsense words are all tasks that require the mediation of this component.

Recent work in neuropsychology that tests for the existence of the phonological buffer involves two methodologically distinct stages (Bub et al., 1986; Caramazza et al., 1986). First, an association of deficits in reading, writing, and repetition is used to localize the lesion in the functional architecture. Next, some theoretical claim is made to infer a specific pattern of performance errors if the component malfunctions at the lesion site. Bub et al. (1986), for example, showed that their patient's mispronunciation of nonsense words resulted in the substitution of phonemes that were very often one, or at most two, distinctive features removed from the target. Thus MUNT was pronounced MUNK, SIFE as SIVE, and so on. A very similar demonstration has been provided independently by Caramazza et al., 1986. Both groups of authors concluded that the errors were the result of a defect in the activation of phonemic codes at the level of the response buffer.

5.2.4. Issues Related to Studies with Brain-Damaged Patients

We have discussed a methodological approach based on the analysis of deficits within the framework of a processing model of discrete, functionally independent components. Damage is localized to a particular *component* or set of components to explain the performance of a patient with a given disturbance. The question of the underlying relationship between the functional "lesion" and the actual brain tissue responsible for a particular function remains, as many authors have pointed out, a separate issue (e.g., Morton, 1984). The goal of explaining impaired performance in terms of damage to a hypothetical functional architecture places important constraints on the kind of logic that can be used to draw inferences about higher cognition from damaged performance. Caramazza (1984, 1986), in a series of articles, has argued that only single cases can provide the data for testing different processing models, because in each patient the initial step must always be one of defining the affected part of the functional architecture by a number of experimental tests. We cannot average the performance of different patients, according to Caramazza, because we can never be sure that they have sustained damage to equivalent components *a priori*. Thus, each case must stand on its own merits, and any generalized conclusions must be drawn with respect to the functional system we are investigating, not with respect to a group of patients.

The position advocated by Caramazza is an extreme one and has been the focus of much controversy (cf Caplan, 1986; Bub and Bub, in press; Zurif et al., in press). Though we cannot review the many arguments pro and con in this chapter, we note that the argument could in principle only have merit when we deal with the evaluation of a *specific* functional architecture. Many, indeed the bulk of neuropsychological experiments that are cognitively relevant are not immediately concerned with this enterprise, and for them, we remain unconvinced that Caramazza's argument would apply.

Finally, we should reemphasize in closing this section that questions of brain–behavior relationships demand both a clearly defined processing model of a particular cognitive mechanism and a coherent notion of the possible ways in which the functional components of the mechanism are represented in the brain. The recent advent of distributed computational systems, where the activity of many different operating units together represent any one memory trace, highlights the pitfalls awaiting any naive attempt to localize aspects of higher cognitive function in the brain (Allport, 1984).

5.2.5. Task-Related Issues

Neuropsychological research, particularly that pertaining to the functional asymmetry of the cerebral hemispheres, has come to rely on several more or less sophisticated techniques (*see* Bradshaw's chapter, this volume). Some of these techniques involve particular stimulus presentations, such as divided visual-field presentations, dichaptic presentations, or dichotic presentations. Other techniques call for measures of various motor asymmetries, such as dowel balancing and finger tapping, or for the recording of lateral eye movements. According to Caramazza (1984), these techniques do not appear to be sufficiently powerful to study the workings of the brain as it performs complex tasks. Even physiological measures, such as electroencephalography, auditory evoked responses, and cerebral blood flow, have yet to provide satisfactory answers as to the functional asymmetry of the cerebral hemispheres (Bryden, 1982). Because of the difficulties inherent to each of these techniques, researchers have been forced to resort to the study of brain-damaged populations, which has provided, and continues to provide, a major source of information on brain–behavior relationships.

6. Conclusion

At the beginning of this chapter, allusion was made as to the multidisciplinary origins of human neuropsychology. We have attempted to expose the origins and the nature of methods issuing from one of these disciplines, namely, psychology. Following a prescientific period dominated by philosophical issues and devoid of any hard-core scientific evidence, the elaboration of strictly controlled experimental paradigms and the development of standardized psychological tests both contributed to the scientific soundness of the study of human behavior. At the same time, progress within the neurosciences provided information concerning the cortical structures underlying human behavior. However, one should not consider the influence of psychology as being unidirectional. Indeed, although the understanding of the human brain may have benefited to a certain extent from the understanding of human behavior, the opposite is equally true.

References

Allport D. A. (1984) Distributed memory, modular subsystems and dysphasia, in *Dysphasia* (Newman S. and Epstein R., eds.), Churchill Livingston, Edinburgh, pp. 1–35.

Anastasi A. (1976) *Psychological Testing* 4th Ed. (Macmillan, New York).

Beauvois M.-F. and Derouesné J. (1979) Phonological allexia: Three dissociations. *J. Neurol. Neurosurg. Psychiatry* **42,** 1115–1124.

Boring E. G. (1950) *A History of Experimental Psychology* 2nd Ed. (Appleton-Century-Crofts, New York).

Bruce D. (1985) On the origin of the term "neuropsychology". *Neuropsychologia* **23,** 813–814.

Bryden M. P. (1982) *Laterality, Functional Asymmetry in the Intact Brain* (Academic Press, New York).

Bub J. and Bub D. (in press) On the methodology of single-case studies in cognitive neuropsychology. *Cogn. Neuropsychol.*

Bub D., Black S., Howell J., and Kertesz A. (1986) Speech output processes and reading, in *The Cognitive Neuropsychology of Language* (Coltheart M., Sartori G. and Job R., eds.), Laurence Erlbaum, N.J., pp. 79–110.

Caplan D. (1986) In defense of agrammatism. *Cognition* **24,** 263–276.

Caramazza A. (1984) The logic of neuropsychological research and the problem of patient classification in aphasia. *Brain Lang.* **21,** 9–20.

Caramazza A. (1986) On drawing inferences about the structure of normal cognitive systems from the analysis of patterns of impaired performance: The case for single-patient studies. *Brain and Cognition* **5,** 41–66.

Caramazza A., Miceli G., and Villa G. (1986) The role of the (output) phonological buffer in reading, writing and repetition. *Cogn. Neuropsychol.* **3,** 37–76.

Coltheart M. (1985) Cognitive neuropsychology and the study of reading, in *Attention and Performance IX* (Posner M. I. and Marin O. S. M., eds.) Laurence Erlbaum, N.J. pp. 3–37.

Diller L. and Gordon W. A. (1981) Rehabilitation and clinical neuropsychology, in *Handbook of Clinical Neuropsychology* (Filskov S. and Boll T. J., eds.), John Wiley and Sons, New York, pp. 702–733.

Funnell E. (1983) Phonological processes in reading: new evidence from acquired dyslexia. *Br. J. Psychol.* **74,** 159–180.

Geschwind N. (1965) Disconnection syndromes in animals and man. *Brain* **88,** 237–294, 585–644.

Golden C. J. (1981) A standardized version of Luria's neuropsychological tests: a quantitative and qualitative approach to neuropsychological investigations, in *Handbook of Clinical Neuropsychology.* (Filskov S. and Boll T. J., eds.), John Wiley and Sons, New York, pp. 608–642.

Goldstein K. (1948) *Language and Language Disturbances* (Grune and Stratton, New York).

Halstead W. C. (1947) *Brain and Intelligence* (University of Chicago Press, Chicago).

Head H. (1926) *Aphasia and Kindred Disorders of Speech* (Cambridge University Press, London).

Hécaen H. and Albert M. L. (1978) *Introduction to Human Neuropsychology* (John Wiley and Sons, New York).

Henderson L. (1985) Issues in the modelling of pronunciation assembly in normal reading, in *Surface Dyslexia* (Patterson K. E., Marshall J. C., and Coltheart M., eds.), Laurence Erlbaum, N.J., pp. 459–508.

Hothersall D. (1984) *History of Psychology* (Temple University Press, Philadelphia).

Kohler W. (1947) *Gestalt Psychology* (Liveright, New York).

Kolb B. and Wishaw I. Q. (1985) *Fundamentals of Human Neuropsychology* 2nd Ed. (W. H. Freeman and Company, New York).

Lecours A. R., Basso A., Moraschini S., and Nespoulous J. L. (1984) Where is the speech area and who has seen it, in *Biological Perspectives on Language* (Caplan D., Lecours A. R., and Smith A., eds.), MIT Press, Cambridge, pp. 220–246.

Lezak M. D. (1983) *Neuropsychological Assessment* 2nd Ed. (Oxford University Press, New York).

Luria A. R. (1970) The functional organization of the brain. *Sci. Amer.* **222,** 66–78.

Marr D. (1976) Early processing of visual information. *Phil. Trans. Roy. Soc. Lond.* **B-275,** 483–524.

Morton J. (1984) Brain-based and non-brain-based models of language, in *Biological Perspectives on Language* (Caplan D., Lecours A. R., and Smith A., eds.), MIT Press, Cambridge, pp. 40–64.

Patterson K. E., Marshall J. C., and Coltheart M. (1985) *Surface Dyslexia.* (Laurence Erlbaum, N.J.

Schiller F. (1979) *Paul Broca, Founder of French Anthropology, Explorer of the Brain* (University of California Press, Berkeley).

Shallice T. (1979) Case study approach in neuropsychological research. *J. Clin. Neuropsychol.* **1,** 183–211.

Shallice T. (1981) Neurological impairment of cognitive processes. *Br. Med. Bull.* **37,** 187–192.

Taylor J. (1932) *Selected Writings of John Hughlings Jackson* (Hodder and Slaughton, London).

Zurif E. G., Gardner H., and Brownell H. H. (in press) The case against the case against group studies. *Brain and Cognition.*

Contributions of Linguistic Approaches to Human Neuropsychology

Aphasia

John Ryalls, Renée Béland, and Yves Joanette

1. Introduction

The portion of human neuropsychology in which linguistics has had its greatest impact is that of aphasiology. It is only logical that aphasia—an impairment in language as a result of neurological damage—is most likely to benefit from linguistics, the science of language. However, it should also be noted that, somewhat differently from psychology, what linguistics has had to offer neuropsychology (and aphasia in particular) is more in terms of theory or frameworks than in terms of methods.

This chapter will attempt to reveal some of the ways in which linguistics has influenced our understanding of aphasia. We have selected only a few studies that we feel to be most exemplary of the manner in which linguistic methods have been most useful in clarifying the nature of aphasia. Needless to say, such a selection is somewhat subjective and certainly limited. Such a sampling cannot hope to give an appreciation of the scope of linguistic influence.[1] Our intention here is to illustrate some of the ways in which linguistic methods have allowed greater insight and precision in defining the language deficit of aphasia.

Although most of the interaction has been in the form of linguistics influencing the study of aphasia, in some ways, aphasia

[1]The reader is referred to Lesser (1978) for a more detailed, although unfortunately already somewhat dated, treatment of the influence of linguistics in aphasia. A more contemporary somewhat more theoretical treatment will be found in Caplan (1987).

does also represent a testing ground for linguistic models. That is, a theoretical model that can both explain normal and pathological language behavior is to be favored over a model that can only account for the facts of normal function. Yet, it is only recently, and then only on a very limited scale, that aphasic behavior has been used to further test linguistic models rather than linguistics being used to test aphasia. One might expect the influence of aphasia on linguistics to grow as more powerful and explicit linguistic models are developed. In fact, Caplan (1987) has recently dealt with the evolving influence of linguistics on aphasia. In his treatment, he distinguishes a branch of neurolinguistics—linguistic aphasiology—which is more concerned with theories of language processing. It is partially, according to Caplan, the advent of the influence of aphasia on theories of normal language processing that distinguishes linguistic aphasiology from its parent discipline, neurolinguistics.

Since contemporary linguistics is a fairly young discipline, one understands that the influence that it has exerted upon aphasiology is relatively recent. Reconsidering the classics of aphasia, it seems as if aphasiology had been waiting for a better understanding of language in order to advance, and just such an advance was offered by the development of linguistics. The influence of linguistics proper can be traced to the early portion of the twentieth century. There are three researchers (and their collaborators in the case of one) who are most salient in this introduction of linguistic methodology into the study of aphasia.

First of all is the contribution of Arnold Pick, whose monograph on agrammatism (1913) can be considered the first linguistic treatment of an aphasic syndrome. Pick's work was originally published in German and was only sporadically translated much later. It is perhaps largely because of this lack of translation that he did not enjoy a wider international audience and more prominent position in early neurolinguistics. However, a contemporary reading of his work shows just how modern his treatment really was. Goodglass and Blumstein (1973) have pointed out how much his hierarchical concept of language organization resembles the formulation in transformational-generative grammar of the late 1950s and early 1960s (Chomsky, 1957, 1964). As Spreen has noted in his chapter in Goodglass and Blumstein (1973), Pick was widely versed in the linguistic treatments of his day, and realized the potential that linguistics had to offer the study of aphasia.

The next contribution to early neurolinguistics was that of Alajouanine et al. (1939). Here was the first collaboration of a neurologist (Théophile Alajouanine), a psychologist (André Ombredane), and a linguist (Marguerite Durand). Their monograph demonstrates the fruitful result that can be derived with the combination of expertise from several disciplines. *Le syndrome de désintegration phonétique dans l'aphasie* is the first example of how a highly developed methodology from linguistics, that of early acoustic phonetics, can be used to arrive at a much more precise conception of a neuropsychological syndrome, that of phonetic disintegration (or "apraxia of speech" to some).

Although the phonetic instrumentation of the day was rather crude by today's standards—essentially sound vibrations traced onto a revolving Rousselot cylinder (either onto smoke-covered glass plate or a wax drum)—these authors were able to quantify such changes in speech production as longer and less precise articulation. This improvement of methodolgy employing objective measures instead of relying on subjective impressions is one that is not always used even today. One often finds theoretical statements based on subjective listener impressions, even though it is a well-known fact that speech is perceived in a categorical manner and differences in acoustic entities that are between categories are largely undetected by listeners (*see* Repp, 1983, for a review).

Finally, the third name that appears in the formative days of neurolinguistics is that of the Russian linguist Roman Jakobson. In a monograph that was first published in the German language in Norway in 1941 (translated into English in 1968). Jakobson formulated one of the first theoretical linguistic claims about aphasic speech: that the phonological dissolution of aphasic speech would follow, in reverse order, that of phonological acquisition. Although there is speculation along almost identical lines in Alajouanine et al. (1939), we have a more well-developed theoretical formulation of phonological acquisition in Jakobson.

Like Pick, Jakobson's influence seems to have been somewhat dampened by the rather late translation of his work into other languages such as French and English. Certainly, World War II did nothing to promote the popularity of texts written in the German language. Jakobson (1956) went on to formulate other theoretical claims about aphasia, such as a linguistic difference between the two main types of aphasia. Jakobson was one of the first linguists widely familiar with the work of neuropsychologists and neurologists, most notably that of his compatriot Luria. Luria himself also knew of Jakobson's work and employed his linguistic formulations.

These are the first three instances of what we shall call truely linguistic influences in the study of aphasia. The next period of influence would have to wait until the revolutionary reformulation of the concept of human language proposed by transformational grammar in the late 1950s by Chomsky. Chomsky's work, which brought about a much more explicit theory of normal language, gave the potential for much more testable proposals about pathological language. Today, we can see Chomsky's main contribution being a hierarchical view of language and an attack on the linear view that language was comprised of a complex set of learned associations, such as advocated by Skinner (1957).

Above, we have alluded to some of the covert influences of modern linguistics on human neuropsychology. We shall now turn our attention to the more overt influences in the form of specific theoretical frameworks that have been derived from linguistics. In order to organize such an enterprise, it will be necessary to divide our presentation into different linguistic aspects or levels. There are different ways of dividing these different levels, but most approaches agree on at least three distinctions: that of **semantics,** the level of meaning; **syntax,** the information conveyed by word order and sentence structure; and **phonology,** the level of individual language sounds (or phonemes). In the present treatment, two further subdivisions will be added: that of **morphology,** or the level of the internal structure of words (which is a level between that of phonology and that of syntax); and **phonetics,** which is the level that deals with the acoustic nature of language sounds, as well as the manner in which they are articulated.

2. Semantics

The area of semantics within linguistics has had somewhat less influence in the study of aphasia than have other linguistic levels. Surely part of the reason for this smaller effect is the lack of a strong theory of semantics, which would allow for generation of testable predictions, such as can be found at the level of syntax or phonology *(see below)*. However, there are some indications that this situation is changing and that stronger semantic theories (Montague, 1974) are being applied.

One concept that has been developed and has been borrowed from linguistics to develop more specific tests for aphasics is that of semantic features (Katz and Fodor, 1963). For an example of the concept of semantic features, we can take the words "mother" and "dog." One of the most salient differences between the meanings of these two words is that the former refers to a human being and the latter does not. Thus, these words are conceived to differ in the semantic feature for "humaneness," with "mother" being (+human) and "dog" being (–human). This notion of semantic features suggests that a group of words could be grouped on the basis of such a criterion. For example, given the words "man," "mother," and "fish" and asked to find the "odd man out," a subject would be expected to group "man" and "mother" together, both being (+human) and to choose "fish" as the odd member being (–human).

In a widely cited article, Zurif et al. (1974) compared the semantic clustering behavior of normal subjects and aphasic patients using such a task with the same test set of words. They found that Broca's aphasics grouped words together in a manner very similar to that of normal subjects, except that they apparently introduced a different feature of "ferocity" to group certain animals together, such as "tiger" and "shark," which was not used by the normal subjects. However, the responses of Wernicke aphasics essentially demonstrated no systematic pattern in their word groupings—a result that these authors take to reflect a semantic impairment in such patients.

Another study that tended to support the notion of a semantic deficit on the part of certain aphasics was that of Whitehouse et al. (1978). In this study, subjects were required to select a name for drawings of prototypical and nonprototypical objects. For example, they were shown a drawing of the form of a typical cup, except that the handle was missing, and asked to select a name for this item from the choice of "bowl, cup, basket." Although most cups do indeed have handles, normal subjects as well as Broca's aphasics still selected the word "cup" to describe such a drawing. Anomic aphasics with posterior lesions, in contrast, did not categorize such "fuzzy" items in the same manner. In this example, they may have chosen "bowl" probably because a bowl also typically does not have a handle, even though the general target form was that of a cup. Such results were interpreted by these authors to reveal an impairment in such patient's underlying semantic organization.

This study shows where a concept borrowed from linguistics, that of semantic features, has been employed in a test that revealed a change in patients' underlying conceptual organization. No previous study had separated a patient's ability to retrieve a name for an item from their underlying perceptual categorization. Such a study tends to demonstrate that semantic deficits in aphasia are probably not simply the result of an inability to retrieve a name for an item, but that it also seems to affect the manner in which perceptual information or semantic concept may be categorized and integrated.

It should be mentioned in closing this section that currently much interesting work is also being conducted on patients with lesions in the right hemisphere (and therefore generally not aphasic), since such patients seem to demonstrate semantic deficits (Hannequin et al., 1987). We can expect that, as linguistic conceptions of semantics are made more explicit in a manner that allows for more testable predictions, more work will be done at this linguistic level with neurologically damaged patients.

3. Syntax

One important contribution of syntax to studies of language pathology is to provide a hierarchical concept of sentence organization. Implicit in this conception is the notion that, beyond the simple left-to-right order of words in a sentence, there is a hierarchical organization that encodes structural information. Let us consider for a moment this hierarchical structure. In a sentence such as "The Dog chased the cat," if we were to make the first logical division in the sentence in the process of dividing it up into its component parts, it would be between "The dog" and "chased the cat." By "hierarchical structure," syntactic theory refers to our intuition that this division between what is referred to as the Noun Phrase (or NP) and the Verb Phrase (VP) is somehow more basic than the division between "The" and "dog." The way in which this hierarchical information is represented is by means of a "Syntactic tree" (which is similar to the sentence diagrams to which many of us may have been exposed in grammar school).

The important insight captured by syntactic trees is that the divisions nearest to the top are more fundamental than divisions near the bottom. Syntactic trees have been especially useful in pointing out structural difference inherent in the two different meanings of ambiguous sentences, such as "Flying airplanes can be dangerous." One meaning is that the act of flying airplanes can be dangerous, and the other is that airplanes that are flying can be dangerous to people on the ground. Such differences cannot be captured by the simple left-to-right order of words, which is the same for both meanings. Syntactic theory allows one to make predictions about linguistic behavior. For example, a sentence, such as "The dog chased the cat and then got punished," can be shown to have a more complex structure in that it is composed of two more basic sentences than an equally long sentence, such as "The big brown dog chased the small black cat." Syntactic theories allow us to predict that the former sentence will be more difficult for aphasic patients than the latter, because it is structurally more complex. Let us turn to some studies that have used syntax to study aphasic behavior.

Perhaps one of the most influential studies of aphasia at the level of syntax is that of Zurif et al. (1972). Up until this time, it was generally conceived that agrammatism, or the problem with word order and missing elements in Broca's aphasic's speech production, was the byproduct of a strategy employed by such patients to get across the content of the message using the least amount of effort by omitting nonessential words (Pick, 1913).[2] Thus, their problems in syntax were blamed on their difficulty in speech production, and not on a difficulty with appreciation of the structural information encoded in word order. The fact that such patients were usually quite proficient in understanding spoken language reinforced the notion that this disability was limited to the production modality.

[2]Agrammatism, more than any other aphasic syndrome, has attracted a linguistic sophistication in its investigation not previously attained in aphasiology. It is with the study of agrammatism that linguistic theory has itself begun to feel the reciprocal influence from the study of aphasia. It is well beyond the scope of this chapter to deal in any detail with the specific linguistic hypotheses that the study of agrammatism has led to. Recently, a volume has been published which deals with several of these linguistic accounts (Kean, 1985). The reader is also referred to Grodzinsky (1984) and LaPointe (1985). Kean's account (1977) will be briefly discussed below under morphology.

However, Zurif et al. (1972) succeeded in demonstrating that such patients may also have trouble comprehending the structural information given by word order. Their methodology employed a relatively easy task that aphasic patients are quite capable of performing. The subjects were requested to indicate which two or three word sentence fragments went together best. The advantage of such a task is that it allows patients the extra time that they might require in making responses. This may not be the case when using acoustically presented sentences where the auditory store may have faded before a patient can respond. It also does not require the patient to use his or her impoverished speech production system, and also therefore spares the patient the frustration of hearing his or her own poor production.

It has been shown that the responses of normal subjects reflect the structural information given by word order in sentences. For example, normal subjects would first of all group subject noun and verb together in the first cluster. However, such was not the case of the judgments of the aphasic patients in this study. As the authors note: "Quite clearly, the relatedness judgments of the control subjects were constrained by the surface syntactic properties of these four utterances, which those of the aphasic subjects were not." (Zurif et al., 1972, p. 411). The aphasic patients tended to violate the normal unity of noun and verb phrases. These and similar findings have been essential is reformulating the concept of Broca's aphasia to be more than a reflection of problems in language outputting. Broca's aphasia is conceived of more recently as more of a central deficit that can affect both production and comprehension.

Zurif et al.'s work raises the possibility that this information is not available or not employed by aphasic patients in the same manner as normal subjects. It demonstrated that the production problems of Broca's aphasics may also reflect a problem in apprehending linguistic structure rather than simply being a problem of meeting the needs of speech production. More recently, however, Linebarger et al. (1983) have shown evidence of some preserved syntactic judgment ability in agrammatic Broca's aphasics. Future work needs to define the limits of such ability.

The results of Zurif et al. (1972), as well as other important studies of syntactic comprehension in aphasia (i.e., Zurif et al., 1976; Caramazza and Zurif, 1978; Goodglass et al., 1979 and references therein) demonstrate that a linguistic formulation of tests offers the possibility of demonstrating deficits in aphasia that are not apparent from simple language evaluations—deficits that were previously ignored. Here we have an important example of how linguistic sophistication has changed the concept of aphasia and, consequently, the concept of language and the brain.

An example of a very recent approach to syntax in aphasia is that of Caplan et al. (1985). Here Caplan and coworkers demonstrate that using carefully constructed linguistic materials, aphasic patients' individual syntactic deficits begin to emerge. In other words, Caplan and collaborators have raised the possibility that patients, even with the same aphasic syndrome, may have isolated impairments with specific syntactic structures. What we are referring to by specific syntactic structures are differences in the "tree diagrams" *(see above)* for sentences. In other words, what Caplan et al. have shown is that certain aphasic patients seem to have problems with some syntactic structures and not with others, and that these structures are not necessarily the same ones that pose problems for another patient even with the same type of aphasia or damage to the same area of the brain. Thus, Caplan et al. did not find a correlation of syntactic impairment with either type of aphasia or lesion site. If, as it is usually construed, syntax is represented in the same manner in all subjects, such results suggest that aphasia is much more diverse and complex than has previously been suspected. Once again, such a study demonstrates that only increasingly sophisticated approaches can hope to gain new insights into the complex nature of aphasia, but the reward to be gained from linguistic sophistication in study of aphasia is also the added benefit of better appreciation of the intact language system.

Let us mention some overviews of the contribution that methods derived from syntax have made on the study of aphasia. One of these is a chapter on syntax in Lesser (1978), which is a good introduction to the different methods found in syntactic approaches and their main results, as well as the advantages and drawbacks of several different testing paradigms. Another more recent and more theoretical overview is to be found in Berndt and Caramazza's chapter in Sarno's (1981) volume. It should also be mentioned that there are several recent challenges to classic theories of syntax (e.g., Gazdar et al., 1985). Although the manner in which these theories would change the concept of language and syntax is not entirely clear, it could still be expected that, as such, theoretical reformulations become more integrated into linguistic theory at large and that they will also be adapted to study aphasia. We shall now turn our attention from the sentential level to the level of words and word formation.

4. Morphology

The place and function of morphology in linguistic grammar have been the matter of much discussion (Anderson, 1982). In fact, it is the relation between morphology and syntax that has proved one of the most recalcitrant problems of modern linguistic theory. It is because of this unclear status of morphology that work on aphasia in this area has often been considered as either phonology or syntax. As theories in morphology become more explicit, we can anticipate some significant contributions to aphasia from work at this level. Below, we shall consider some research that we feel to be important at the level of morphology.

Kean (1977, 1982) has formulated a linguistic hypothesis to explain the difference between items retained in agrammatic speech production vs those items that are omitted based on agrammatic speech corpora previously published in the literature. To summarize what is proposed in Kean's analysis, let us note that morphological processes affected in agrammatism (such as in inflectional processes) and function words that are omitted form a homogeneous class of words called "phonological clitics" at the phonological level of representation. The class of phonological words corresponds to the content words or "open-class words"; and clitics correspond grossly to function words or "closed-class words." A general definition of content and function words is given by Clark and Clark (1977):

> Content words are those that carry the principal meaning of sentence. They name objects, events and characteristics that lie at the heart of the message the sentence is meant to convey . . . Function words, in contrast, are those needed by the surface structure to glue the content words together, to indicate what goes with what and how. (p. 21.)

Bradley (1978) has opposed open-class and closed-class words in a recognition task. She found that for open-class words, normal subjects demonstrated a sensitivity to their relative frequency of occurrence, and their initial segments were playing a preeminent role in guiding recognition. Neither of these effects were observed with closed-class words. She concluded that these two word classes were recognized via two distinct systems.

Bradley et al. (1980) investigated the recognition of these two word classes by Broca's aphasics. According to this study, Broca's aphasics do not perform like normal subjects in the recognition task, since they show a sensitivity to frequency and the initial portion of the words for both word classes. Gordon and Caramazza (1982) have called the results of Bradley et al. (1980) into question, showing that the difference between the open- and closed-class vocabularies may be one of differences in distributional frequencies between the items of the two classes and not one of differences in the way they are processed.

Another means of demonstrating the difference between closed- and open-class vocabularies is to use nonword interference. That is, it takes longer to reject a nonword that begins with an actual word than it does to reject a simple nonword. Taft and Forster (1976) found an interference effect for open-class headed nonwords (e.g., "footmilge"), but not for closed-class headed nonwords (e.g., "thenmilge"). However, this result has been questioned by Kolk and Blomert (1985). These later authors attribute such effects to poor control of the word list, namely the fact that there were no real closed-class items included in the stimuli.

As can be seen, studies focusing on the difference between open- and closed-class items encounter a great deal of methodological problems. Another area of study in morphology has been in the problems that agrammatic aphasic patients have both in derivational and inflectional processes. Goodglass and Berko (1960) found that phonological complexity does not predict correct use of morphemes in aphasia. Moreover, this study suggested that syntactic and inflectional aspects of grammar could be selectively impaired. In analyzing a group of morphemes in their obligatory context in spontaneous aphasic speech, De Villiers (1974) found that some morphemes were more difficult than others. Thus, there was a hierarchy of difficulty, but it was not related to the one found in acquisition.

Turning to derivational processes in aphasia, Eling (1986) has raised the question whether derivational word forms are represented as independent lexical items. That is, are they recognized separately from their stem form in the lexicon? Since all subjects were sensitive to surface-form frequency (the frequency of the item itself) rather than base frequency (the frequency of the base from which the complex form is derived), he concluded that both normal subjects and Broca's aphasics recognize derivational word forms by construing them as separate items.

The study of aphasia at the level of morphology has been somewhat limited. As alluded to above, this is probably because of the ambiguous status of morphology within a linguistic grammar. Recent models, such as lexical phonology (Mohanan, 1982; Pulleyblank, 1983), provide an effective integration of the lexicon, phonology, and the morphology, but the interaction of these three components with syntax is still far from clear. As the role of morphology and the manner in which it relates to syntax and phonology become more clearly delineated in linguistic theory, we can expect more studies in aphasia to ensue. We shall now turn our attention to the phonological level—an area of linguistic inquiry that has undergone very rapid theoretical expansion over the last two decades.

5. Phonology

In linguistics, phonology is concerned with the organization of phonemes and syllables into words. This level is to be distinguished from the phonetic level, which is concerned with the set of articulatory gestures required in oral production *(see below)*. In aphasiology, a phonological deficit is characterized by the presence of phoneme substitutions, syncopations, and additions in the production of a subject not resulting from arthric difficulties. These phonological errors interest linguists as well as aphasiologists. From a linguistic point of view, these errors should be predictable on the basis of phonological models. In addition, aphasiologists are interested in establishing the deviant psycholinguistic processing responsible for these errors. In this section, we will concentrate on some of the phonological models that have been used for the description and understanding of phonological errors in aphasia.

French-language aphasiologists first turned their attention to "functionalism" because of a model developed by Martinet (1960) and Buyssens (1967), whereas early English-language work, such as Blumstein (1973), was directly inspired by Jakobson. Both models issued from Trubetzkoy (1958) and put emphasis on distinctive features, thereby providing an evaluation procedure for phoneme substitutions. Lecours and Lhermitte (1969), Blumstein (1973), and Nespoulous et al. (1984) have evaluated the distance, in number of distinctive features, between the target phoneme and the phoneme produced by the aphasic patient in a phonological error. They have all found that the distances tend to be small, generally no more than one or two distinctive features, even though there are differences in the feature system used and some differences according to the type of aphasia.

The model of generative phonology proposed by Chomsky and Halle (1968) has provided aphasiologists with new hypotheses and means of analyzing phonological errors. This model makes use of distinctive features, but also assumes two levels of phonological representations: the underlying representation (UR) and the surface representation (SR). The mapping from the UR to the SR is achieved by application of phonological rules. Schnitzer (1971) applied this theoretical framework to the phonological errors produced by an aphasic patient. Essentially, he attempted to infer the incorrect UR used by the patient from the observed incorrect SR. In some cases, Schnitzer attributed errors to a mistaken or a nonapplication of phonological rules. For this author, aphasic errors bear evidence of generative processing in phonological production.

Generative phonology (GP) was followed by a model called "Natural Generative Phonology" (NGP) (Hooper, 1976). The major differences between GP and NGP can be summarized by the following: (1) NGP refuses abstract URs and abstract derivations, and (2) NGP reintroduces the syllable as a phonological unit. Aphasiological studies conducted by Dressler (1979) and Kilana-Schoch (1982) were based on NGP. They emphasized the similarity between aphasic errors and phonological processes that were observed diachronically in the evolution of natural languages.

Perhaps the most important change in contemporary phonological theory is to be found in the important role assigned to the syllable, which is now considered an autonomous phonological unit. Although other approaches to phonology have also considered syllabic units, the syllable was a linear object as opposed to the tridimensional representation suggested in metrical phonology (Liberman and Prince, 1977). In Levin (1984), Grignon (1984), Archangeli (1985), and Halle and Mohanan (1985), a phonological representation lies on two planes: the syllabic plane and the melodic plane. These two planes are linked together by the so-called "skeleton," which might be viewed as a number of timing or prosodic units to which segments (vowel and consonant phonemes) are associated. The melodic plane encodes featural information about the segments, whereas the syllabic plane gives the hierarchical organization of segments into syllables. This new nonlinear (or tridimensional) system for phonological representations has been applied to the analysis of aphasic errors in Béland (1985).

One of the most important objectives in Béland (1985) was to investigate if such a three-dimensional model was needed to provide an adequate description of phonological errors. Indeed, the nonlinear representational model turned out to be essential to both the description and understanding of phonological errors made by aphasic patients. First, phonological error type is predictable in such a model on the basis of the syllabic organization of the target word. Secondly, aphasic errors respect syllabic constraints, which determine the nature of the segments that can intrude, be omitted, or substituted. For example, in the English word "blue," the "l" can only be substituted by an "r," which is the only other consonant that can play the same syllabic role. Here the "l" is the second member of a branching constituent called the "onset," and this position strictly limits potential substitutions.

In previous work reported in the first part of this section, aphasiologists considered a word as a linear segmental string and, thus, considered phonological errors as the result of simple concatenative operations. The possibility offered by this new theoretical framework is also to take into account the number of segments in a word, the featural content of these segments, and their syllabic role in the word as three independent variables. This gives rise to new concepts with regard to the origin of phonological errors in aphasic speech.

Phonology, as mentioned, has been undergoing a great deal of evolution in the past five years. We shall not end this section without mentioning the most recent development in nonlinear phonology—the theory of "charm and government" proposed by Kaye et al. (1985). In this model, there are no rules and no distinctive features—only elements and government relations between segments. Work is presently being conducted in our laboratory to apply this new theory to aphasiological data (Béland

and Lecours, 1987; Valdois, 1987). We now turn our attention to phonetics, or the sound level of speech.

6. Phonetics

A classical problem in the study of aphasia is that of distinguishing the level of error involved in literal paraphasias. That is, when an aphasic patient replaces the target word "dog" with a production that we, as listeners, perceive as "bog," is the error one of improper selection of basic speech elements of "phonemes," or one of faulty articulation of a properly organized response? This distinction is usually referred to as the one between the phonological or phonemic level, and the motor or phonetic level (*see* Blumstein, 1981 for discussion).

One approach to providing evidence about which level is involved is to make a detailed study of aphasic speech and compare it to that of normal speakers. The most direct comparison would be that of articulation itself. However, since such methods often involve X-rays or implantation of electrodes, little research has actually made such direct comparisons. More research has been conducted making acoustic comparisons of recorded pathological speech. Surely recording speech is much more likely to be accepted by aphasic patients who are already ill than is implantation of electrodes or cineradiography, for example. It should be pointed out that some less disruptive techniques have been developed, such as ultrasound (Keller and Ostry, 1983) and surface electrode myography (Shankweiler et al., 1968), which offer less invasive means for direct comparison of speech production. Although acoustic data has the advantage of being much easier to collect, it is still difficult to analyse and interpret. One large problem is that there are not well-defined standards for what constitutes "normal speech" production. There is a great deal of acoustic variation both between and within speakers.

In spite of such limitations, there has been a substantial number of acoustic–phonetic investigations of aphasic speech, which have greatly improved our understanding of aphasia.[3] We will

[3]A contemporary overview of phonetic approaches to aphasia entitled *Phonetic Approaches to Speech Production in Aphasia and Related Disorders* can be found in Ryalls, 1987.

review just a very few here. As mentioned above, Alajouanine et al. (1939) were the first researchers to use a comparative phonetic approach to speech production in aphasia. They were quite successful in giving a more precise descriptive characterization of what they called "phonetic disintegration" in aphasia. They provided some of the first data demonstrating acoustic differences between aphasic and normal speech production.

One of the rare studies to use the more direct comparison of speech production of normal and aphasic patients using electrogmyography (EMG) or recordings from speech articulator muscles is that of Shankweiler, et al. (1968). These authors demonstrated that muscle recordings from aphasic patients were both abnormal in form and highly variable compared to those of a normal speaker. Such results indicate that the problem in at least anterior or Broca's type aphasia is not simply one of confusing phonemes, but also one of disintegrated articulation.

Another important study of speech production in aphasia is that of Blumstein and her colleagues (1980). This study is interesting in that the authors were able to arrive at an operational definition of what would constitute a phonetic vs a phonetic error in the timing of vocal cord vibration ("voicing" or V.O.T.—voice onset time—which is the delay from the release of the consonant to the onset of periodicity of the following vowel) for stop consonants. These authors reasoned that productions of the wrong target phoneme that respected normal acoustic boundaries would be "phonemic" in nature, whereas productions that violated normal acoustic boundaries would be "phonetic." Notice that such a definition is conservative in its estimation of "phonetic" errors, because extreme phonetic deviations that in fact end up in the correct timing for another category will still be counted as "phonemic." Even though their definition has some such limitations, this study may be regarded as one of the first to try to effectively titrate out the contribution of the conceptual organization from that of the motoric realization in aphasic speech.

The authors found that, although both Broca and Wernicke aphasics had errors of both the phonemic and phonetic type, the Broca's aphasics were characterized by a significantly greater number of phonetic-type errors. This is an interesting result in that it seems to confirm the classic, but descriptive, notion that Broca's aphasics's problem is one of motor realization (as Alajouanine et al.'s (1939) term "phonetic disintegration" suggests).

Blumstein et al.'s (1980) results for stop consonant production can be compared with another acoustic-phonetic study that considered vowel production (Ryalls, 1986). This study attempted to eliminate obvious "phonemic" errors from the aphasic data and then extract the relevant acoustic information for vowels and compare them to those for normal subjects or the same task. Again both Broca-type and Wernicke-type aphasics were included in order to compare their respective performances.

Results showed very few significant differences in the acoustic characteristics of aphasic vs normal vowels. In fact, greater variability of these characteristics from repetition to repetition was the only significant difference between vowel production of the normal control subjects and that of both aphasic groups. There seem to be two important points that result from comparing this study to that of Blumstein et al. for stop consonant production. These are: (1) vowel production does not seem to lead to the same type or degree of acoustic disintegration in aphasia as that found for at least VOT in consonants, and (2) the type of speech disintegration that is found in vowels is not significantly different for Broca-type aphasics than it is for Wernicke-type aphasics. Recall that VOT characteristics of consonant production does distinguish these two aphasic groups. Of course, it should also be pointed out that it was essentially timing characteristics that were measured in the consonantal study, whereas it was spectral or frequency characteristics that were addressed in the vowel study. In fact, it may indeed be that poor timing integration is an important descriptive factor in Broca-type aphasics.

Such an interpretation receives some additional support from an additional study of consonant production in aphasia focusing on more static spectral characteristics (Shinn and Blumstein, 1983). In this study, the static spectral characteristics of consonant production were found to be essentially preserved and were not different in the Broca-type aphasics than in the Wernicke-type control.

These three studies taken together seem to indicate that the critical problem of Broca's speech production may be the precise timing requirements imposed by fluid speech production. Again we have an example of how methodology derived from linguistics, here that of acoustic-phonetics, has lead to both a more precise conception of the impairment entailed by aphasia, and to development of experimental hypotheses of increasing strength.

7. Conclusion

In conclusion, we hope that this chapter has been successful in demonstrating the diversity of the influence of linguistics on aphasiology in particular and on human neuropsychology in general. As has already happened in linguistics proper, each level of linguistics has resulted in its own fairly autonomous specialization in neurolinguistics as well. We can expect this specialization to continue, but hold a hope both for interaction of researchers working on each of the different levels, and between their disciplines of neurology, psychology, and linguistics. For not only is human language behavior hierarchical and highly specialized, but it is also greatly interactive. It seems that only a multifaceted and yet somehow eventually integrated approach can hope to understand the complexity of human language behavior and the nature of its representation in the brain.

References

Alajouanine Th., Ombredane A., and Durand M. (1939) *Le Syndrome de Désintégration Phonétique Dans L'aphasie* (Masson, Paris).

Anderson S. (1982) Where's morphology? *Linguistic Inquiry* **13**, 571–612.

Archangeli D. B. (1985) Yokuts harmony: Evidence for coplanar representation in nonlinear phonology. *Linguistic Inquiry* **16**, 335–372.

Béland R. (1985) Contraintes syllabiques sur les erreurs phonologiques dans l'aphasie. Unpublished Ph.D. dissertation, Université de Montréal.

Béland R. and Lecours A. R. (1987) Analyse phonologique des séquences d'approximations. Paper presented at ACFAS: Ottawa, May. Association Canadian Francais pour l'Avancement de la Science.

Berndt R. and Caramazza A. (1981) Syntactic aspects of aphasia, in *Acquired Aphasia* (Sarno M. T., ed) New York, Academic Press.

Blumstein S. (1973) *A Phonological Investigation of Aphasic Speech*. Mouton, The Hague.

Blumstein S. (1981) Phonological aspects of aphasia, in *Acquired Aphasia* (Sarno M. T., ed.) Academic Press, New York.

Blumstein S., Cooper W., Goodglass H., Statlender S. and Gottlieb J. (1980) Production deficits in aphasia: A voice-onset time analysis. *Brain and Language* **9**, 153–170.

Bradley D. (1978) Computational distinctions of vocabulary type. unpublished Ph.D. dissertation, M.I.T.

Bradley D., Garrett M. and Zurif E. (1980) Syntactic deficits in Broca's aphasia, in *Biological Studies of Mental Processes* (Caplan D., ed.) M.I.T. Press, Cambridge, MA.

Buyssens E. (1967) *La Communication et L'articulation Linguistique* (Presses Universitaires de France, Paris).

Caplan D., Baker C., and Dehaut F. (1985) Syntactic determinants of sentence comprehension in aphasia. *Cognition* **21,** 117–175.

Caplan D. (1987). *Neurolinguistics and Linguistic Aphasiology.* Cambridge University Press, New York.

Caramazza A. and Zurif E. (1978) Disociation of algorithmic and heuristic processes in language comprehension. *Brain and Language* **3,** 572–582.

Chomsky N. (1957) *Syntactic Structures* (Mouton, The Hague).

Chomsky N. (1964) *Aspects of the Theory of Syntax* (MIT Press, Cambridge, MA).

Chomsky N., and Halle M. (1968) *The Sound Pattern of English* (Harper and Row, New York).

Clark H. and Clark E. (1977) *Psychology and Language: An Introduction to Psycholinguistics* (New York, Harcourt Brace and Jovanovich).

De Villiers J. (1974) Quantitative aspects of agrammatism in aphasic. *Cortex* **10,** 36–54.

Dressler W. U. (1979) Experimentally induced phonological paraphasias. *Brain and Language* **8,** 19–24.

Eling P. (1986) Recognition of derivations in Broca's aphasics. *Brain and Language* **28,** 346–356.

Friederici A. and Schonle P. (1980) Computational dissociation of two vocabulary types: Evidence from aphasia. *Neuropsychologia* **18,** 11–20.

Gazdar G., Klein E., Pullum G., and Sag I. (1985) *Generalized Phrase Structure Grammar* (Harvard University Press, Cambridge Mass.).

Goodglass H. and Berko J. (1960) Agrammatism and inflectional morphology in English. *J. Speech Hearing Res.* **3,** 257–267.

Goodglass H. and Blumstein S. (1973) *Psycholinguistics and Aphasia* (Johns Hopkins University Press, Baltimore).

Goodglass H., Blumstein S., Gleason J., Hyde M., Green E., and Statlender S. (1979) The effect of syntactic encoding on sentence comprehension in aphasia. *Brain and Language* **7,** 201–209.

Gordon B. and Caramazza A. (1982) Lexical decision for open- and closed-classwords: A failure to replicate differential frequency sensitivity. *Brain and Language* **15,** 143–160.

Grignon A.-M. (1984) Phonologie tri-dimensionelle du Japonais. unpublished Ph.D. dissertation, Université de Montréal.

Grodzindsky Y. (1984) The syntactic characterization of agrammatism. *Cognition* **16,** 99–120.

Halle M. and Mohanon K. P. (1985) Segmental phonology in modern English. *Linguistic Inquiry* **16,** 57–116.

Hannequin D., Goulet P., and Joanette Y. (1987) *La Contribution de L'hemisphere Droit à la Communication Verbale* (Masson, Paris).

Hooper J. B. (1976) *Introduction to Natural Generative Phonology* (Academic Press, New York).

Jakobson R. (1941) *Kindersprache, Aphasie und Allegemeine Lautgesetze* (Uppsala, Sprakvetenskapliga Sallskapets Forhandlingar).

Jakobson R. (1968) *Child Language, Aphasia and Phonological Universals* (The Hague, Mouton).

Jakobson R. (1956) Two aspects of language and the two types of aphasic disturbances, in *Fundamentals of Language* (Jakobson R. and Halle M., eds.) Mouton, The Hague.

Katz J. and Fodor J. (1963) The structure of a semantic theory. *Language* **39,** 170–210.

Kaye J., Lowenstamm J., and Vergnaud J. R. (1985) The internal structure of phonological elements: A theory of charm and government. *Phonology Yearbook* **2,** 305–328.

Kean M.-L. (1977) The linguistic interpretation of aphasic syndromes, in *Explorations in the Biology of Language.* (Walker E., ed.) Bradford Books, Mongomery, Vt.

Kean M.-L. (1982) Grammatical representations and the description of language processing, in *Biological Studies of Mental Processes.* (Caplan D., ed.) M.I.T. Press, Cambridge MA.

Kean M.-L. (1985) *Agrammatism* (Academic Press, New York).

Keller E. and Ostry D. (1983) Computerized measurement of tongue dorsum movements with pulsed-echo ultrasound. *J. Acoustical Soc. America* **73,** 1309–1315.

Kilani-Schoch M. (1982) *Processus Phonologiques Processus Morphologiques et Lapsus dans un Corpus Aphasique.* Publications Universitaires Européennes, Série XXI Linguistique, Bern.

Kolk H. and Blomert L. (1985). On the Bradley hypothesis concerning agrammatism: the nonword-interference effect. *Brain and Language* **26,** 94–105.

LaPointe S. (1985) A theory of verb form use in the speech of agrammatic aphasics. *Brain & Language* **24,** 100–155.

Lecours A. R. amd Lhermitte F. (1969) Phonemic paraphasia: linguistic structures and tentative hypotheses. *Cortex* **5,** 193–228.

Lesser R. (1978) *Linguistics Investigations of Aphasia* (Elsevier, New York).

Levin J. B. (1984) Conditions on syllable structure and categories in Klamath phonology. Internal manuscript, M.I.T.

Liberman M. and Prince A. (1977) On stress and linguistic rhythm. *Linguistic Inquiry* **8(2)**, 249–236.

Linebarger M., Schwartz M., and Saffran E. (1983) Sensitivity to grammatical structure in so-called agrammatic aphasics. *Cognition* **13,** 361–392.

Martinet A. (1960) *Elements de Linguistique Générale* (A. Colin, Paris).

Mohanan K. (1982) Lexical phonology. unpublished Ph.D. dissertation, M.I.T.

Montague R. (1974) Formal philosophy: Selected papers of Richard Montague (Thomason R., ed.) Yale U. Press, New Haven.

Nespoulous, J. L., Joanette Y., Béland R., Caplan D., and Lecours A. R. (1984) Phonological disturbances in aphasia: is there a "markedness effect" in aphasic phonemic errors? in *Advances in Neurology, vol. 42., Progress in aphasiology.* (Rose C. F., ed.) New York, Raven Press.

Pick A. (1913) *Die Agrammatischen Sprachstorungen* (Studien zur psychologischen Grundlegung der Aphasielehre, Berlin).

Pulleyblank D. (1983) Tone in lexical phonology, unpublished Ph.D. dissertation, M.I.T.

Repp B. (1983) Categorical perception: issues, methods, findings. In *Speech and Language: Advances in Basic Research and Practice* (Lass N., ed.) New York, Academic Press.

Ryalls J. (1986) An acoustic study of vowel production in aphasia. *Brain and Language* **29,** 48–67.

Ryalls J., ed. (1987) *Phonetic Approaches to Speech Production in Aphasia and Related Disorders* (San Diego, College Hill Press).

Sarno M. T. (ed.) (1981) *Acquired Aphasia* (Academic Press, New York).

Schnitzer M. (1971) *Generative Phonology: Evidence from Aphasia.* (Penn. State University Press, University Park, PA).

Shankweiler D., Harris K., and Taylor M. (1968). Electromyographic studies of articulation in aphasia. *Arch. Physical Med. Rehabilitation* **49,** 1–8.

Shinn P. and Blumstein S. (1983) Phonetic disintegration in aphasia: acoustic analysis of spectral characteristics for place of articulation. *Brain and Language* **20,** 90–114.

Skinner B. F. (1957) *Verbal Behavior* (Appleton-Century-Crofts, New York).

Spreen O. (1973) Psycholinguistics and aphasia: The contributions of Arnold Pick, in *Psycholinguistics and Aphasia* (Goodglass H. and Blumstein S., eds.) Baltimore, John Hopkins Press.

Taft M. and Forster K. (1976) Lexical storage and retrieval of polymorphemic and polysyllabic words. *J. Verbal Learn. Verbal Behav.* **15,** 607–620.

Trubetzkoy N. S. (1958) *Grundzuge der Phonologie* (Vandenhoek & Ruprecht, Gottingen).

Valdois S. (1987) Analyse des perturbations phonémiques dans l'aphasie. Ph.D. dissertation, Université de Montréal

Whitehouse P., Carammaza A., and Zurif E. (1978) Naming in aphasia: Interacting effects of form and function. *Brain and Language* **6,** 63–74.

Zurif E., Caramazza A., and Myerson R. (1972) Grammatical judgments of agrammatic aphasics. *Neuropsychologia* **10,** 405–417.

Zurif E., Caramazza A., Myerson R., and Galvin J. (1974) Semantic feature representations for normal and aphasic language. *Brain and Language* **1,** 167–187.

Zurif E., Green E., Caramazza A., and Goodenough C. (1976) Grammatical judgements of aphasic patients: sensitivity to functors. *Cortex* **12,** 183–186.

Techniques for Imaging Brain Structure

Neuropsychological Applications

Terry L. Jernigan

1. Introduction

One of the principal goals of neuropsychologists has always been to establish relationships between the discernible qualities of brain and those of behavior. One avenue for this pursuit has been clinico-anatomic correlation, i.e., the search for brain–structural abnormalities occurring in association with specific behavioral aberrations. In the not-too-distant past, this search relied almost entirely on the neuropathological examination of autopsy material for information about the brain. There are at least two drawbacks to this approach. First, the psychologist is at a distinct disadvantage if, in order to answer his or her experimental question, concurrent behavioral and neuroanatomical assessments are required. Also, as most candid neuropathologists would agree, this literature has been characterized by considerable inconsistency, much of which may be attributed to the problems of representing and quantifying the complex morphological data that emerge from brain-cuttings.

Today, following over a decade of experience with in vivo tomographic medical imagers, we can obtain remarkably high-resolution images of the living brain, and the images produced reflect a growing number of neurobiological dimensions. Notwithstanding the still large discrepancy between the spatial resolution of magnified brain sections and that of in vivo brain images, these developments will certainly enhance the fortunes of neuropsychologists. In principle, at least, it is now possible, not only to study the behavior of the subject concomitantly with the imaging, but also with repeated examinations to study the development of changes in brain and behavior, and even in some instances to intervene, and then to assess the results of such interventions in both domains. The study of primary disorders of cognition, such as mental retardation and the dementias, are key areas of current neuropsychological research. The ability to observe how brain abnormalities evolve in relation to the evolving behavioral abnormalities is especially important in the study of these disorders, since they have very long courses, they are known to interact strongly with development and normal involution, and they often change very dramatically in character between the time of their emergence and the time of the patient's death (i.e., autopsy).

Intelligent exploitation of the exciting opportunities offered by in vivo brain imaging requires that we face some of the previously mentioned goblins that have harassed our colleagues, the neuropathologists and neuroanatomists, for many years. We must find ways to define, detect, and accurately measure the morphology present in these images. The aim of this chapter is to describe briefly the technical bases of the two major structural brain imaging methods: X-ray computed tomography (CT) and magnetic resonance (MR) imaging* and then to discuss some of the methodological issues and strategies relevant to their interpretation. Finally, some exciting prospects for the future are outlined.

2. X-Ray Computed Tomography of the Brain

A cranial CT image is a two-dimensional map of estimated X-ray attenuation for a sectional volume from the head. In order to compute such an image, a narrow X-ray beam is transmitted through the head, and the X-ray photons emerging on the other side are detected and counted. The reduction in the number of photons emerging relative to the number of photons emitted is the attenuation value. In practice, a row of separate attenuation estimates is collected, each of the values estimating the attenuation in a particular narrow strip of the volume. This row of values, collected from a set of adjacent strips, is called a projection. In order to estimate the attenuation values of areas deep inside the head

*The discussion below of the basis of computed tomographic imaging techniques was strongly influenced by lucid descriptions by William H. Oldendorf in his excellent primer *The Quest for an Image of Brain* (1980).

separately from those of peripheral areas, multiple projections are obtained at different angles through the object.

The reconstruction of two-dimensional distributions from such one-dimensional projections is the basis for all computed tomography. In order to gain an intuitive understanding of how this works, consider the following analogy: Imagine that the sectional volume one is measuring is simply from a cube of water, except that a single glass rod is held upright in the cube and runs through the water. If one measured the attenuation values in a projection along one side of the cube, confining the X-ray beam to a narrow slab of the cube, almost all of the attenuation values in the projection would be the same, i.e., the attenuation value produced by a strip of water. However, from the few strips through the cube that contained the rod, the attenuation measured would be quite different, because glass attenuates X-ray to a different degree than water. In this case, it would be easy to tell something about the contents of the section from the very first projection; it would appear that an object was present within specific strips, but it would not be clear whether the object was present throughout the whole length of the strips or, if not, where in the strips it was. Now imagine that one took another such projection along another side of the cube. The projection would be much the same as the first, most strips would yield attenuation values corresponding to water, but a few would show the alteration of attenuation caused by the presence of the rod. Combining the information in this projection with that from the first provides considerably more information about the position of the rod, since from the second angle the rod "casts its shadow", so to speak, in a different place.

One way of combining projections such as these is called back-projection. First, a matrix is constructed to represent the two-dimensional space imaged. Next, each attenuation value from the first projection is assigned to the whole row of matrix values corresponding to the strip that gave rise to it. Then, each attenuation value from the second projection (new angle) is added to the whole row of values corresponding to the strip that gave rise to it, and so on. In the case of the rod analogy above, since the strips containing the rod would always have elevated attenuation values, regardless of the angle from which the projection was collected, these successive back-projections would result in very high values in the matrix cells representing the position of the rod. The more projections used, the larger the contrast between the summed values in the rod area and those in the surrounding water. After all of the projections had been back-projected in this way, the two-dimensional matrix could be visualized and would appear as a blurry picture of the rod in the cube of water. CT reconstructions work essentially in this way, although the mathematical techniques for accomplishing this and the algorithms for deblurring and otherwise processing the images are very complex.

In the case of CT sections of brain, the images reveal the structure of the brain because of differences in attenuation of X-rays by different tissues. In Fig. 1, a mid-ventricular CT section of brain is shown through the ventricles. Skull bone and calcifications within the brain have very high attenuation values and appear white on the image, fluid is very much less attenuating, and the cerebrospinal fluid (CSF) in the ventricles and in the cortical sulci is very dark, whereas the soft tissues yield intermediate values with gray matter somewhat higher (brighter) than white matter. Modern CT scanners produce images with spatial resolution in the plane of section approximately 0.8 mm, full-width at half maximum. Section thickness can be varied, but sections thinner than about 5 mm show a noticeable reduction in the signal-to-noise ratio.

One unfortunate artifact in CT images is called spectral shift, or beam-hardening, artifact. It occurs because the X-ray beam is not monochromatic; that is, it has energy at more than one frequency. One can accurately compute the X-ray attenuation of a volume of tissue only when the frequency of the X-ray beam is known. Although the spectrum of the beam that is originally emitted can be determined, tissues attenuate X-rays of different frequencies to different degrees, so the spectrum of the beam will change as it passes through the tissue, and this change in the spectrum will be different, depending on which tissues are in the path of the beam. For this reason, the exact spectrum of the beam as it passes through the tissue is indeterminate. In practice, this results in artifactual attenuation values, especially at the interface of high with low density materials. CT images show an artifactual elevation of brain values near the skull, and on higher sections, where the skull is effectively thicker, all tissue CT values are higher. This artifact reduces the accuracy of qualitative and quantitative measurements of CT images, especially measurements of structures near the skull, such as the cerebral cortex.

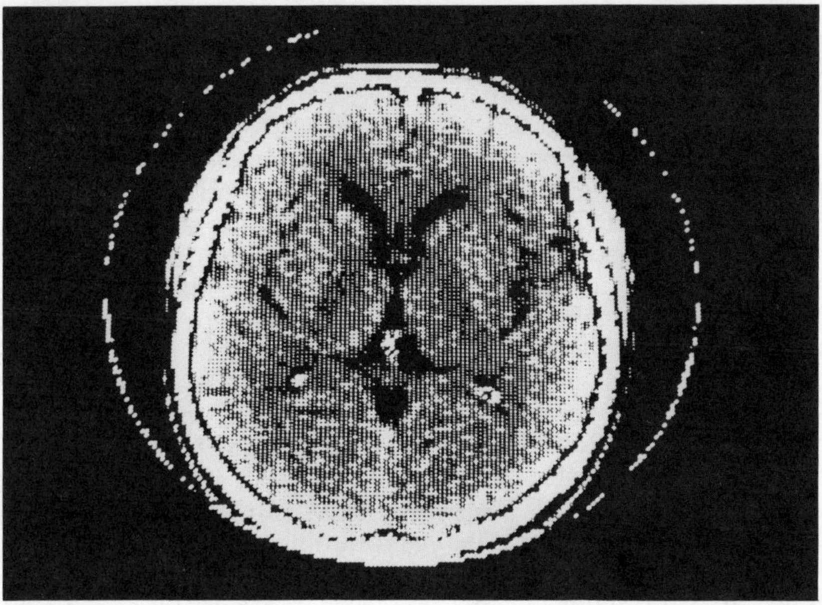

Fig. 1. A midventricular CT section at the level of the thalamus. Black areas are CSF in ventricles, Sylvian fissure, and cortical sulci. Darker gray areas are white matter, light gray areas are gray matter, and white areas are bone. Note the elevation of brain pixel values near the skull.

Because of the high signal contrast between bone and soft tissues, CT has always been an excellent method for visualizing cranial defects and for detecting the abnormal calcification sometimes present in certain types of brain tumors. This use of CT, however, is rarely important in neuropsychological studies. Much more relevant is the sensitivity of the technique to fluid increases. Because of this sensitivity, and because virtually all damage to the brain, either directly or secondarily, results in increased intracranial fluid, CT has very often been used to examine cerebral fluid spaces and abnormal accumulations of fluid.

In studies of patients with neurological disorders, CT has been used to confirm the presence of a focal abnormality, such as an area of infarcted tissue, and to aid in the more precise localization of the damage. This has led to increased information about the role of damage to specific brain structures in the development of aphasia

(Naeser, 1983; Kertesz, 1983b; Rubens and Kertesz, 1983), memory disorders (Ladurner et al., 1982; Ross, 1980a,b), and other cognitive dysfunctions (Heilman et al., 1983a,b; Kertesz, 1983c; Alexander and Albert, 1983).

In many disorders with prominent psychological symptoms, such as advanced aging, the dementias, chronic alcoholism, and the major psychiatric disorders, focal brain abnormalities are rarely observed. In these cases, degenerative brain changes, when they have been established to be present at all, are diffuse, and it is unclear what relationships exist between specific brain changes and behavioral abnormalities. Since diffuse brain degeneration is often reflected in enlarged cerebral ventricles, widened cortical sulci, or both, it has been possible to use CT to describe such changes in aging (Zatz et al., 1982a; Pfefferbaum et al., 1986; Barron et al., 1976; Gyldensted, 1977; Gonzales et al., 1978; Jacobs et al., 1978; Earnest et al., 1979; Meese et al., 1980; Jacoby et al., 1980), Alzheimer's Disease (Bird, 1982), Huntington's Disease (Barr et al., 1978; Terrence et al., 1977; Sax et al., 1983), alcoholism (Jernigan et al., 1986), and major psychiatric disorders (Pearlson et al., 1981; Weinburger, 1982; Scott, et al., 1983). In all of these disorders, statistically significant correlations have been reported between psychological or cognitive measures and measures of cerebral fluid spaces, but these correlations have in almost all cases been very modest.

Studies such as these, of ventricles and sulci, are often unsatisfying to the neuropsychologist, because no one presumes that behavioral effects are mediated by fluid changes *per se*. The assumption is that specific changes in the brain alter the function of the structures affected and, in doing so, produce behavioral symptoms. Increased fluid is assumed to be a secondary change, and only in rare cases does the location of the fluid increase strongly implicate a location for the parenchymal changes. Partly for this reason, several investigators have attempted to measure the CT attenuation values in specific brain regions directly. The rationale here is that cellular changes will result in an alteration of the density of the tissue that will be detectable in a controlled study. It is hoped that, if such abnormalities can be detected in a group of psychologically impaired subjects, perhaps regional patterns within the brain will implicate specific systems in the development of the symptoms.

Reduction of tissue CT values has been observed in normal aging (Zatz et al., 1982b). Although most studies of demented elderly have suggested reduced tissue CT values, especially in periventricular white matter, some reports are of increased values or no difference (Jernigan, 1986; Naeser et al., 1980; Bondareff et al., 1981; Wilson et al., 1982; Bird, 1982; Steingart et al., 1987; Rezek et al., 1987; McQuinn and O'Leary, 1987). Rarely have local changes been related to specific deficits; however, Jernigan (1986) reported a dissociation between the cognitive impairments associated with frontal lobe tissue changes and those associated with temporal lobe changes in a group of patients with gradual cognitive deterioration.

In a recent study of local CSF volumes and tissue CT values in amnesia (Shimamura et al., 1988), Korsakoff patients were compared to alcoholic and nonalcoholic controls. Both amnesic and nonamnesic alcoholics had increased frontal and peri-sylvian sulcal fluid, but the Korsakoff patients showed abnormalities beyond that seen in alcoholism alone. These more specific abnormalities were observed in third ventricle and sylvian fissure size, and in caudate and thalamus. When the memory scores of the Korsakoff patients were correlated with tissue abnormalities, the thalamic, but not the caudate, measure showed significant association.

Unfortunately, CT studies such as this one have limited utility for establishing specific neuro–behavioral relationships. Many structures of interest are not well visualized with CT, and the technique is not particularly sensitive to subtle abnormalities, even in the visible structures. The brain-imaging method described below, MR, has numerous advantages over CT.

3. Magnetic Resonance Imaging of the Brain

Although chemical analysis of biological tissues using nuclear magnetic resonance began in the 1940s, attempts to use the technique to obtain regional maps, or images, of chemical properties began only a couple of decades ago. Since that time, advances in high-speed computation and three-dimensional reconstruction techniques have contributed to the development of high-resolution imaging of magnetic resonance signals from the human body.

Theoretically, one can use MR to study the distribution and behavior of many of the magnetized nuclei in the body, and at present, several nuclei are being imaged experimentally. Because hydrogen atoms (protons) are extremely plentiful in the body, are strongly magnetized, and have low mass, they are an ideal source of MR signals, and most MR imaging of the human body today is proton MRI. The body is placed in a strong static magnetic field, so that the magnetized protons align with the magnetic moment of the field. In the presence of this field, the protons tend to precess, or wobble, like a tilted, spinning top, in a circular course about the longitudinal axis of the magnetic field. For a given field strength, the proton has a characteristic frequency of precession. This is called the resonant frequency of the nucleus. If a weak, rapidly alternating electromagnetic (RF) signal at this frequency is then passed through the field, the protons will absorb energy and precess through a wider circle. Having absorbed energy as a result of their perturbation by the RF pulse, they emit the energy at the resonant frequency when the pulse is discontinued. At high magnetic field strengths, the proton emits energy at a frequency in the short-wave radio spectrum.

The spatial information in MR imaging is obtained by producing magnetic field gradients within the volume to be imaged. Since the proton's resonant frequency is a function of field strength, this means that protons in different parts of the field will absorb and emit energy at different frequencies. Only protons along a particular line in a plane in the imaged volume will emit energy at a given frequency, so the signal strength at that frequency is a strip measurement much like the strip attenuation values in CT. By measuring the signal strength at different frequencies, a projection of signal strengths from adjacent strips may be used to reconstruct two-dimensional maps.

The strength of the signal emanating from the nuclei is related to the concentration of protons in the volume of tissue within the controlled magnetic field. When the pulse is discontinued and the protons emit their signals, they revert from a high-energy state to their equilibrium, low-energy state. The rate at which this return to equilibrium, or *relaxation,* occurs can be measured and reveals additional information about the composition of the tissue within the field. Actually, there are two measureable components to this return to equilibrium: one exponential time constant describes the return to magnetic equilibrium in the longitudinal plane of the magnetic field, and a second describes the return to equilibrium in the transverse, or x-y plane. These parameters are generally referred to as T1 and T2, respectively. Fortunately, these rate constants are influenced differently by such tissue characteristics as temperature, viscosity, protein content, and the magnetic effects of neighboring atoms.

Images of the proton signals from the human head contain remarkable anatomical detail, as can be seen in Fig. 2. Never before has it been possible to examine so closely the structure of the living human brain, undistorted by the obscuring effects of the surrounding bony cranium. The anatomical detail in the images is the result of the technique's sensitivity to tissue variations in proton concentration (mostly in water molecules) and to its sensitivity to the changes in the proton's behavior in different biochemical environments. Images like those in Fig. 2 are essentially maps of the distribution of water in the brain, but they contain much information about alterations of the tissue by disease processes and the precise locations of such alterations. As an added bonus, the technique, unlike its predecessor, CT scanning, involves no ionizing radiation and has no known biological hazards, so it may be repeated.

Images produced by MR are usually based on both T1 and T2 values, but they vary in terms of the weighting of the two parameters. The images labeled B and E in Fig. 2 are heavily T2 weighted, whereas C and D have relatively more T1 weighting and A has even more. This "relaxation" information provides much of the anatomical detail and sensitivity to tissue abnormalities observed with the technique. To estimate these parameters accurately, however, specific pulse sequences are required and adequate imaging time is critical. In practice, the MR examination usually must be tailored to provide as much of this information as possible in the amount of time that can reasonably be allotted or that can be tolerated by the patient, who must remain very still during the exam. It is rarely possible to obtain all of the information that the technique could provide in any examination or even in several.

Although MRI has been available for clinical studies of patients for only a few years, several interesting observations have been made already. One group of studies focuses on the morphology of the brain. Because of the dramatic increase in anatomical detail afforded by MR brain images, it is now possible to delineate accurately the borders of brain structures. This makes it possible to

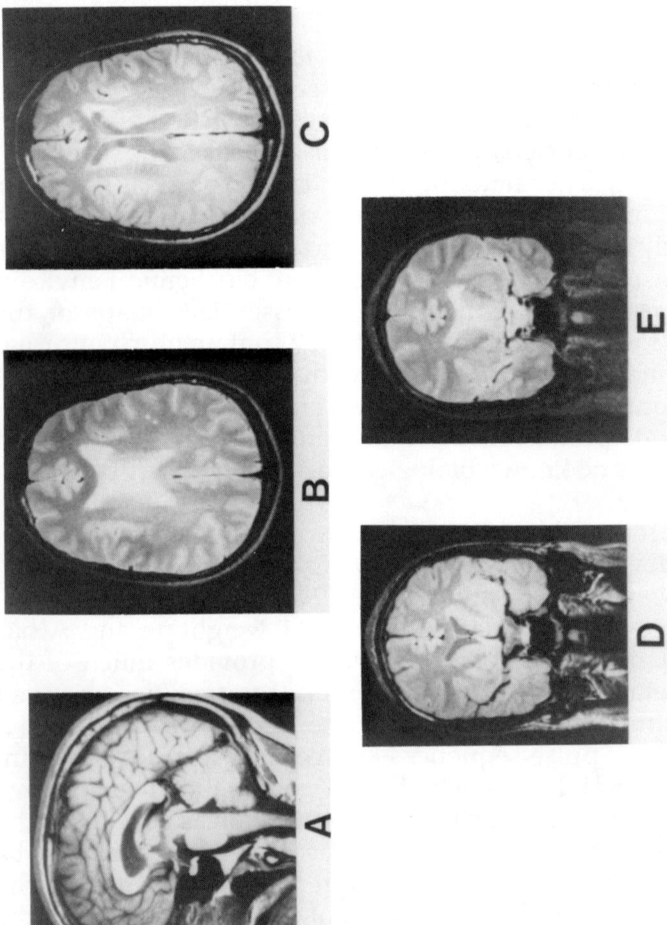

Fig. 2. Representative MR images in different imaging planes. A: Sagittal section, SE 600/20. B: Axial section, SE 2000/70 (T_2 weighted). C: Axial section, SE 2000/25 (proton density weighted). D: Coronal section, SE 2000/25 (proton density weighted). E: Coronal section, SE 2000/70 (T_2 weighted). All sections 5-mm thick, matrix 256 × 256.

examine the structures for evidence of abnormal size or shape. Several investigators have begun to examine the brains of schizophrenics, and preliminary studies suggest that the cerebrum, and particularly the frontal lobes, may be reduced in size (Andreasen et al., 1986), and that abnormalities in the shape of the corpus callosum may also be present in some schizophrenics (Nasrallah et al., 1986). These early findings suggest that some alteration, possibly in development, has changed the organization of cerebral subsystems within the brain in schizophrenics.

Another important morphological observation has been made in a group of autistic individuals (Courchesne et al., 1987, 1988). A reduction in the size of a part of the cerebellar vermis has been measured in a large proportion of these subjects. The abnormality appears to be the result of hypoplasia of the region rather than shrinkage, and differs from cerebellar abnormalities observed in several other disorders known to affect the cerebellum. The particular part of neocerebellum implicated has been linked in animal studies to many of the behavioral functions affected in autism. One of the most exciting clues from this finding may emerge from a description of the pattern of affected and unaffected portions of the cerebellum. The pattern narrows the point in development during which the abnormal event or events acted to disturb the growth of the affected structures. This finding could help to focus on the critical point in brain maturation when some process, such as toxicity, virus, or injury, could produce this syndrome.

Other studies have exploited the sensitivity of MR to detect subtle tissue changes. T2 weighted images are especially likely to reveal subtle changes in proton relaxation. Pathology studies have confirmed that signal abnormalities on these images are often associated with focal ischemia, demyelination, or gliosis (Awad et al., 1986a). These changes occur as a result of vascular damage, multiple sclerosis, old injury, inflammation, or viral infection. Many of these abnormalities are not discernible in life by any other means.

Since the advent of MR, such abnormalities have been reported frequently in many clinical groups, including nonsymptomatic older people (Brant-Zawadzki et al., 1985; Awad et al., 1986b; Agnoli and Feliciani, 1987). After much initial confusion, we are now beginning to understand them better. Within normal volunteers, they appear to be very rare in individuals under 50 yr of age, and to be more and more common after that age. Also, within normal elderly, they are quite strongly associated with vascular risk factors, suggesting that they may reflect some latent cerebrovascular abnormality. There is still considerable uncertainty about the significance of these signal abnormalities in individual cases, especially when they occur in association with other central nervous system abnormalities.

Such signal abnormalities have recently been detected in a large proportion of patients with bipolar affective disorder (Dupont et al., 1987). These patients averaged only 38 yr of age, and none of their age-matched controls had such brain abnormalities. This was a very surprising result, and its implications are not yet understood. The patients with the abnormalities were no older that the other patients, but they did seem to have more severe illnesses, as reflected in a larger number of hospitalizations. Perhaps what has been detected is some degenerative process manifesting as an emotional dyscontrol syndrome, or maybe the remnants of old injuries to the nervous system that prevent the normal function of certain brain systems. It is also possible that the abnormalities reflect treatment effects. Only longitudinal studies are likely to reveal the specific relationship of these abnormalities to the symptoms and etiology of bipolar affective disorder.

In an interesting study of patients with histories of Wernicke's encephalopathy, MR imaging revealed an apparent reduction in the size of the mamillary bodies (Charness and De LaPaz, 1987). Unfortunately, only limited information was provided about the cognitive function of the patients, so the relationship of this abnormality to memory impairments could not be determined.

In the following sections, a number of methodological problems relevant to this research area are discussed. When possible, suggestions are made for addressing these problems, or at least reducing their effects on the results.

4. Image Artifacts

The matrices of values underlying CT and MR images are subject to numerous artifacts. The beam-hardening artifact of CT, described above, is just one of these. In MR imaging, imperfections in the structure of the magnetic field and in the radiofrequency pulses give rise to distortion of image values. In practice, even when the imaging protocol is carefully standardized, substantial fluctuations occur in the signal values, such that the signal strength in CSF, for example, will vary in different parts of the image, and will be even more variable from one examination to the next.

In most cases, such artifacts increase the "noisiness" of measures of brain morphology. The resulting loss of measurement reliability reduces the sensitivity of the techniques for detecting morphological abnormalities. In some cases, however, such artifacts may even act to generate spurious "findings." The following examples underscore the importance of understanding and considering the effects of such artifacts in neuropsychological research.

As mentioned above, CT values near the skull are artifactually higher than those farther removed, and values in sections nearer the vertex are higher than those in lower cerebral sections. This complicates the comparison of signal values from different brain regions. Some investigators have attempted to detect abnormalities in brain tissue by measuring the CT attenuation values of the tissue. One commonly used method is to sample CT values in white matter areas adjacent to the ventricles. If, however, samples are located by reference to the ventricular borders, such samples will tend to be located nearer to the skull in patients with enlarged ventricles than in those with small ventricles. Since CT values are higher near the skull, the patients with enlarged ventricles will seem to have increased tissue values. This may account for some findings of increased tissue CT values in groups of demented patients with large ventricles. Although the group difference in this case occurs because of a real morphological abnormality in the patients, the interpretation of the result as evidence of altered tissue composition is incorrect.

A similar mistake sometimes occurs when the effects of partial voluming are not adequately considered. Partial voluming, which occurs in both CT and MR, refers to the effect on image values of the presence of different tissues within the volume element, or voxel. Since the voxel is usually about 1 mm square in the image plane and several mm deep, it frequently contains more than one tissue. The different tissues within the voxel will contribute to the summed signal value in the approximate proportion of their quantities (strictly speaking, some nonlinearity occurs for signals emerging from different depths within the voxel). Thus, the CT value from a voxel wholly within a ventricle will be characteristic of CSF, and a nearby voxel in the adjacent white matter will have a value characteristic of white matter, but a voxel on the edge of the ventricle will have a value in between the two since it will contain both tissues. Some investigators have attempted to measure the signal value of brain tissue by averaging the values of all voxels not in fluid spaces, the fluid voxels being removed by elimination of all those with fluid values. The problem occurs because voxels at the edges of fluid spaces will have altered signal values because of the presence of some fluid, but will not meet the criterion for elimination. The larger the fluid spaces, the greater the number of such altered "brain" voxels and the greater the contamination of the "brain" average resulting from the presence of partially volumed fluid. Thus, again, patients with more intracranial fluid will appear to have altered brain tissue relative to those with less.

Both CT and MR imagers are subject to drift in signal values because of varying calibration. Also, hardware and software updates may result in subtle, but significant changes in the image values. For this reason, it is important that the investigator attend to possible confounds between the experimental variables in the study and the time of imaging. Controls should be scanned concurrently with patient groups, right hemisphere patients concurrently with left hemisphere patients, older subjects concurrently with younger subjects, and so on, depending on the study. Already, studies have appeared in which patients were studied longitudinally and changes in signal values at followup were attributed to the course of the illness or intervening therapy. Possibly, the changes in signal values at followup resulted from calibration drift in the imager. It is critical that such studies include appropriate control measurements at followup.

Even when care has been taken to avoid confounds in the design, such signal variations from one examination to the next can represent a large source of irrelevant variation in the brain measurements. In some cases, correction of signal values can be accomplished by collecting standardization values at each examination. Some investigators have accomplished this by imaging a standardization object during every brain imaging session and then using the values from this object as a reference for correcting tissue values. In the CT study of Korsakoff's patients described above (Shimamura et al., 1988), CT values from fully volumed samples of CSF were used to correct tissue CT values taken from different brain structures.

When measuring MR image values, such calibration is especially important, since signal values exhibit dramatic spatial distortions and even greater variation across examinations. In fact, this variation in the absolute values obtained in the brain is generally prohibitive of study designs calling for comparisons of MR signal strengths. Even studies using more time-intensive procedures for estimating absolute T1 and T2 values have suggested disappointing instability in these computed values. Perhaps with continued improvements in the instrumentation, direct comparisons of MR signal values will be feasible in the future.

5. Correlation and Localization

Several general problems arise when attempting to correlate the results of brain imaging with neuropsychological variables. The first of these is the definition of the "result" itself. The human brain is exceedingly complex, and not easily described with a small number of variables. Some studies focus on a particular aspect of the brain's morphology, such as ventricular enlargement, but even for these studies there is little consensus about the best, or most sensitive, measurements to take. It is not unusual for 8 or 10 separate measurements of the ventricular system to be made in a single study of ventricular enlargement, perhaps as an attempt to "cover all the bases." Studies with more descriptive aims, or with broader hypotheses, may attempt to characterize many other dimensions of brain morphology as well, and it is easy to imagine how literally dozens of brain measurements could emerge from a single CT or MR examination, each to be correlated with a set of neuropsychological variables. The statistical problems associated with so many dependent variables are substantial, to say the least. This problem, although not specific to brain-imaging research, seems to be endemic in this area.

Neuropsychologists, with their multiple test instruments and computed indices, are no strangers to the problem of statistical test proliferation. There is continuing controversy about how to handle statistical analysis of results from test batteries, for example. To narrow the focus here, however, the discussion will emphasize attempts to relate a single functional variable to measures of brain morphology. Assume that an investigator wants to determine the relationship between cerebral atrophy and a measure of recognition memory in amnesics. Investigators interested in cerebral atrophy commonly make measurements of the ventricular system and of the cortical sulci and fissures. The form of the measurements may be subjective ratings, linear distances, or computed area or volume estimates. Often, as mentioned above, at least 10 separate brain measures are obtained for each subject. Frequently, the investigator simply computes 10 correlation coefficients to test his or her hypothesis that recognition memory is impaired in those subjects with cerebral atrophy. Any correlation with $p < .05$ under the null hypothesis is reported as confirming the hypothesis. This practice is clearly unacceptable. The probability of obtaining this "confirmatory" result is about .50 if the variables are all independent random variables. Although the brain variables probably are not independent of each other, their statistical relationships to each other are rarely known in such studies.

Why are mistakes like the one described above so common? Perhaps the most important answer has to do with construct validity. Although psychologists are well aware that a construct such as "recognition memory" must be clearly defined and operationalized if hypotheses about it are to be tested, they may ignore these considerations with a construct, such as "cerebral atrophy." The latter, although it may seem relatively concrete, is, in fact, a weakly validated construct by psychological standards. Let us assume that an investigator decides to define cerebral atrophy in terms of ventricular enlargement and to estimate such "enlargement" from the present ventricular volume. Then the question of how to compute an accurate and reliable measurement arises. The current consensus is that volume estimates from areas are more accurate than those from distances or ratings. Although the tasks of defining the construct clearly, designing a measurement likely to be sensitive to it, and establishing reliability for that measure are arduous, they are more likely to be scientifically fruitful than is the practice of collecting lots of inferior measures.

A second issue relating to the design of correlative studies is the problem of mismatch between hypotheses and tests. In the study described, the *a priori* hypothesis may, at least implicitly, be more complex than it at first seems. Often, multiple measures, for example of the ventricles, are collected, not only because they are different ways of assessing overall size, but because they are believed to measure different aspects of the enlargement, one of which may be more important for recognition memory than the others. A common example is the inclusion of a third ventricular measure as a more specific index of diencephalic atrophy. These more complex hypotheses, encompassing localization as well as correlation, may sometimes not be uncovered until the discussion section of a paper when, to the reader's surprise, the single "significant" correlation is interpreted as confirming both that ventricular enlargement is associated with poor recognition memory, and that the relationship is specific to third ventricular changes (diencephalic atrophy). The problem, of course, is that the analysis conducted is no more adequate as a test of this more complicated hypothesis than it was of the simpler one, and the result does not confirm either. The point here is that, as in all other psychological experiments, the actual hypothesis must be clearly stated and then formalized in a specific test that meets the necessary conditions for statistical inference. Hypotheses about localization should be explicitly stated at the outset, and the tests conducted should specifically test "localization."

In truth, proving localization, i.e., demonstrating that there is a specific relationship between a brain structure or group of structures and a cognitive function is an extremely difficult task. Kertesz (1983a) has provided a very helpful discussion of some of the conceptual issues involved. Suggested here is an approach that begins by scaling down the explicit hypothesis of a localization study to one that can reasonably be tested within a single study. This should help to reduce the confusion that often sets in when neuropsychologists try to relate their results to their own models and to those of others.

To illustrate the approach, let us take another example. Again, the construct of interest is recognition memory. The working hypothesis of the investigator may be that a critical role in this function is played by the hippocampus. A test of this hypothesis is beyond the scope of a single (feasible) neuropsychological study. Suppose, however, that the investigator recruited a group of patients with mild to severe deficits on a test of visual recognition memory in whom he or she suspected atrophy of the hippocampus might be present. A study of these patients with MR might provide some indication about whether poor scores were associated with hippocampal damage. Since the study can only provide evidence relevant to those neurobehavioral variables assessed, the selection of variables for the study is a critical decision. Measuring only hippocampal volume for correlation with the memory scores actually provides no evidence relevant to localization. If the correlation is *not* significant, the questions raised by all negative findings, i.e., about measurement sensitivity and statistical power, prevent any localization inferences. If it *is* significant, the possibility remains that many psychological tests unrelated to memory may have also shown correlation with the hippocampal measure, and/or that reduced volume of other brain structures, had they been measured, would have shown an association with poor visual recognition. Given these inferential constraints, it would seem that proof of localization would require measuring all possible structures and all possible functions, a study that, if not unthinkable, is certainly not practical. In fact, localization is at best only meaningful relative to some particular standard of equipotentiality. Correspondingly, a meaningful localization hypothesis must state the standard against which localization is being measured. A practical solution is to partition the functional and structural domains into a few relatively separable parts and define localization in terms relative to these parts.

As an example, the investigator could define the visual recognition function as distinct from visual discrimination functions. Now the localization hypothesis can be formalized: It is that hippocampal atrophy will have a stronger association with poor visual recognition than with poor visual discrimination. Note that it is the difference between the correlations that is critical to the test. Unfortunately, even if the test is passed, little can be concluded. It could be argued that visual discrimination is simply an easier test than visual recognition, and that atrophy anywhere in the brain would impair recognition performance more than discrimination. The best defense of localization is a double dissociation. If discrimination is a separately localizable function, it should be more vulnerable to damage in another part of the brain. A double dissociation hypothesis might be the following: Although atrophy of infero-temporal cortex will be more strongly associated with discrimination than recognition scores, hippocampal atrophy will show the opposite pattern of association. Although this is a stringent test, if passed, it demonstrates that there definitely exist functional differences between the two tests that relate to functional differences between the two brain structures.

One weakness of the double dissociation hypothesis as stated is that the conditions of the hypothesis will only be met if the tests and brain measures used produce relatively independent variables with simple factor structure. In other words, each variable must measure what it is supposed to measure and little else. Sometimes it is a sensible and more powerful strategy to use a technique, such as canonical correlation, to construct independent variables from the original, less "pure" variables. Complex multivariate methods, such as canonical analysis, should be used with caution, but a double dissociation established with this method offers the additional advantage of providing clues about how more sensitive tests of the underlying functional variables, or more sensitive measures of the neuropathological processes, might be devised. The use of this method in neuropsychology is described in more detail in Jernigan (1986).

6. Future Prospects

New MR techniques under development provide some of the most exciting near-future prospects for studying neurobehavioral disorders. One of these is in vivo spectroscopy. MR spectroscopy has long been used in vitro to provide biochemical analysis of tissue. This is possible because, although the resonance frequencies of nuclei of different elements (or even different isotopes) are quite distinct, for a given nucleus, small differences in the frequency are induced by variations in the chemical environment of the nuclei. For this reason, a spectrum of the energy emitted by the resonant phosphorus-31 nucleus, for example, has several peaks. Biochemists have been able to identify the peaks as corresponding to phosphorus contained in certain compounds. The relative sizes of the peaks in the spectrum reflect the concentrations of the different compounds in the tissue sample. This method can be used to monitor metabolic processes and detect metabolic changes associated with biochemical interventions.

By using surface coils near the skull, spectra may be obtained to assay the phosphorus compounds in a volume of tissue under the coil. In this way, actual chemical analysis of internal tissues can be obtained noninvasively. Advances in MR technology designed to make these methods more practical include the development of higher field strength imagers (because high-field strengths are needed to obtain well-defined spectra) and improved methods for shaping the magnetic field, so that spectra can be obtained from specific, localized regions in the brain.

This method is already being used in studies of brain–behavior relationships. In a recent investigation, in vivo P-31 spectroscopy was used to show that lithium-induced inhibition of a brain enzyme results in an increase in the level of a phosphorus-containing metabolite in the brain. This change is considered a possible mediator of the poorly understood therapeutic action of lithium in bipolar affective disorder (Renshaw et al., 1986).

MR contrast agents represent another promising development. Nontoxic metal ions have been adapted for injection into the body for providing magnetic contrast. They work by changing the local magnetic environment, thus altering local proton relaxation. Such agents may be used to examine the local integrity of the blood–brain barrier, for example.

Newer agents under investigation, however, may go beyond these early applications. Recently, researchers interested in the benzodiazepine GABA receptor linked the ligand clonazepam with a complex that lengthens T1 and T2 values. They demonstrated with in vivo experiments in rabbits that parenteral administration of the compound led to a regionally variable alteration of brain signal values, with greater alterations in regions where specific benzodiazepine binding is expected to occur (Coffman et al., 1986). If further developed, this technique could lead to the use of MRI for receptor labeling in humans. Such experiments could yield important information about possible alterations in receptor density in neurobehavioral disorders and about the action of drugs in these disorders. For example, such studies might help to resolve some issues surrounding the role of reduced cortical monoamines in producing the various cognitive deficits of primary dementias. The advantage of such studies over PET studies would be the improved localization made possible by higher spatial resolution in MRI.

7. Conclusion

It is hoped that the preceding discussion of neuropsychological brain-imaging research will underscore both the exciting possibilities and the critical need for experimental rigor. The relevant technologies in this field are progressing so rapidly that the prospect of staying abreast of new developments can be quite intimidating. An unavoidable consequence of the growing complexity of the techniques is that sound research in this area cannot be accomplished without the active collaboration of several disciplines. When different experts are each contributing a "piece" of the study, the more "holistic" aspects may go unattended. Several of the methodological points raised in this chapter relate to these more global properties of the research: the design, the inferential process, and the generation of testable hypotheses. It is one of the challenges in this new field to achieve high methodological standards in a research milieu within which no member of the team is expert on all aspects of the study. In this regard, it is important that each member keep a vigilant eye on the overview, as well as the methodological details, of the research.

References

Agnoli A. and Feliciani M. (1987) Nuclear magnetic resonance imaging in the aging brain. *Gerontology* **33,** 247–252.

Alexander M. P. and Albert M. L. (1983) The anatomical basis of visual agnosia, in *Localization in Neuropsychology* (Kertesz A., ed.), Academic, New York, pp. 393–418.

Andreasen N. C., Nasrallah H. A., Dunn V., Olson S. C., Grove W. M., Ehrhardt J. C., Coffman J. A., and Crossett J. H. W. (1986) Structural abnormalities in the frontal system in schizophrenia. *Arch. Gen. Psychiatry* **43,** 136–144.

Awad I. A., Johnson P. C., Spetzler R. F., and Hodak J. A. (1986a) Incidental subcortical lesions identified on magnetic resonance imaging in the elderly. II. Postmortem pathological correlations. *Stroke* **17,** 1090–1097.

Awad I. A., Spetzler R. F., Hodak J. A., Awad C. A., and Carey R. (1986b) Incidental subcortical lesions identified on magnetic resonance imaging in the elderly. I. Correlation with age and cerebrovascular risk factors. *Stroke* **17,** 1084–1089.

Barr A. M., Heinze W. J., Dobben G. E., Valvassori G. E., and Sugar O. (1978) Bicaudate index in computerized tomography of Huntington's disease and cerebral atrophy. *Neurology* **28,** 1196–1200.

Barron S. A., Jacobs L., and Kinkel W. R. (1976) Changes in size of normal lateral ventricles during aging determined by computerized tomography. *Neurology* **26,** 1011–1013.

Bird J. M. (1982) Computerized tomography. Atrophy and dementia: A review. *Prog. Neurobiol.* **19,** 91–115.

Bondareff W., Baldy R., and Levy R. (1981) Quantitative computed tomography in dementia. *Arch. Gen. Psychiatry* **38,** 1365–1368.

Brant-Zawadzki M., Fein G., Van Dyke C., Kiernan R., Davenport L., and de Groot J. (1985) MR imaging of the aging brain: Patchy white-matter lesions and dementia. *American Journal of Neuoradiology Associated with American Journal of Radiology* **6,** 675–682.

Charnes M. E. and DeLaPaz R. L. (1987) Mamillary body atrophy in Wernicke's encephalopathy: Antemortem identification using magnetic resonance imaging. *Ann. Neurol.* **22,** 595–600.

Coffman J. A., Barfknecht C. F., Hunter W., Chakers D., Neff N., Dunn V., and Kleeman B. (1986) Benzodiazepine receptor labeling for study by MRI, in *Society of Magnetic Resonance in Medicine, Book of Abstracts,* Vol. 2, (Society of Magnetic Resonance in Medicine, Inc., Berkeley), p. 547.

Courchesne E., Hesselink J. R., Jernigan T. L., and Yeung-Courchesne R. (1987) Abnormal neuroanatomy in a nonretarded person with autism: Unusual findings with magnetic resonance imaging. *Arch. Neurol.* **44,** 335–341.

Courchesne E., Yeung-Courchesne R., Press G., Hesselink J., and Jernigan T. L. (1988) Hypoplasia of cerebellar vermal lobules VI and VII in infantile autism. *New Eng. J. Med.* **318(21),** 1349–1354.

Dupont R. M., Jernigan T. L., Gillin J. C., Butters N., Delis D. C., and Hesselink J. R. (1987) Subcortical signal hyperintensities in bipolar patients detected by MRI. *Psychiatry Res.* **21,** 357–358.

Earnest M. P., Heaton R. K., Wilkinson W. E., and Manke W. F. (1979) Cortical atrophy, ventricular enlargement, and intellectual impairment in the aged. *Neurology* **29,** 1138–1143.

Gonzales C. F., Lantieri R. L., and Nathan R. J. (1978) The CT scan appearance of the brain in the normal elderly population: A correlative study. *Neuroradiology* **16,** 120–122.

Gyldensted C. (1977) Measurements of the normal ventricular system and hemispheric sulci of 100 adults with computed tomography. *Neuroradiology* **14,** 183–192.

Heilman K. M., Rothi L., and Kertesz A. (1983a) Localization of apraxia-producing lesions, in *Localization in Neuropsychology* (Kertesz A., ed.), Academic, New York, pp. 371–392.

Heilman K. M., Watson R. T., Valenstein E., and Damasio A. R. (1983b) Localization of lesions in neglect, in *Localization in Neuropsychology* (Kertesz A., ed.), Academic, New York, pp. 471–492.

Jacobs L., Kinkel W. R., Painter F., Murawski J., and Heffner R. R., Jr. (1978) Computerized tomography in dementia with special reference to changes in the size of normal ventricles during aging and normal pressure hydrocephalus, in *Alzheimer's Disease: Senile Dementia and Related Disorders* (Katzman R., Terry R. D., and Bick K. L., eds.), Raven, New York, pp. 241–260.

Jacoby R. J., Levy R., and Dawson J. M. (1980) Computed tomography in the elderly: 1. The normal population. *Br. J. Psychiatry* **36,** 249–255.

Jernigan T. L. (1986) Anatomical validators: Issues in the use of computed tomography, in *The Handbook of Memory Assessment of Older Adults* (Poon L. W., ed.), American Psychological Association, Washington, D.C., pp. 353–358.

Jernigan T. L., Pfefferbaum A., and Zatz L. M. (1986) Computed tomography correlates in alcoholism, in *Neuropsychiatric Correlates of Alcoholism* (Grant I., ed.), American Psychiatric Association, Inc., Washington, D.C., pp. 21–36.

Kertesz A. (1983a) Issues in localization, in *Localization in Neuropsychology* (Kertesz A., ed.), Academic Press, New York, pp. 1–20.

Kertesz A. (1983b) Localization of lesions in Wernicke's Aphasia, in *Localization in Neuropsychology,* (Kertesz A., ed.), Academic, New York, pp. 209–230.

Kertesz A. (1983c) Right-hemisphere lesions in constructional apraxia and visuospatial deficit, in *Localization in Neuropsychology,* (Kertesz A., ed.), Academic, New York, pp. 455–470.

Ladurner G., Skvarc A., and Sager W. D. (1982) Computer tomography in transient global amnesia. *Eur. Neurol.* **21,** 34–40.

McQuinn B. A. and O'Leary D. H. (1987) White matter lucencies on computed tomography, subacute arteriosclerotic encephalopathy (Binswanger's disease), and blood pressure. *Stroke* **18,** 900–905.

Meese W., Kluge W., Grumme T., and Hopenmuller W. (1980) CT evaluation of the CSF spaces of healthy persons. *Neuroradiology* **19,** 131–136.

Naeser M. A. (1983) CT scan lesion size and lesion locus in cortical and subcortical Aphasias, in *Localization in Neuropsychology,* (Kertesz A., ed.), Academic, New York, pp. 63–120.

Naeser N. A., Gebhardt C., and Levine H. L. (1980) Decreased computerized tomography numbers in patients with presenile dementia. *Arch. Neurol.* **37,** 401–409.

Nasrallah H. A., Andreasen N. C., and Coffman J. A. (1986) A controlled magnetic resonance imaging study of corpus callosum thickness in schizophrenia. *Biol. Psychiatry* **21,** 274–282.

Oldendorf W. H. (1980) *The Quest for an Image of Brain: Computerized Tomography in the Perspective of Past and Future Imaging Methods* (Raven Press, New York).

Pearlson G. D., Veroff A. E., and McHugh P. R. (1981) The use of computed tomography in psychiatry: Recent applications to schizophrenia, manic-depressive illness and dementia syndromes. *Johns Hopkins Med. J.* **149,** 194–202.

Pfefferbaum A., Zatz L. M., and Jernigan T. L. (1986) Computer-interactive method for quantifying cerebrospinal fluid and tissue in brain CT scans: Effects of aging. *J. Comput. Assist. Tomogr.* **10(4),** 571–578.

Renshaw P. F., Summers J. J., Renshaw C. E., Hines K. G., and Leigh J. S., Jr. (1986) Changes in the 31-P-NMR spectra of cats receiving lithium chloride systemically. *Biol Psychiatry* **21,** 691–694.

Rezek D. L., Morris J. C., Fulling K. H., and Gado M. H. (1987) Periventricular white matter lucencies in senile dementia of the Alzheimer type and in normal aging. *Neurology,* **37,** 1365–1368.

Ross, E. D. (1980a) Sensory-specific and fractional disorders of recent memory in man: I. Isolated loss of visual recent memory. *Arch. Neurol.* **37,** 193–200.

Ross E. D. (1980b) Sensory-specific and fractional disorders of recent memory in man: II. Unilateral loss of tactile recent memory. *Arch. Neurol.* **37,** 267–272.

Rubens A. B. and Kertesz A. (1983) The localization of lesions in transcortical aphasias, in *Localization in Neuropsychology* (Kertesz A., ed.), Academic, New York, pp. 245–268.

Sax D. S., O'Donnell B., Butters N., Menzer L., Montgomery K., and Kayne H. L. (1983) Computed tomographic, neurologic, and neuropsychological correlates of Huntington's disease. *Int. J. Neurosci.* **18,** 21–36.

Scott M. L., Golden C. J., Ruedich S. L., and Bishop R. J. (1983) Ventricular enlargement in major depression. *Psychiatry Res.* **8,** 91–93.

Shimamura A. P., Jernigan T. L., and Squire L. R. (1988) Korsakoff's syndrome: Radiological (CT) findings and neuropsychological correlates. *J. Neurosci.* **8(11):** 4400–4410.

Steingart A., Hachinski V. C., Lau C., Fox A. J., Fox H., Lee D., Inzitari D., and Merskey H. (1987) Cognitive and neurologic findings in demented patients with diffuse white matter lucencies on computed tomographic scan (leukoaraiosis). *Arch. Neurol.* **44,** 36–39.

Terrence C. F., Delaney J. F., and Alberta M. C. (1977) Computed tomography for Huntington's disease. *Neuroradiology* **13,** 173–175.

Weinburger D. R. (1982) Computed tomography (CT) findings in schizophrenia: Speculation on the meaning of it all. *J. Psychiatr. Res.* **18,** 477–490.

Wilson R. S., Fox J. H., Huckman M. S., Bacon L. D., and Lobick J. J. (1982) Computed tomography in dementia. *Neurology* **32,** 1054–1057.

Zatz L. M., Jernigan T. L., and Ahumada A. J., Jr. (1982a) Changes on computed cranial tomography with aging: Intracranial fluid volume. *American Journal of Neuradiology* **3,** 1–11.

Zatz L. M., Jernigan T. L., and Ahumada A. J., Jr. (1982b) Changes on computed cranial tomography in white matter with aging. *J. Comput. Assist. Tomogr.* **6,** 19–23.

Functional Neuroimaging in Neurobehavioral Research

Frank Wood

1. Introduction

Functional neuroimaging techniques are to the second century of neurobehavioral research what the clinicopathological method was to the first century—the ultimate empirical method by which theoretical speculations are to be tested. Thus, from Charcot's day until our own, the "gold standard" criterion for local brain damage—if such damage is offered as an explanation for behavioral deficit—has been the careful postmortem examination of the brain, both grossly and through the microscope. That this method is not yet exhausted is illustrated by the recently rich and fruitful cytoarchitectural studies of dyslexic brains by Galaburda (1983) (*see also* Geschwind and Galaburda, 1985a–c for a fuller review of the theoretical neurobehavioral context to which such cytoarchitectural studies have been related).

For the first time in the history of neuroscience, however, it has become possible to investigate the functioning of localized areas of the brain by direct measurements of markers of local blood flow or glucose metabolism in the living brain while that brain is engaged in a particular behavioral or cognitive task. Such measurements allow a different correlation: between experimentally isolable features of the behavioral task and localized intensities of neuronal activation in the brain (instead of correlations between behavioral dysfunction and site of lesion). Our charge in this chapter is to consider the prospect of this new method making as great a contribution to neurobehavioral theory as the clinicopathological method has already made.

As would be expected in the initial stages of any sustained scientific inquiry, not all the problems and issues are known. Still, the effort has been proceeding for more than a decade, so certain obstacles and opportunities have become clear, and further progress depends on understanding them well. The issues group themselves logically into approximately three categories: technical, statistical, and experimental (adequacy of an experimental design to answer a specific question).

Technical issues are naturally the farthest advanced: in this area, as in many areas of science, technological development forces theoretical and empirical progress. Having a telescope stimulates astronomy. With respect to particular issues in the technology of regional cerebral blood flow measurements, several good reviews are available. For the xenon-133 method (geographically coarse and limited to the exterior cortical surface, but still the one giving the most accurate separation of gray and white matter flow estimates), *see* Stump and Williams (1980). For cerebral blood flow measurements by positron emission tomography (PET) (allowing much finer temporal and spatial resolution, but some relative "blurring" of the gray vs white matter boundary) *see* Frackowiak et al. (1980), Herscovitch et al. (1983), and Raichle et al. (1983).

Statistical problems have also received increasing attention. They apply, of course, to PET glucose studies as well as to regional cerebral blood flow studies. *See* Wood (1983) for a general review of the range of issues. Among the issues considered in that review are variance differences (between groups or between activation conditions); the ubiquitous correlations between means and variances; the commonly nonnormal, sometimes bimodal distributions; and the problem of differentially strong intercorrelations among separate subsets of brain locations. In recognition that traditional ANOVA and MANOVA approaches are inadequate and based upon faulty assumptions, some researchers have proposed specific new approaches. Of these, the "scaled subprofile model" of Moeller et al. (1987) is the most carefully considered and thorough. Readers consulting this proposal will also find references to most earlier proposals, but *see* especially those by Clark et al. (1984) and by Clark and Stoessl (1986).

The Moeller et al. proposal seeks explicitly to separate three different sources of variance: "global" (variance between subjects that is independent of regions, hence a scaling or normalizing factor); group mean profile variance (reflecting variance between sites that is common to the group); and subject residuals (comprising not only unique variance but also that variance that can be accounted for by patterns or factors of interregional covariation).

One particularly noteworthy specific feature of the proposal is the ratio scaling: error is considered greater in profiles with high metabolic activity, so raw value departures from the mean profile by subjects with lower profiles essentially receive greater weight than arithmetically identical departures by high metabolic (or flow) subjects. This assumption is often, perhaps usually, correct. In special cases, however, it will fail—as when there is a ceiling effect or when some activation particularly constrains metabolism or flow values at a certain region to be uniformly high with little variance. *See* Wood (1987) for a more extended discussion. Nonetheless, the Moeller et al. proposal is quite helpful as it is, and it is adjustable for the special-case exceptions noted.

Technological and statistical issues aside, this review concentrates largely on questions of experimental design and inference. They are, as always, the caboose on the scientific train—the last to arrive, often in the least tidy condition, but carrying the essential tools for effective operation and use of the machine. We consider especially the relation between experimental strategies and the assumed models of brain functioning to which they are referred.

The earlier studies, though limited and sometimes even naïve in retrospect, are the essential foundations for the later progress. Those using normal subjects and behavioral activation paradigms are particularly instructive (*see* Wood, 1983, for a review). These have confirmed expected topographical representations, such as those involving tactile and motor functions along the banks of the central sulcus (Roland, 1977), or those involving auditory stimulation and the temporal lobes (Knopman et al. 1980). They have also developed newer findings, including the now-familiar notion of hyperfrontality (Ingvar, 1979; Prohovnik et al., 1980; Ingvar, 1985) whereby most states of rest or activation are accompanied by relatively higher frontal than posterior flows. Expected cognitive laterality phenomena have also been demonstrable (Risberg et al., 1975; Gur and Reivich, 1980), as have interactions between stimulus or response laterality and attention or effort (Halsey et al., 1979, 1980; Maximilian et al., 1980; Prohovnik et al, 1981). Studies such as these have set the stage for the newer investigations.

For concreteness, I shall review in detail three particular experiments, each in its own theoretical context. They each represent the "second generation" of brain activation studies, inasmuch as they go beyond the simpler approaches that characterized the earliest investigations. Accordingly, each experiment offers one or more new insights to refine and sharpen forthcoming research; together they cover many of the major but less obvious points that should be considered in a manual of methodology for the field.

2. The Verbal Fluency Study of Parks et al. (1988): Inverse Correlations Between Glucose Utilization and Task Performance

Two groups of normals were studied. The larger (N = 35) underwent PET scans of glucose utilization during rest; the smaller (N = 16) group carried out the traditional neuropsychological task of verbal fluency throughout the glucose uptake period. In a formal sense, then, the study was initially a straightforward rest vs cognitive activation experiment, in which the activation variable was completely between groups.

One feature of interest is the duration of the activation itself: for 30 min, subjects produced as many words as they could think of that began with a certain letter, the particular letter being changed by the experimenter every minute. Anyone familiar with this task will recognize the considerable sustained effort this required from the subjects! PET glucose studies require this duration of activation, but few have used a task this intense: unlike most studies (continuous performance, for example) this task required subjects to perform at their maximum speed throughout the 30 min. There can be no doubt that this represented a strong and significant activation of verbal generative processes.

The two groups were reasonably well-matched for age and Wechsler IQ, both verbal and performance, and exquisitely well-matched also on a 3-min version of the activation task itself—a control that lifts the experiment out of the ordinary context, to a level that permits sharper and more focused conclusions. By this control, the experimenters allow us to conclude that—if permitted to do so—the N = 35 resting group would have performed the cognitive activation task at the same level of accuracy as did the N = 16 activation group. Differences in the glucose utilization levels and profiles will not, therefore, be attributable to differences between the groups in underlying ability to perform the task (for example, less able subjects possibly having lower flows in general, regardless of task conditions).

Differences were found: the activation condition generated significantly higher flows than the rest condition in the frontal, temporal, and parietal "compartments" of the analysis—thus excluding only the occipital lobe from a general effect (even there, the trend was certainly suggestive at $p = .058$). There was also a main effect of hemisphere, the right being significantly higher in glucose utilization than the left. Detailed analyses using values that were normalized to the occipital lobe (in effect, controlling for overall activity levels) showed the most substantial effects in the temporal lobes, bilaterally.

The less obvious finding related to the correlation between task performance and glucose utilization, within the experimental group. The correlation was negative: in all regions, the higher a subject's glucose utilization, the lower the task performance. Note well (again representing careful experimental and statistical control) that these correlations were independent of age and IQ. Thus, the within-group variance in glucose utilization that was significantly inversely associated with task performance was not variance in either IQ or age.

The authors interpreted their findings in terms of an effort model: subjects performing the task less well might be expected to find it more difficult and therefore more effortful—assuming there was authentic task engagement in the first place. An early example of this finding of inverse correlation between task performance and brain activity level was reported by Wood et al. (1980), who discussed the relation between recognition-memory accuracy and regional cerebral blood flow, especially in the occipital areas. General reviews of these inverse correlations are also found in Wood (1983, 1987). The fundamental question concerns the kind of brain activity model that is implied by these types of inverse correlations.

Let us note in the first place that, in the absence of these intra-group correlations, we would assume a straightforward activation model whereby a task (or some component of the task) simply "engages" a certain brain region. Once effort is allowed into the model, however, that assumption fails: there is no basis for assuming that effort would be exerted only by structures that are actually "doing" the task. Indeed, part of the notion of excess effort during difficult tasks is precisely that inefficient effort will spill over into regions that would not be involved at all if the task were efficiently performed.

This "spillover" notion has at least three specific subcategories: (1) the "widening" of effort beyond its normal limits, so as to recruit additional resources that actually help the performance; (2) the mobilization of inhibitory activities to suppress competing stimuli or responses; and (3) the inability—likely constitutional—to limit activation to a circumscribed region, hence a type of neural imprecision. The first possibility is illustrated by skilled vs clumsy use of the hammer: an experienced carpenter recruits minimal arm muscle activation; because the blows are accurately aimed, they drive the nail with only moderate impact. The novice, however, is less precise: the blows are only approximately accurately directed, so greater impacts are required to drive the nail. This particular analogy invites a second, more refined question: is the extra effort located in precisely the same muscle or group of muscles that is used by the skilled carpenter? Is strength itself the *only* difference? Alternatively, the extra effort might recruit muscle systems not ordinarily used for the discrete level of strength expended by the experienced carpenter: the novice perhaps does use his or her shoulder or trunk muscles, in particular, more than the expert, either because the stronger blows require such use for overall body balance, or because the stronger blows simply cannot be delivered with the limited muscle group employed by the skilled carpenter. In brief, this first analogy raises the question of whether or under what circumstances a more intense activation inherently requires a more widespread one.

The second mechanism—recruitment of inhibitory processes—is illustrated by the general fact that people who are trying hard often spend effort to reduce distractions (turn off the TV, take the phone off the hook, and so on). A more particular example is the carpenter again: in fine sawing, especially of curved lines, it is common to see a carpenter purse the lips so as to continually blow the sawdust off the line being sawn. One who is highly familiar with the particular shape being sawn, however, may need to do this quite a bit less than a novice would. Notice that, in any comparison of rest to activation in the sawing of finely curved lines, one focus of activation would be the lips. One would not wish to conclude from that activation, however, that the lips were really involved in the sawing. Nor is the analogy to brain activity fanciful: it is indeed generally assumed that much, perhaps most, of the brain's activity is inhibitory; certainly it is a behavioral fact that states of high arousal with focused attention are routinely and inherently accompanied by a reduction of general motor activity. The posture of thought is a quiet one.

The third mechanism, spillover, is familiar in clinical neurology as "overflow movements," as when a child is asked to perform an exercise with one hand, but performs all or some parts of the movement with both hands. Usually considered a sign of developmental immaturity, the finding connotes imprecision of activation, but does not reveal the mechanism whereby the more mature precision is developed. Bilateral "overflow" to homologous structures, as when finger tapping in one hand is accompanied by "mirror" finger tapping in the other hand, may be different from the nonspecific shoulder shrugging or "throwing up your hands" that accompanies frustration or puzzlement. Consider, however, the baby who executes the tonic neck synergism that includes head and trunk turning, unilateral grasping with one hand (usually the right), and discrete vocalization at pleasant levels. This is obviously an approach or appetitive synergism: it terminates in eating or in the attempt to eat the grasped object. The contrasting synergism is that of avoidance or aversion, and it is accompanied by bilateral strong thrusting of arms and legs, head turned up or alternating side to side, and unpleasantly loud crying. In turn, approach requires both perceptual and motoric operations that isolate the to-be-approached target from its surrounding field.

It could be, therefore, that the imprecision of neural or motoric activation seen in spillover or overflow movements is a natural consequence of an insufficiently goal-directed approach—regardless of whether the insufficiency is a constitutional unreadiness or a motivational unwillingness. To be sure, only a little is gained by substituting the notion of goal directedness for the less complicated notion of imprecision of activation. What is gained, however, is a recognition of the possibility that excess activation, by this spillover mechanism, may represent a truly different state of task or goal orientation—whatever the reason, whether constitutional or motivational. Differences in goal orientation, or in some similar internal state, could be what is signaled by excess neuronal activity in the PET scans of subjects who are performing a cognitive task relatively poorly. Obviously, anxiety—not unrelated to effort or to goal orientation—is one possibility that springs readily to mind.

In general, this first experiment, showing inverse relations between neuronal activation learning and task performance, relates to a model of brain functioning that is biologically as well as psychologically plausible (which is why it is easy to analogize to carpenters and their movements, babies and their crying, and the like). It naturally directs attention to system-wide patterns of responding (as should be expected from any system that deserves the name of organism"); it properly directs our attention away from purely modular or componential models—the familiar "boxes in the head" models of information-processing psychology. Flow-charts assigning operations in boxes and describing transfers between boxes do not naturally or readily accommodate inverse correlations between performance and activation.

For all that, it must also be said that the type of experiment reported here leaves hanging an almost poignant question of specificity. Can we not do better than to say that a verbal fluency task engages almost all areas of the brain, the more so in brains that find it difficult? Surely there is at least some localization of function; lesion evidence certainly suggests so. We look in vain through the results of this experiment for any help on legitimate questions of localization and lateralization of function. Even if we grant the limitations of a componential or modular model of the brain, must we forsake all notions of specificity of brain activation? The next experiment certainly purports to give a clear answer in favor of specificity and localization.

3. The Single-Word Processing Study of Peterson et al. (1988): The Ultimate in Modularity and Specificity

Seventeen normals underwent a series of blood flow PET scans using oxygen-15 labeled water. This procedure requires only 40 s of activation time, and the same subject can repeat several scans, each using a different cognitive task. This obviously allows within-subject comparisons to be made relatively easily. A classical subtraction logic is employed to create comparisons between tasks that differ only by the addition of a single processing component: when the two scans are subtracted, the difference is taken to represent the additional processing evoked by the additional task component. Thus, in the first level of comparison, the subtraction is between the passive viewing of single words and the passive viewing of a single fixation point. In this case, the hypothesized "additional" operations relate to the difference between these two tasks. Passive viewing of single words differs from passive viewing of a fixation point, by this logic, in the specificity of the visual stimuli (real words vs fixation point) without regard to motoric or other response. (Though not discussed explicitly, the conditions apparently differ also in total luminosity, degrees of visual angle subtended, and in the cyclical vs persisting nature of the stimulus display.)

Additional subtractive contrasts in the visual modality are between oral pronunciation of the presented words vs passive viewing of them (on the one hand) and between spontaneous generation of words vs oral pronunciation of printed words (on the other hand). In addition, this three-tiered series of subtraction contrasts was also presented in the auditory modality, again contrasting: passive listening to words vs no auditory stimuli; active oral repetition of words vs purely passive listening; and generation of semantic associates to auditorally presented words vs oral repetition of auditorally presented words.

In general, this approach identified specific contrasts believed to represent three processes: (1) modality-specific auditory or visual processing of word stimuli without response requirements; (2) oral repetition of auditorally or visually presented words; and (3) oral generation of words that are semantic associates of either auditorally or visually presented words.

By subtraction, the specific brain regions associated with these processes were identified and averaged across subjects. These were as follows:

1. Generally bilateral extrastriate occipital activation for passive visual words;
2. Bilateral superior temporal and anterior cingulate activation for passive auditory words;
3. Generally bilateral perirolandic, perisylvian, and medial superior frontal activation for oral repetition of spoken or written words;
4. Anterior cingulate and left inferior premotor frontal activation for generation of words that are semantic associates of spoken or written words.

A number of familiar neurobehavioral concepts are supported by these findings. These include not only the accepted mapping of visual and auditory processing onto occipital and temporal cortex, respectively, but also the identification of an involvement of left premotor cortex with the generation of semantic associates. An attentional role for the cingulate cortex was suggested, on the basis that the activation there was greatest when there were more rather than fewer targets in the semantic association condition.

In addition to these classical questions, however, the study addressed specific issues in reading theory. Most particularly, the authors concluded that parallel sensory-to-motor pathways had been demonstrated for the auditory and visual presentation of words: there was no temporal or temporoparietal activation associated either with repetition of visual words or with the generation of semantic associates to words in either modality. This suggested that there was no obligatory recording of visually presented words into a phonological code (on the understandable assumption that phonological recoding would be superior temporal or temporoparietal in its locus).

Obviously, this study is grist for a localizationist mill. When a method so clearly demonstrates accepted localizations of visual and auditory sensory processing, it does command respect when the same method demonstrates the localization of cognitive or attentional components. Of these, the left frontal involvement with generation of semantic associates is particularly compelling, since it fits with many generally accepted notions of lateralized frontal activity. Indeed, verbal generative fluency itself—employed in the Parks (1988) study above—is supposed to elicit activation in this locus, so the present study clearly succeeds in the localization arena, precisely where the former study seemed to fail.

The localizing success of this study appears to result mainly from the more modular and componential nature of the task comparisons. Whereas the Parks study employed only a rest vs verbal-generation comparison, the present study identifies at least three separate components within that range of comparison. It may also be that the capability of restricting the activation to only 40 s, rather than 30 min, is helpful in obtaining a more focused and circumscribed activation pattern.

Notwithstanding that this study seems at first glance to resolve some important localization questions, it also has serious limitations that must be faced. Obviously, some of these limitations are precisely those raised by the positive findings in the Parks study and reviewed above. Thus, we must ask whether the implicated cognitive loci represent the activation of those who are doing the task well or poorly. Admittedly, it is perhaps difficult or strained to imagine gradations of task performance or effort when the task is simply passive listening or viewing. It is not at all difficult, however, to imagine a variety of emotional or cognitive states during such stimulation. Which of these states, if any, are necessary conditions for the demonstration of the expected loci of sensory activation? For example, does anxiety intensify or minimize the foci of sensory activation? With respect to the foci of cognitive activation, do they differ in size or intensity, or even in location, with differing accuracies of task performance?

The Petersen et al. study requires averaging across subjects in order to get stable subtraction images. The above questions of individual differences are inherent in such averaged data, and the Parks et al. study clearly suggests that within-group correlations between task performance and size, intensity, or location of the subtraction foci would be necessary for any strong theoretical conclusions. In a separate review, Posner et al. (1988) acknowledge the possibility that their subtraction method would leave open the question of within-subject strategy differences emerging as the task becomes more complex. (Such strategy differences might apply even to the simple components.) They contend, however, that since the most extreme comparison—between the passive nonword sensory condition and the most active semantic generation condition—yields a "subtraction image" that is essentially the sum of the images obtained by individual comparison at successive stages of the hierarchy, then no evidence for strategy alterations is provided. That argument, however, does not seriously refute the long history of the numerous state variables that have been shown to contaminate the subtraction method—a history that goes all the way back to the classical structuralist vs functionalist arguments of the last century. It is, indeed, an ironic pun: against Kulpe and the Wurzburg school, and against the functionalists generally, the present authors contend that there can be no "imageless" thought since there is no PET image!

Consider William James' beautiful argument in this regard. A sentence behaves like a bird (note again the ease with which a functionalist can recruit a biological analogy). Birds do perch, sometimes, and when they do so they can be considered in a substantive state (well localized). Sentences also have their stopping places, which are substantive words. Indeed, such substantives take up the majority of the space in a sentence. However, the truly interesting thing about birds is not that they perch at various places, but that they fly from place to place; likewise, sentences are interesting not so much for their content as for their direction, directions that are marked by transitive, nonsubstantive words, such as "if," "then," or "furthermore." Might it not be the same with localized brain activity as with localized bird or word activity? Might the "transfers" or "recodings" from modular sensory to semantic generative activity not be relatively inconspicuous?

Consider yet another analogy, of a baseball pitcher who catches (modality-specific reception) and then throws the ball (generation). A scan of his musculature would show little evidence of the transfer (recoding) from the catching hand to the throwing hand, even if he sometimes caught with his throwing hand, and transferred first to his glove or catching hand, then back to his throwing hand. Certainly the throwing arm could be expected to show activation attributable to the output phase, but so would both of the legs and the trunk. A consideration of the relative scope and intensity of the activation might well lead to the conclusion that the major work of pitching a baseball is done by the leg and trunk muscles—not an inaccurate conclusion, especially for fastball pitching (high effort), but nonetheless misleading about where the ball was actually "handled."

Appealing as it is for its seeming precision, the Petersen et al. approach must be seen as inherently limited and in need of broadening—especially at the point of individual differences. (Obviously, if transitive processes in neural information processing are really nonimageable, there is little we can do; but certainly we must study the impact of functional state and trait variables.) The next study has individual differences as its major focus.

4. The Dyslexia Study of Flowers et al.: Individual Differences in Brain Organization

Two experiments were performed. In the first, 72 normals underwent xenon-133 blood flow measurements during an auditory-orthographic task that required subjects to signal whether an auditorally presented word was exactly four letters long. Words were presented by earphones at the rate of one every 2.5 s. In the second experiment, 73 subjects did the same task. These were subjects who had been referred for dyslexia evaluations in childhood, and all had achievement and IQ scores available from their childhood records. The current blood flow testing was done an average of 25 years after the childhood evaluations. This population obviously affords a unique opportunity to study persisting residual deficit from a chronic developmental disorder that was documented in childhood, instead of retrospectively.

In the first experiment, a positive correlation was found between left Wernicke's area (superior temporal lobe) activation and accuracy of task performance. Specifically, the analysis was a multiple regression of task accuracy by flows at three brain sites: left Wernicke's area, left angular gyrus, and left inferior temporooccipital junction. (A corresponding, separate analysis was made to predict task accuracy from the three homologous right hemisphere sites.) Note that the inclusion of all three sites in the multiple regression means that the other two sites—angular gyrus and inferior temporooccipital junction—are functioning as control sites or statistical covariates. The finding really means that, holding angular gyrus or inferior temporooccipital flow constant, Wernicke's area flow is positively correlated with task accuracy. Hence, it is really the slope or difference between Wernicke's area and these other two sites that is predictive of task accuracy. The group mean profile showed that Wernicke's area had higher flow than the other two areas. The relation of flow to accuracy was independent of age or sex. No significant relations were found involving the right hemisphere sites.

These three sites were chosen to test a theoretical notion (Ojemann, 1983; Wood, 1985) that certain types of cognitive or learning disability might represent situations in which there was a posterior displacement of language activation from its usual site in Wernicke's area to more posterior sites in the angular gyrus or the inferior temporooccipital junction.

The second experiment, with subjects who had been assessed in childhood, employed a stratification of the cases into normal, borderline, and severe categories (with respect to the presence and severity of dyslexia in childhood). This variable history of dyslexia predicted task accuracy (not surprisingly). Task accuracy was also significantly predicted by left angular gyrus flow in the same type of analysis as before (in which the three sites jointly predicted childhood history). Thus, with Wernicke's area flow and temporooccipital flow controlled, it was angular gyrus flow that predicted history of dyslexia: holding the other two sites constant, the higher the angular gyrus flow, the greater the likelihood of severe dyslexia in childhood. The group mean profile for the dyslexic group showed angular gyrus flows to be higher than the other two sites; however, the borderline and normal groups showed the normal profile of higher Wernicke's area flow than either angular gyrus or temporooccipital flow. It is of interest that the relation between dyslexia and angular gyrus flow is present even when task accuracy and state anxiety are controlled.

This study was interpreted by its authors as showing frank displacement of the activation focus, from the left superior temporal region of Wernicke's area to an immediately posterior angular gyrus activation site. In turn, this confirmed a theoretical expectation that true dyslexia—already believed from other evidence to involve a congenital lesion in the temporal planum or Wernicke's area—involves an actual relocation or redistribution of function. It was interpreted as not simply a greater spread of activation from the same Wernicke's area focus, since the activation in Wernicke's area was actually less in true dyslexics than in normals.

As an illustration of a particular research strategy, this study brings individual differences to the forefront, with the explicit expectation and finding that such differences can involve actual redistribution of functional localization. It may properly caution us that we should not assume that all populations, or even all normals, have the same functional neuroanatomical map. (Nor is it plausible in the slightest that dyslexia is the only or the major disorder that might show such differences; far more likely is the prospect that most groups intended as normal controls, to say nothing of frankly abnormal populations, will have individual differences in functional localization.) In turn, once this assumption of a universal "map" of the brain is forsaken, only some careful consideration of topographical geometry is likely to preserve order in this domain: if certain activation foci are displaced from one subject to the next, such foci may still retain their relative location (above, behind, etc.) with respect to other foci.

Clearly, this study lacks what the Petersen et al. study so richly possesses—a series of discrete contrasts that might disclose relatively narrow and limited components of processing. What it provides instead is nevertheless an interesting caution and corrective about distortion or displacements in the underlying anatomical map. It also illustrates the power of larger-N studies allowing statistical correction for a variety of factors, such as intelligence, anxiety, age, sex, and the like.

5. Conclusions

The studies reviewed in detail offer a compendium of suggestions and hints for researchers using functional neuroimaging techniques to study basic brain–behavior relations in humans. These can conveniently be enumerated as follows:

1. Attempt large-N studies that allow for the control of as many subject variables as possible; important relationships in the areas of interest may be obscured by subject variance caused by factors such as age, sex, intelligence, and anxiety. Small-N studies squander the power of the methods: in their quest for plainly visible phenomena, such studies override the known complexity of brain function and so tend to disclose trivial rather than important findings.
2. Manipulate or measure task performance, as speed, accuracy, or both. Correlations between task performance and flow or metabolism will often, indeed usually, be highly instructive. In this connection, however, consider two specific points:
    a. Task accuracy or speed can be a measure of some underlying ability, so it would be good to attempt other measures of the presumed ability. If the other measures correlate less strongly with brain activation than does task performance itself, then specific task activation may indeed be partly what is measured by the task. Otherwise, the correlation with task accuracy may really be with an underlying ability.
    b. Lack of correlation with task accuracy does not mean that a brain region thus lacking is unrelated to the task. It may simply indicate an obligatory set or similar mechanism that is activated whenever the task is attempted.

3. Seek discrete, deconfounded comparisons between one task and another, so that specific operations can be at least heuristically isolated for study. In doing so, however, consider the above two points—without which the strategy will fail.
4. Explore state variables as thoroughly as trait variables, both by measurement (as of state anxiety) and by direct manipulation (as of reward value of the task). More broadly, notwithstanding the inherently structuralist "slant" of these techniques, keep the historic functionalist critiques readily at hand, for they will serve well to call attention to variables otherwise overlooked.
5. Recognize, at least at the point of data analysis, that loci of task-specific activation may shift as a result of either state or trait variables.
6. Do not consider individual sites in isolation. Sometimes there will be a significant relation involving a particular site only if other, usually adjacent, sites are held constant. In other words, slopes, or gradients of change of activation between adjacent regions, may often be the fundamental indices that represent activation. To use one last analogy, it is sometimes the "definition" of muscles—how well they "stand out"—rather than their sheer size that indicates the level of training or skill.
7. Finally, expect to need converging experiments to isolate a particular mechanism. There is no *experimentum crucis* in this field, no definitive settling of arguments by a single experiment. At this stage, each result raises further questions and invites converging operations to validate it.

References

Clark C. M. and Stoessl A. J. (1986) Glucose use correlations: A matter of inference. *J. Cereb. Blood Flow Metab.* **6,** (letter) 511–513.

Clark C. M., Kessler R., Buchsbaum M. S., Margolin R. A., and Holcomb H. H. (1984) Correlational methods for determining regional coupling cerebral glucose metabolism: Pilot study. *Biol Psychiatry* **19,** 663–678.

Flowers D. L., Wood F. B., and Naylor C. E. Regional cerebral blood flow in adults diagnosed as reading disabled in childhood. (Under editorial review).

Frackowiak R. S. J., Lenzi G. L., Jones T., and Heather J. D. (1980) Quantitative measurement of regional cerebral blood flow and oxygen metabolism in man using ^{15}O and positron emission tomography: Theory, procedure and normal values. *J. Comput. Assist. Tomogr.* **4,** 727–736.

Galaburda A. M. (1983) Histology, architectonics, and asymmetry of language, in *Neuropsychology of Language, Reading and Spelling* (Kirk U., ed.), Academic, New York.

Galaburda A. M., Sherman G. F., Rosen G. D., Aboitiz F., and Geschwind N. (1985) Developmental dyslexia: Four consecutive cases with cortical anomalies. *Ann. Neurol.* **18,** 222–233.

Geschwind N. and Galaburda A. M. (1985a, b, c) Cerebral lateralization, biological mechanisms association and pathology I, II, III. *Arch. Neurol.* **42,** I(a) 428–459, II(b) 521–552, III(c) 634–654.

Gur R. C. and Reivich M. (1980) Cognitive task effects on hemispheric blood flow in humans: Evidence for individual differences in hemispheric activation. *Brain Lang.* **9,** 78–92.

Halsey J. H, Blauenstein U. W., Wilson E. M. and Wills E. L. (1979) VCBF comparison of right and left hand movement. *Neurology* **29,** 21–28.

Halsey J, Bauenstein U, Wilson E, and Willis E. (1980) Regional cerebral blood flow activation in a patient with right homonymous hemianopia and alexia without agraphia. *Brain Lang.* **9,** 137–140.

Herscovitch P., Markham J., and Raichle M. E. (1983) Brain blood flow measured with intravenous $H_2{}^{15}O$. I. Theory and error analysis. *J. Nucl. Med.* **24,** 782–789.

Ingvar D. H. (1979) Hyperfrontal distribution of the cerebral grey matter flow in resting wakefulness; on the functional anatomy of the conscious state. *Acta Neurol. Scand.* **60,** 12–25.

Ingvar D. H. (1985) Memory of the future: An essay on the temporal organization of conscious awareness. *Hum. Neurobiol.* **4,** 127–136.

Knopman D. S., Rubens A. B., Klassen A. C., Meyer, M. W., and Niccum N. (1980) Regional cerebral blood flow patterns during verbal and nonverbal auditory activation. *Brain Lang.* **9,** 93–112.

Maximilian V. A., Prohovnik I., Risberg J., and Hakansson K. (1980) Regional cerebral blood flow changes in the left cerebral hemisphere during word pair learning and recall. *Brain Lang.* **6,** 22–31.

Moeller J. R., Strother S. C., Sidtis J. J., and Rottenberg D. A. (1987) Scaled subprofile model: A statistical approach to the analysis of functional patterns in positron emission tomographic data. *J. Cereb. Blood Flow Metab.* **7,** 649–658.

Ojemann G. A. (1983) Brain organization for language from the perspective of electrical stimulation mapping. *Brain Behav. Sci.* **20,** 189–230.

Parks R. W., Loewenstein D. A., Dodrill K. L., Barker W. W., Yoshii F., Chang J. Y., Emran A., Apicella A., Sheramata W. A., and Duara R. (1988) Cerebral metabolic effects of a verbal fluency test: A PET scan study. *J. Clin. Exper. Neuropsychol.* **10,** 565–575.

Petersen S. E., Fox P. T., Posner M. I., Mintun M., and Raichle M. E. (1988) Positron emission tomographic studies of the cortical anatomy of single-word processing. *Nature* **331,** 585–589.

Posner M. I., Petersen S. E., Fox P. T., and Raichle M. E. (1988) Localization of cognitive operations in the human brain. *Science* **240,** 1627–1631.

Prohovnik I., Hakansson K., and Risberg J. (1980) Observations on the functional significance or regional cerebral blood flow in "resting" normal subjects. *Neuropsychology* **18,** 203–217.

Prohovnik I., Risberg J., Mubrin Z., Bolmsjo M., and Von Sabsay E. (1981) Further improvements of the 133-Xe inhalation method. *J. Cereb. Blood Flow Metab.* 1(Suppl. 1): 108–109.

Raichle M. E., Martin W. R. W., Herscovitch P., Mintun M. A., and Markham J. (1983) Brain blood flow measured with intravenous $H_2{}^{1}5O$. II. Implementation and validation. *J. Nucl. Med.* **24,** 790–798.

Risberg J. L., Halsey J. H., Wills E. L., and Wilson E. M. (1975) Hemispheric specialization in normal man studied by bilateral measurements of the regional cerebral blood flow: A study with the 133-Xe technique. *Brain* **98,** 511–524.

Roland P. E., Skinhoj E., Larsen B., and Lassen N. A. (1977) The role of different cortical areas in the organization of voluntary movements in man. A regional cerebral blood flow study. *Acta Neurol. Scand.* **56,** 542, 543.

Stump D. A. and Williams R. (1980) The noninvasive measurement of regional cerebral circulation. *Brain Lang.* **9,** 35–46.

Wood F. (1983) Cortical and thalamic representation of the episodic and semantic memory systems converging evidence from brain stimulation, local metabolic indicators and human neuropsychology. *Behav. Brain Sci.* **6,** 189–230.

Wood F. (1987) Focal and diffuse memory activation assessed by localized indicators of CNS metabolism: The semantic-episodic memory distinction. *Hum. Neurobiol.* **6,** 141–151.

Wood F., Armentrout R., Toole J., McHenry L., and Stump D. (1980) Regional cerebral blood flow during rest and memory activation in a patient with global amnesia. *Brain Lang.* **9,** 124–136.

Intracarotid Sodium Amobarbital Procedure

Rebecca Rausch and Michael Risinger

1. Background

1.1. Historical Perspective

Intracarotid injections of amobarbital have been performed for clinical purposes since 1949, when Wada described a method for determination of hemispheric language dominance. It was noted that the intracarotid injection of amobarbital, performed in an attempt to investigate the interhemispheric spread of epileptiform discharges, produced a transient ipsilateral paralysis of hemispheric function without eliciting unacceptable sedation or interruption of vital functions. It was reasoned that this method would be useful for determination of hemispheric language dominance in patients who were to undergo neurosurgical procedures on the language dominant hemisphere (Wada, 1949). Eighty patients were evaluated by Wada between 1948 and 1954 without major complications (Wada and Rasmussen, 1960).

These observations were enlarged upon by Wada and Rasmussen in 1960. They described 20 additional patients from the Montreal Neurological Institute (MNI) who were similarily tested. Wada and Rasmussen also reported animal studies that documented the safety of dilute concentrations of amobarbital for intracarotid injection. Branch et al. (1964) subsequently reported experience with an additional 103 patients, and the safety and efficacy of the technique for determination of language dominance was established.

A new indication for the intracarotid sodium amobarbital procedure (IAP) was proposed in 1962 by Milner et al., who described their study of memory function after intracarotid injection of sodium amobarbital in 50 consecutive patients at the MNI. Preceding reports by Scoville and Milner (1957) and Penfield and Milner

(1958) had described a syndrome of severe anterograde memory dysfunction in two types of neurosurgical patients: (1) those with bilateral surgical removal or destruction of mesial temporal structures and (2) those with unilateral mesial temporal lobe excision (performed to provide relief from medically intractable seizures of temporal lobe origin), and evidence of physiological or structural abnormality affecting the contralateral mesial temporal lobe. It was hypothesized that those patients at risk for amnesia after unilateral temporal lobe excision (i.e., those with additional contralateral temporal lobe dysfunction) could be identified by the emergence of a transient amnesia following pharmacologic inactivation of the hemisphere containing the identified seizure focus. In the initial series of patients, memory dysfunction was seen in 12/50 patients, but always after injection of the hemisphere *contralateral* to the known seizure focus. No instance of transient amnesia was noted after injection of the hemisphere ipsilateral to seizure origin, and no postoperative amnestic syndrome resulted. After providing these negative findings, the authors proposed that the IAP was a suitable procedure for assessing the risk of postoperative amnesia in patients undergoing unilateral temporal lobectomy. Since adoption of this technique for evaluation of memory function, no cases of global amnesia have been reported in temporal lobectomy patients at the MNI (Penfield and Mathieson, 1974). Others (Kløve et al., 1970; Blume et al., 1973; Rausch et al., 1984) have described their modifications of the IAP for evaluation of memory function, and the IAP is currently in use in the large majority of centers that provide comprehensive presurgical evaluations for patients with medically refractory complex partial seizures (Rausch, 1987).

1.2. Evolving Indications

1.2.1. Prognostic Value

1.2.1.1. Hemispheric Language Dominance. The IAP is the definitive method for determining hemispheric language dominance. The contribution of each hemisphere to language functioning can be directly assessed and independently evaluated. Methods of indirectly assessing language dominance are not sufficiently reliable to have predictive value for an individual case. The earliest indirect method of assessing hemispheric language dominance was by determination of handedness. In 1865, Broca proposed a direct relationship between handedness and hemispheric language dominance. Although the majority of right-handers have been shown (by the IAP and by lesion studies) to have left hemispheric language dominance (Rasmussen and Milner, 1975; Gloning et al., 1969; Penfield and Roberts, 1959; Zangwill, 1960), some right-handers have right hemispheric language dominance. Rasmussen and Milner (1975) reported that 13% of their right-handed patients with clinical evidence of early left hemispheric damage demonstrated right hemispheric language dominance with the IAP. Similarly, Rausch and Walsh (1984) reported that 15% of their right-handed patients with seizures of left temporal origin and no strong evidence of early brain damage showed right hemispheric language dominance. Speech representation in the left-hander has been found to be even more variable (Rasmussen and Milner, 1975; Rausch and Walsh, 1984). Specialized techniques, such as the dichotic listening procedure and visual half-field tasks, have also shown a significant relationship to cerebral dominance as assessed by the IAP (Kimura, 1961; Strauss et al., 1985), but these procedures, also, are not sufficiently reliable to predict laterality in the individual case.

Electrical stimulation is an alternative direct method for determination of hemispheric language dominance (Penfield and Roberts, 1959). Stimulation studies can be performed in the operating room with local anesthesia or outside the operating room if the patient has implanted cerebral electrodes. In either instance, a small amount (up to 15 mA) of current is applied to a specific pair of electrodes and evidence of language interruption is noted (Ojeman, 1983; Lesser et al., 1986). This method is reliable and safe in experienced hands, but obvious reservations exist:

1. cranial surgery is necessary
2. the presence of negative findings (i.e., no interruption of language functions) does not necessarily indicate that the hemisphere being stimulated is not language dominant; the stimulation may be insufficient or the electrode may not be optimally placed and
3. only with specific preparation can both hemispheres be tested (i.e., only when bilateral implants or bilateral craniotomies are performed).

Patients most likely to undergo the IAP for solely determining language dominance are neurosurgical candidates in whom the planned excision may encroach upon critical language areas. These patients include either right hemisphere or left hemisphere surgical candidates in whom there is a possibility of abnormal speech representation. Indications of such a possibility would be: (1) non-right-handedness; (2) unusual neuropsychological test patterns (i.e., abnormal lateralizing findings with dichotic listening or tachistoscopic studies, or lateralizing findings on the neuropsychological evaluation contralateral to the planned surgery); and (3) history of early insult to the left hemisphere.

1.2.1.2. Memory Assessment. Clinical validation of IAP for prediction of a potential amnestic syndrome in the temporal lobe surgical candidate has been based primarily on negative findings reported from the MNI (Penfield and Mathieson, 1974). The recently reported combined experience of 15 epilepsy centers provides similar negative data. Patients who did not become transiently amnestic with amobarbital perfusion of the hemisphere containing a known temporal lobe seizure focus (the "epileptic" hemisphere) did not subsequently become amnestic after unilateral temporal lobectomy (Rausch, 1987). Positive validation of the IAP for prediction of postlobectomy amnesia is difficult to establish. Demonstration of such would require the identification of patients in whom a *predicted* amnestic syndrome was *documented* after a unilateral mesial temporal resection. Such cases are rarely encountered. The prediction of a postlobectomy amnestic syndrome is, for practical and ethical reasons, rarely put to the test. One case of predictable global amnesia following selective amygdalohippocampectomy has recently been reviewed by Rausch et al. (1986). There have been a few reports of patients with "failing" performance on memory tasks following IAP injection of the "epileptic" hemisphere who were not amnestic following unilateral temporal lobe resection. These reports are difficult to evaluate, since no universally accepted criteria for behavior assessment after intracarotid amobarbital injection exist. A "failing" performance at one center might be considered "passing" at another center using different criteria (Rausch, 1987).

Evidence has recently been presented that shows that intact memory performance during IAP may indeed require the anatomical integrity of critical memory structures in the contralateral hemisphere. Rausch et al. (1989) reported that 5 of 6 patients with severe hippocampal sclerosis had poor memory performance following sodium amobarbital injection of the contralateral hemisphere. Noteworthy, the one patient whose IAP memory performance was intact did not have the ipsilateral posterior cerebral artery perfused with the injection. This finding indicates that the validity of the IAP to detect patients at risk for amnesia may be compromised if the hemispheric perfusion is restricted.

Use of the IAP to predict a potential amnestic syndrome is most relevant to candidates for temporal lobe resection. In some surgical centers, the IAP is performed prior to temporal lobe surgery of any kind. In others, it is performed only prior to temporal lobe surgery for treatment of epilepsy, whereas in other centers, it is performed prior to temporal lobe surgery only if there is clinical evidence (i.e., EEG or neuropsychological) of dysfunction of the contralateral hemisphere (Rausch, 1987). The variations in application as well as procedure of the IAP have made comparisons of results across centers difficult.

1.2.2. Diagnostic Value

1.2.2.1. HEMISPHERIC LANGUAGE REORGANIZATION. Patients with clinical evidence of early damage to the left hemisphere, with or without accompanying handedness change, have an increased probability of right hemispheric language dominance (Rasmussen and Milner, 1975). The presence of dysfunction of the left hemisphere early in life (as evidenced by epileptiform activity in the absence of structural damage) also increases the probability of right hemispheric language dominance. In the UCLA series, 15% of right-handed patients with seizures of left temporal lobe origin demonstrated right hemispheric language dominance, whereas *none* of the right-handed patients with seizures of right temporal origin had right hemispheric language dominance (Rausch and Walsh, 1984). The finding of abnormal hemispheric language representation has been diagnostically useful in a population of patients with intractable epilepsy of unknown etiology in whom surgical treatment is being considered. These patients have no known structural lesion, and surgery is contingent upon identifying the primary seizure focus. In diagnostically difficult cases (in which seizures of left temporal lobe origin were ultimately documented), the presence of bilateral or right hemispheric language dominance has provided additional confirmatory evidence of left hemisphere dysfunction (Engel et al., 1983, 1981).

1.2.2.2. HEMISPHERIC DYSFUNCTION INDICATOR. In an epilepsy surgery center where the IAP is routinely performed for evaluation of memory function, it has been reported that 63% of patients with seizures of temporal lobe origin demonstrated poor memory function when the hemisphere contralateral to the proposed surgery (the "nonepileptic" hemisphere) is injected (Rausch et al. 1989). Also, poor performance contralateral to the seizure focus occurred regardless of the perfusion pattern or degree of contralateral hippocampal damage. Some clinicians use memory performance during the IAP as a diagnostic indicator of the functional integrity of the noninjected hemisphere; such information may provide the confirmatory evidence of dysfunction in the suspect hemisphere (Engel et al., 1981; Rausch, 1987).

It has been reported that memory problems following injection of the "nonepileptic" hemisphere occur primarily when the contralateral lesion is localized to the temporal lobe (Milner et al., 1962). However, this finding requires confirmation. Current standard practice allows for, at best, lateralization of dysfunction. More information will be required before the IAP can be used for intrahemispheric localization.

2. Methodological Considerations

2.1. *Factors Affecting Assessment*

The IAP requires assessment of behavioral changes following injection of a centrally active drug. It is critical that stable baseline behavioral characteristics be clearly documented. The patient should be cooperative, attentive, and well-rested. Lack of cooperation or attentiveness makes evaluation of language function difficult and evaluation of memory function impossible.

The patient's age and intellectual capability must be taken into consideration when planning an IAP. The patient should have a basic understanding of the required tasks and should be able to perform during a stressful situation. In order to accommodate the younger child or individual with a lower intellectual capacity, the testing procedure may be modified. An IAP with children is best performed with a pediatric nurse or assistant who can devote their time providing psychological support to the child. In suboptimal situations, an adequate assessment of language function can usually be made. A reliable assessment of memory function, which requires greater patient cooperation, is more difficult to obtain.

It may be difficult to assess behavioral changes in patients who are acutely anxious, depressed, or psychotic. Preexisting psychiatric dysfunction may be accentuated following drug injection. In such instances, the procedure should be postponed until the patient has regained his/her baseline behavioral state. In the highly anxious patient, additional education and exposure to the testing protocol may be beneficial.

In the patient with intractable epilepsy, a recent generalized or partial seizure may interfere with the behavioral assessment. Postictal effects may be manifest as sedation, confusion, psychosis, or selectively depressed cognitive functioning. Ideally, the patient should be in a fairly stable interictal state. As a guideline, if a generalized seizure has occurred within 24 h or a partial seizure has occurred within 3 h, postponement of a scheduled IAP procedure is recommended. Patients with postictal psychosis pose a special problem. The authors have had considerable difficulty assessing behavioral changes during the IAP with several patients whose postictal psychosis had apparently resolved 1 wk prior to the procedure. These patients experienced a transient reemergence of psychiatric symptoms during the testing period, and full evaluation was felt to be unreliable. Several weeks are recommended before attempting an IAP following recovery from a postictal psychosis.

It is conceivable that anticonvulsant medication levels could affect performance during the IAP. No data are currently available to confirm or deny this possibility. The authors' anectdotal experience suggests that patients receiving sedative anticonvulsants or multiple anticonvulsants may have difficulty maintaining attentiveness during the IAP.

2.2. Neuroradiological Procedures

As initially described by Wada (1949), the IAP required percutaneous puncture of the common carotid artery, the same as was required for cerebral angiography. The IAP is now routinely performed after transfemoral catheterization of the internal carotid artery. This catheterization technique is well described in standard references (Osborn, 1980; Rumbaugh et al., 1983).

The patient is not sedated prior to the catheterization procedure. After careful local anesthesia, puncture of the femoral artery (usually the right) is performed under sterile conditions, and a guide wire and catheter are successively advanced. The catheter is connected, via a closed, sterile system, to a source of normal saline that is maintained under approximately 300 psi. The neuroradiologist can regulate, via an adjustable valve, the delivery of saline through the catheter. Small amounts of iodinated contrast are administered and, under fluroscopic control, either the right or left internal carotid artery is catheterized. In most patients, the internal carotid artery has its origin at approximately the C3-C4 level, and the catheter is placed slightly distal to this point. The timing of the procedure should be coordinated so that the sodium amobarbital injection can be performed expeditiously after documentation of correct catheter placement. Limiting the time that the catheter remains in place in the internal carotid artery will minimize the possibility of vessel wall injury or vessel occlusion. Immediately following the sodium amobarbital and subsequent flush injections, the catheter is withdrawn from the internal carotid artery and remains in place in the aorta while the testing period elapses. If further selective catheterizations are not to be performed, the catheter may be withdrawn completely at the earliest opportunity. If contralateral selective internal carotid catheterization is to be performed during the same testing session, the catheter will remain in place in the aorta for approximately 30 min, and careful attention to sterile technique will be continued.

Cerebral angiography is a necessary prerequisite for safe and efficient performance of the IAP. Prior angiography will document abnormal or anomalous vessel patterns that may influence the distribution of injected drug. Rarely, anomalous connections between the carotid and basilar arterial systems may be encountered. Failure to recognize such anomalous connections could result in inadvertent perfusion of the brainstem with sodium amobarbital intracarotid injection, causing an unexpected respiratory arrest.

Ideally, cerebral angiography and the IAP technique would be performed under identical circumstances, and with similar volumes and injection speeds. One could then reasonably assume that flow patterns noted after injection of iodinated dye would be replicated after injection of the amobarbital solution. On the contrary, if the angiographic study is performed by automated injection (at approximately 7 mL/s) the flow pattern noted may be different from that obtained with a hand injection of sodium amobarbital solution (at approximately 2 mL/s). In most cases, practical considerations make duplication of circumstances difficult.

However, if a hand injection angiogram is not performed immediately prior to the injection of sodium amobarbital, variation in the ipsilateral or bilateral distribution of the injected drug may be difficult to predict and the validity of the procedure may be compromised (Rausch et al., 1989).

Complications, both major and minor, of the IAP are similar to those reported after cerebral angiography (Mani et al., 1978; Rausch, 1987). Major complications (strokes, major arterial occlusion, respiratory arrest, death) occur at a rate of less than 1.0%. Minor complications (local hematoma formation, arterial spasm with transient neurological deficit, minor allergy to dye or drug) occur more frequently, but are of limited consequence.

2.3. Pharmacology

Amobarbital is a di-alkyl substituted oxybarbiturate with the structural formula: $C_{11}H_{18}N_2O_3$. The sodium salt used for parenteral injection has the formula: $C_{11}H_{17}N_2NaO_3$. Amobarbital is highly lipid-soluble and readily crosses the blood/CSF barrier. The kinetics of amobarbital metabolism after intravenous administration have been described (Balasubramaniam et al., 1970), but these kinetic measurements have little relevance to the particular circumstance of intracarotid injection.

Jacobs et al. (1962) have described the behavioral effects of intracarotid amobarbital in conscious intact cats, and have compared these effects to those produced by thiopental, phenobarbital, and other agents. They described two distinct syndromes that may be produced by intracarotid drug injections: a *lateralized* syndrome with prominent unilateral neurological signs and little obtundation; and a *generalized* syndrome with prominent sedation and obtundation and less prominent lateralized signs. The type of syndrome produced after intracarotid injection depends on four factors:

1. the relative permeability of the drug across the blood/CSF barrier
2. the cerebral extraction ratio (that fraction of the drug in the arterial blood extracted by the brain)
3. the systemic persistence of the active drug in the general circulation and
4. the proportion of drug bound to plasma proteins and thus not available for diffusion into the brain.

An "ideal" agent for transient hemispheric inactivation would produce an instantaneous onset of strictly unilateral dysfunction of sufficient duration to allow for full clinical testing. The agent would penetrate the blood/CSF barrier easily and would be extracted to a nearly complete extent during the initial pass through the unilateral carotid circulation. The ideal agent would then be rapidly deactivated or eliminated from the systemic circulation, thus producing only minimal *generalized* signs of sedation and obtundation. Amobarbital satisfies these criteria to a reasonable extent although not perfectly. The onset of clinical signs is often transiently bilateral, and persistence of active drug in the systemic circulation produces some sedation.

Other sedative/hypnotic agents are not used for intracarotid injection. The safety of alternative agents has not been established. There is limited evidence to suggest that thiopental may produce vascular damage when injected intra-arterially (Ghersi et al., 1954).

In Wada and Rasmussen's 1960 study, it was determined that concentrations of amobarbital above 10% produced an unacceptable rate of CNS damage in experimental animal models. Thus, concentrations of 10% or less are recommended for intra-arterial injection in human subjects.

Opinions differ as to optimal dosage, concentration, or injection speed for the IAP. Recommended dosages vary from 75–200 mg and concentrations vary from 1.25–10%. Variation in injection speed is less, with most examiners injecting the drug (by hand) over 2–6 s (Rausch, 1987). The authors have found that 125 mg of amobarbital in 10cc of normal saline injected over 4 seconds produces a fairly consistent effect.

2.4. EEG Monitoring

Most, but not all, epilepsy centers utilize simultaneous EEG monitoring during the IAP (Rausch, 1987). The EEG responses to intracarotid amobarbital have been described in detail elsewhere (Serafetinides et al., 1965; Terzian, 1964; Werman et al., 1959). The usual response is a pattern of high amplitude semirhythmic δ activity, which appears within 2 s of bolus injection. The initial scalp EEG response is frequently bilateral in its distribution, but in most cases, becomes clearly lateralized over the hemisphere ipsilateral in injection within 10 s. Less prominent responses are noted with slower or incremental injections. The immediate EEG response is determined by the speed of injection, the amount of drug injected, and the concentration of drug injected. The speed of injection is the major determinant when total dosages in the range of 100–200 mg/injection are used (Serafetinides et al., 1965; Terzian, 1964).

Simultaneous EEG monitoring during the IAP is potentially advantageous for three reasons.

1. EEG monitoring allows differentiation of partial seizures from unusual tonic responses, both of which may be occasionally encountered during performance of the IAP
2. EEG monitoring may help in evaluating obtunded states that may be seen after intracarotid amobarbital injection. Obtundation following intracarotid injection of amobarbital may indicate either an unexpected bilateral spread of injected drug or a unilateral drug effect in combination with preexisting damage in the contralateral hemisphere. Bilateral changes on the EEG suggest that obtundation is the result of bilateral drug effect. Unilateral EEG changes in this clinical situation suggest unilateral drug effect and preexisting contralateral damage
3. EEG monitoring also allows for a rough estimate of duration of drug effect and provides an independent confirmation of hemispheric recovery after intracarotid drug injection.

EEG changes after intracarotid amobarbital injections may provide other information in particular instances. Duration of unilateral slowing may have a relationship to unilateral preponderance of cerebral damage (Rausch et al., 1984; Serafetinides et al., 1965). Changes in epileptiform discharges (over the lateral surface or in the temporal depth) may be seen after intracarotid amobarbital injection (Rovit et al., 1961; Perez-Borja and Rivers, 1963; Coceani et al., 1966; Garretson et al., 1966). These changes are currently of little diagnostic use. The mechanism by which intracarotid amobarbital perfusion produces changes in epileptiform discharges in mesial temporal lobe structures is poorly understood (Perez-Borja and Rivers, 1963).

Montage selection for routine EEG monitoring is largely a matter of professional preference as long as a bilaterally symmetric array of scalp or depth electrodes is chosen. It is generally not useful to employ sphenoidal electrodes because of their susceptibility to artifact induced by jaw movements. The environment of the neuroradiological suite is often "electrically hostile," and EEG artifacts similar to those seen in an ICU setting are frequently encountered. Sixty Hz activity is a commonly seen artifact, and careful attention to balanced electrode impedences will minimize this problem.

Patients undergoing catheterization of the great vessels must be considered electrically "sensitive," and strict electrical safety standards are mandatory. Leakage current of the electroencephalograph and its connections must not exceed 20 μA. The use of extension electrical supply cords is forbidden. All electrical equipment attached to the patient should be connected to a common group of electrical outlets that share a single pathway to ground. Double grounding should be avoided (Seaba, 1980). The use of an isolated or "floating" ground on the scalp is permissable provided that the integrity of the isolation device is documented.

2.5. Behavioral Assessment

2.5.1. General Protocol

The general schema of the neuroradiology suite is shown in Fig. 1. Prior to each injection, practice tasks are given, and assessment of baseline language and memory function is made. The patient is then given two items to commit to memory. Prior to the injection, the patient is asked to slowly count aloud, while the grip strength of both hands is monitored. Within seconds of the injection, behavioral changes occur. With injection of the dominant hemisphere, counting stops. Following injection of the nondominant hemisphere, counting may either continue or stop temporarily. A profound hemiparesis occurs with maximal weakness in the contralateral upper extremity. In approximately 40% of the cases, hemianopsia is present (Kløve et al., 1970). Conjugate eye deviation (toward the injected hemisphere) frequently occurs. Continued assessment of the presence and extent of neurological deficits provides an indication of the duration of the drug effect, but these gross measures of dysfunction are sometimes subject to unpredictable fluctuation. Thus, additional independent indicators of drug effect are helpful. The presence of unilateral EEG slowing is one such additional indicator. The degree of language recovery may be

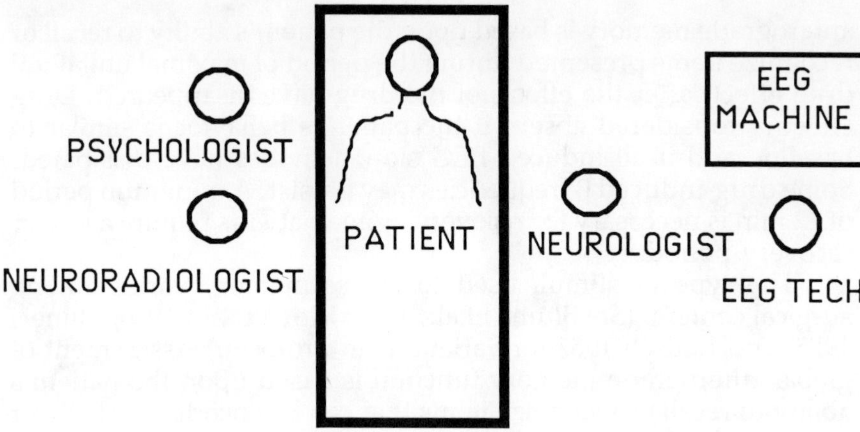

Fig. 1. Schema of IAP testing suite.

used to assess drug effect when the dominant hemisphere is injected. The duration of the maximal drug effect varies among patients and may last from 90 to 300s.

2.5.2. Language Evaluation

Language is assessed immediately after the injection, following orientation of the patient and prior to presentation of the memory items. This is usually during the first 90 s after injection. The patient is asked to continue counting (if she/he has stopped), repeat words after the examiner, read simple words and/or sentences, name pictures of common objects, abstract verbally (such as defining words), and perform simple motor commands. Testing of sequential language and word fluency is performed if time permits. Different language functions recover at different rates following amobarbital injection of the language dominant hemisphere (Rausch, 1985). Time elapsed since injection is, therefore, a potentially important variable in evaluating language function.

2.5.3. Memory Evaluation

Both retrograde and anterograde memory function are assessed. Memory items are presented prior to injection and during the period of maximal unilateral drug effect following injection. Assessment of retrograde memory is based upon the patient's ability to recall or recognize the items presented prior to injection, after the effects of the injection have disappeared. Assessment of anterograde memory is based upon the patient's ability to recall or recognize items presented during the period of maximal unilateral drug effect, after the effects of the drug have disappeared. Drug effect is considered absent if the patient's behavior is similar to baseline and if all induced EEG slow activities have dissipated. Some drug-induced β frequencies may persist. A minimum period of 12 min is necessary for recovery. Some patients require a longer recovery period.

The type of stimuli used to assess memory varies among surgical centers. (*See* Blume et al., 1973; Kløve et al., 1970; Milner, 1975, and Rausch 1987 for variations in protocol.) Assessment of global anterograde memory function is based upon the patient's ability to recall or recognize items that can be encoded verbally or nonverbally. The stimuli can be actual objects or pictures of common objects. Complex line drawings are not recommended, since they are sensitive to selective hemispheric functions (Levine and Banich, 1982; Warrington and James, 1967). Verbal items, such as words read or repeated by the patient, may be used to assess memory function of the dominant hemisphere selectively (following injection of the nondominant hemisphere). Care must be taken to present visual items in the intact visual field to avoid the possible confounding effect of a temporary visual field deficit. As time permits, five to ten memory items are shown to the patient during the period of maximal unilateral drug effect. Memory for these items, assessed following return to baseline as defined above, is first attempted by recall. If the patient cannot spontaneously recall the items, he/she is asked to recognize the items among a set that contains the items shown as well as a sufficient number of matched foils. The items shown and the foils should be matched in difficulty (such as ease of recognition) and in frequency of exposure to the name of the item (Francis and Kucera, 1982). Generally, the patient is not penalized by failure to recall an item spontaneously if he/she can recognize the item correctly among a set. Figure 2 shows items that may be used to assess memory functioning.

2.6. Interpretations

In order to assess language and memory function reliably, the mental status of the patient during the procedure must be considered. The cooperation of the patient should be assessed, and the ability of the patient to perceive the presented items should be

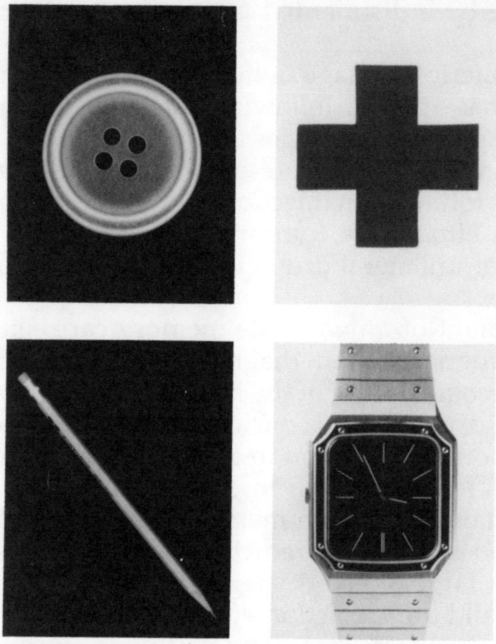

Fig. 2. Examples of items that may be shown during the memory component of the IAP. Memory for these items is assessed after baseline is obtained. Following testing of spontaneous recall, recognition is by forced-choice from a set of 15 items, which contains matched foils. The probability of correctly guessing 3 out of the 15 items would be 0.002. Also shown is a simple geometric shape; recognition memory is similarly assessed from an array.

documented. Following injection of the nondominant hemisphere, this can be accomplished by noting the patient's verbal responses. Assessment of cooperation and perception following injection of the dominant hemisphere is more difficult, since the patient is aphasic and apraxic. The authors rely upon eye movements to indicate whether or not the patient is attending to the individual items. Visual fixation and tracking provide evidence that the dysphasic subject is orienting to presented items. If disorientation occurs, but language skills can be fully elicited and memory is assessed as intact, the test results are reliable. However, when the patient is unable or unwilling to respond, it is difficult to determine whether or not a selective deficit is present. A poor performance by a disoriented patient cannot be considered reliable.

Hemispheric language dominance is determined by the presence of global aphasia following injection of one hemisphere, when no language deficiencies are seen following injection of the other. Bilateral language representation can be inferred when language errors follow both injections or when no errors follow either injection. In either case, care must be taken that there are clear indications of unilateral drug effect after each hemispheric injection.

Assessment of anterograde memory capability is the critical issue for prognostic and diagnostic purposes. Failure to subsequently recognize 67% of the items presented during the period of unilateral drug effect is considered by the authors as an indication of memory dysfunction in the contralateral hemisphere. [*See* Rausch (1987) for variations in criteria among surgical centers.]

Performance scores on memory tasks during the IAP should not be used in isolation to predict postlobectomy memory function. Although a large body of negative results suggests that this method is valid for prediction of postlobectomy amnesia, positive validating evidence is scarce, and questions remain concerning the reliability of the technique. Thus, a "passing" recognition score of 67% in an individual case does not ensure that global memory function will be adequate after unilateral temporal lobectomy, particularly if there are independent indications of bilateral temporal lobe impairment and if the ipsilatural posterior cerebral did not fill with the amobarbital injection. Similarly, a "failing" recognition score of less than 67% does not indicate with certainty that a unilateral temporal lobe resection will produce a severe amnestic syndrome. A battery of investigations designed to identify structural and functional CNS deficits should be performed before a patient is considered for temporal lobe resection (Engel et al., 1981). This information must be considered along with the results of the IAP before a realistic assessment of risk can be formulated.

3. Summary

The intracarotid injection of sodium amobarbital produces a transient and unilateral suppression of hemispheric function. A systematic evaluation of behavior during the transient period of hemispheric suppression makes possible: (1) the identification of deficits resulting from drug effect, and (2) the evaluation of the functional integrity of the contralateral hemisphere. This technique is the definitive method for determination of hemispheric language dominance. It is widely used for assessment of memory capability in patients with intractable complex partial epilepsy of temporal lobe origin. The results may be used to predict a potential amnestic syndrome in patients being considered for temporal lobe resections. It may also provide indirect diagnostic evidence of focal cerebral dysfunction. The technique that is supported by a large anectodal experience is simple in concept and relatively safe in practice.

References

Balasubramaniam K., Lucas S. B., Mawer G. E., and Simons P. J. (1970) The kinetics of amylobarbitone metabolism in healthy men and women. *Br. J. Pharmacol.* **39,** 564–572.

Blume W. T., Grabow J. D., Darley F. L., and Aronson A. E. (1973) Intracarotid amobarbital test of language and memory before temporal lobectomy for seizure control. *Neurol.* **23,** 812–819.

Branch C., Milner B., and Rasmussen T. (1964). Intracarotid sodium amytal for the lateralization of cerebral speech dominance. Observations in 123 patients. *J. Neurosurgery* **21,** 399–405.

Broca P. (1865) Sur la faculté du langage articulé. *Bull. Soc. d' Anthropol.* (Paris), **6,** 337–393.

Coceani F., Libman I., and Gloor P. (1966) The effect of intracarotid amobarbital injections upon experimentally induced epileptiform activity. *Electroenceph. Clin. Neurophysiol.* **20,** 542–558.

Engel J., Jr., Crandall P. H., and Rausch R. (1983) The Partial Epilepsies, in *The Clinical Neurosciences,* Vol. 2. (Rosenberg R. N., Grossman R. G., Schochet S., Heinz E. R., and Willis W. D., eds.), Churchill Livingstone, New York, pp. 1349–1380.

Engel J. Jr., Rausch R., Lieb J. P., Kuhl D. E., and Crandall P. H. (1981) Correlation of criteria used for localizing epileptic foci in patients considered for surgical therapy of epilepsy. *Ann. Neurol.* **9,** 215–224.

Francis W. N. and Kučera H. (1982) *Frequency Analysis of English Usage.* (Houghton Mifflin Company, Boston).

Garretson H., Gloor P., and Rasmussen T. (1966) Intracarotid amobarbital and metrazol test for the study of epileptiform discharges in man:

A note on its technique. *Electroenceph. Clin. Neurophysiol.* **21,** 607–610.

Ghersi J. A., Costales A., and Mayo F. (1954) Posibilidades de la anastesia con Pentothal por via de la arteria canalizada durante las angiografias cerebrales. *Prensa med argent.* **41,** 800–803.

Gloning I., Gloning K., Haub G., and Quatember R. (1969) Comparison of verbal behavior in right-handed and non-right-handed patients with anatomically verified lesion of one hemisphere. *Cortex* **5,** 43–52.

Jacobs G. B., Rothballer A. B., Coppola F. C., and Jarvik M. E. (1962) Effects of intracarotid and intravertebal thiopental, amobarbital, phenobarbital, chlorpromazine and diphenylhydantoin in concious, intact cats. *Int. J. Neuropharmacol.* **1,** 323–332.

Kimura D. (1961) Cerebral dominance and the perception of verbal stimuli. *Can. J. Psychol.* **15,** 166–171.

Kløve H., Trites R. L., and Grabow J. D. (1970) Intracarotid sodium amytal for evaluating memory function. *Electroenceph. Clin. Neurophysiol.* **28,** 418–419.

Lesser R. P., Lüders H., Morris H. H., Dinner D. S., Klem G., Hahn J., and Harrison M. (1986) Electrical stimulation of Wernicke's area interferes with comprehension. *Neurol.* **36,** 658–663.

Levine S. C. and Banich M. T. (1982) Lateral asymmetries in the naming of words and corresponding line drawings. *Brain Lang.* **17,** 34–45.

Mani R. L., Eisenberg R. L., McDonald E. J., Jr., Pollack J. A; and Mani J. R. (1978) Complications of catheter cerebral arteriography. Analysis of 5,000 procedures. I. Criteria and incidence. *Am. J. Roentgenol.* **131,** 861–865.

Milner B. (1975) Psychological aspects of focal epilepsy and its neurosurgical management, in *Advances in Neurology* Vol. 8 (Purpura D. P., Penry J. K., and Walter R. O., eds.), Raven Press, New York, pp. 299–321.

Milner B., Branch C., and Rasmussen T. (1962) Study of short-term memory after intracarotid injection of Sodium Amytal. *Trans. Am. Neurol. Assoc.* **87,** 224–226.

Ojemann G. A. (1983) Brain organization for language from the perspective of electrical stimulation mapping. *Behav. Brain Sci.* **6,** 189–230.

Osborn A. G. (1980) Technical aspects of cerebral angiography, in *Introduction to Cerebral Angiography.* (Harper and Row, Philadelphia).

Penfield W. and Mathieson G. (1974) Memory: Autopsy findings and comments on the hippocampus in experiential recall. *Arch. Neurol.* **31,** 145–154.

Penfield W. and Milner B. (1958) Memory deficits produced by bilateral lesions in the hippocampal zone. *AMA Arch. Neurol. Psych* **79,** 475–497.

Penfield W. and Roberts L. (1959) *Speech and Brain-mechanisms.* (Princeton University Press, Princeton, New Jersey).

Perez-Borja C. and Rivers M. H. (1963) Some scalp and depth electrographic observations on the action of intracarotid sodium amytal injection on epileptic discharges in man. *Electroenceph. Clin. Neurophysiol.* **15,** 588–598.

Rasmussen T. and Milner B. (1975) Clinical and surgical studies of the cerebral speech areas in man, in *Cerebral Localization* (Zülch K. J., Creutzfeldt O., and Galbraith G. C., eds.), Springer, Berlin, pp. 238–257.

Rausch R. (1985) Recovery rates of selective behaviors following intracarotid sodium amobarbital injections. *International Neuropsychological Society Abstracts,* San Diego.

Rausch R. (1987) Psychological evaluation, in *Surgical Treatment of the Epilepsies* (Engel J., Jr., ed.), Raven Press, New York, pp. 181–195.

Rausch R. and Walsh G. O. (1984) Right-hemisphere language dominance in right-handed epileptic patients. *Arch. Neurol.* **41,** 1077–1080.

Rausch R., Babb T. L., and Brown W. J. (1985) A case of amnestic syndrome following selective amygdalohippocampectomy. *J. Clin. Exp. Neuropsychol.* 7(6), 643.

Rausch R., Babb T. L., Engel J. Jr., and Crandall P. H. (1989) Memory following sodium amobarbital injection contralateral to hippocampal damage. *Arch. Neurol.* 46, 783–788.

Rausch R., Fedio P., Ary C. M., Engel J., Jr., and Crandall P. H. (1984) Resumption of behavior following intracarotid sodium amobarbital injection. *Ann. Neurol.* **15,** 31–35.

Rovit R. L., Gloor P. and Rasmussen T. (1961) Intracarotid amobarbital in epileptic patients. *Arch. Neurol.* **5,** 42–62.

Rumbaugh, C. L., Kido, D. K., and Baker, R. A. (1983) Cerebral angiography: technique, indications and hazards, in *Angiography: Vascular and Interventional Radiology* Vol. 1. (Abrams H. L., ed.), Little Brown, Boston.

Scoville W. B., and Milner B. (1957) Loss of recent memory after bilateral hippocampal lesions. *J. Neurol. Neurosurg. Psychiatr.* **20,** 11–21.

Seaba P. (1980) Electrical Safety. *American J. EEG Tech.* **20,** 1–13.

Serafetinides E. A., Driver M. V., and Hoare R. D. (1965) EEG patterns induced by intracarotid injection of sodium amytal. *Electroenceph. Clin. Neurophysiol.* **18,** 170–175.

Strauss E., Wada J., and Kosaka B. (1985) Visual laterality effects and cerebral speech dominance determined by the carotid Amytal test. *Neuropsychologia* **23 (4),** 567–570.

Terzian H. (1964) Behavioural and EEG effects of intracarotid sodium amytal injection. *Acta Neurchir (Wien)* **12,** 230–239.

Wada J. (1949) [A new method for the determination of the side of cerebral speech dominance. A preliminary report on the intracarotid injection of sodium Amytal in man.] *Igaku to Seibutsugaku (Medicine and Biology)* **14,** 221–222 (Japanese)

Wada J. and Rasmussen T. (1960) Intracarotid injection of sodium amytal for the lateralization of cerebral speech dominance. *J. Neurosurgery* **17,** 266–282.

Warrington E. K. and James M. (1967) Disorders of visual perception in patients with localised cerebral lesions. *Neuropsychologia* **5,** 253–266.

Werman R., Anderson P. J., and Christoff N. (1959) Electroencephalographic changes with intracarotid megimide and amytal in man. *Electroenceph. Clin. Neurophysiol.* **11,** 267–274.

Zangwill O. L. (1960) *Cerebral Dominance and its Relation to Psychological Function.* (Oliver and Boyd, Edinburgh).

Testing the Commissurotomy Patient

Eran Zaidel, Dahlia W. Zaidel, and Joseph E. Bogen

1. Introduction

Testing of split-brain patients over the last 25 years has involved a myriad of procedures, both clinical and experimental. Some were adapted from animal testing, others from clinical neurology, and more from experimental psychology. Of these procedures, some proved cumbersome, others unrewarding, and still others misleading. Those that survived the test of time have been progressively improved by simplification, by the introduction of technological advances, and, above all, by increasing sophistication on the part of the examiners. Experienced split-brain experimenters are often surprised when noted and unquestionably competent neuropsychologists who have a rich experience in testing hemisphere-damaged patients or in assessing laterality effects in normal subjects show themselves initially unequal to the task of testing the commissurotomy patient. The chronic disconnection syndrome is dramatic, widely known, and readily explicable. However, the arsenal developed to assess it is complex, sometimes subtle, and often based on implicit assumptions. This chapter aims to describe that arsenal and make explicit those assumptions. The chapter addresses three interrelated questions: how to find out whether a patient exhibits the disconnection syndrome, how to test hemispheric functions in such a patient once diagnosed correctly, and how to explore the current frontier of extra-callosal communication.

1.1. Disconnection Syndrome

Patients who have had complete cerebral commissurotomy for intractable epilepsy, including sectioning of the corpus callosum, anterior commissure, hippocampal commissure, and massa intermedia (when visualized), are generally unable to transfer high-level information from one cerebral hemisphere to the other (Sperry et al., 1969; Gazzaniga, 1970; Sperry, 1974; Sperry, 1982; Bogen, 1985). The right-handed subject with speech in the left hemisphere (LH) cannot read words or name pictures shown in the left hemifield (left hemialexia); this subject is unable to name objects palpated by the left hand (unilateral tactile anomia), and is unable to compare stimuli between the two hands or across the two visual hemifields. Cross-modal transfer across the midline similarly fails. For example, the patient is unable to retrieve with the left hand an object whose picture had been shown in the right hemifield. The same tasks are performed normally when there is no crossing of the midline, so that the same hemisphere perceives the stimuli and controls the responses. The acute disconnection syndrome includes left-handed apraxia to verbal command together with good left-hand imitation of the same actions. As ipsilateral motor control of the left hand develops in the LH, unilateral apraxia subsides. During the early postoperative period, some patients exhibit intermanual conflict that subsides within several months or even weeks, as compensatory noncallosal integrative mechanisms take over (Bogen, 1987).

The acute disconnection syndrome often includes short-term mutism that persists in a few cases, perhaps those with discordant manual and speech dominance (Bogen, 1987). In the chronic syndrome, however, personality and character remain remarkably unchanged. More or less subtle and persisting deficits often include a poor short- and long-term memory (*see* D. W. Zaidel, in press, for a review), impoverished linguistic description of the patient's emotions (alexithymia; TenHouten et al., 1986), and poor execution of certain pragmatic linguistic functions, including the appreciation of emotion in sentence prosody, the appreciation of metaphor, and discourse processing in auditory presentations and, even more, in reading (E. Zaidel, in press). These deficits presumably reflect failure of normal right hemisphere (RH) contribution to language functions.

1.2. Clinical Evaluation

Visual disconnection can best be demonstrated using half-field tachistoscopy *(see below)*. Visual stimuli can be presented selectively to a single hemisphere by having the patients fix his/her gaze on a screen onto which pictures are projected to either half-field, using exposure times of 150 ms or less. The split-brain patients can read and describe material in the right visual half-field (RVF) essentially as they could before surgery. When stimuli are presented to the left visual half-field (LVF), however, the patients usually report that they see "nothing" or "a flash of light." The disconnection can sometimes be demonstrated with simple confrontation testing. The patient is allowed to have both eyes open, but does not speak and is allowed to use only one hand (sitting on the other, for example). Using the free hand, the subject indicates the onset of a stimulus, such as the wiggling of the examiner's fingers. With such testing, there may appear to be an homonymous hemianopia contralateral to the indicating hand. When the patient is tested with the *other* hand, there seems to be an homonymous hemianopia in the *other* half-field. Occasionally, a stimulus in the apparently blind half-field (on the left when the right hand is being used) will produce turning of the head and eyes towards the stimulus, and *then* the hand will point. This situation must be distinguished from extinction or hemi-inattention deficits following a hemispheric lesion. In the latter case, the patient tends to indicate only one stimulus when the stimuli are in fact bilateral. The double hemianopia is a symmetrical phenomenon, whereas extinction or hemi-inattention is typically one-sided, more commonly to the left.

One can elicit the disconnection syndrome in the clinic by showing failure of intermanual cross-retrieval of small test objects, of cross-replication of hand postures, or of cross-localization of finger tips (*see below:* tactile testing). One of the most convincing ways to demonstrate hemispheric disconnection is by unilateral (left) tactile anomia. The examiner asks the patient to feel with one hand, and then to name various small, common objects, such as a button, safety pin, paper clip, rubber band, key, or the like. It is essential that vision be occluded. A blindfold is notoriously unreliable. It is better to hold the patient's eyelids closed, to put a pillowcase over the patient's head, or to use an opaque screen. The split-brain patient is generally unable to name or describe objects in the left hand, although the patient can readily name the same objects in the right hand. This deficit has been present and persistent in *every right-handed* patient with complete cerebral commissurotomy.

To establish hemisphere disconnection, it is necessary to exclude other causes of unilateral anomia, particularly astereognosis, which may occur with a right-parietal lesion. One can often reasonably exclude astereognosis simply by observing the appropriate manipulation of an object. The most certain proof that the object has been identified is for the subject to retrieve it correctly from a collection of similar objects. Such a collection is most conveniently placed on a paper plate about 12–15 cm in diameter, around which the subject can shuffle the objects with one hand while exploring for the test object. In testing for anomia, one must be aware of strategies for circumventing the defect. For example, the patient may manipulate the test object in some way to produce a characteristic noise, or the subject may identify it by a characteristic smell, and thus circumvent the inability of the LH to identify, by palpation alone, an object in the left hand.

Memory deficits are apparent in the patients' conversations. The same stories and jokes are repeated in separate encounters, and the patients are selectively poor at recalling the recency of events and ordering recent events in the correct chronological order (D. Zaidel and Sperry, 1974; E. Zaidel, in press). More formally, the memory deficit can be demonstrated by comparing the memory quotient on the Wechsler Memory Scale to the intelligence quotient on the Wechsler Adult Intelligence Scale, pre- and postoperatively (D. W. Zaidel, in press).

Pragmatic deficits in conversation are difficult to quantify and to distinguish from a personality disorder (E. Zaidel, in press). Quantifiable deficits have been observed in the "prosody," "pictorial metaphor," and "story recall" subtests of the Right Hemisphere Communication Battery (Gardner and Brownell, 1986). These subtests reflect impaired receptive intonation, indirect speech acts, and discourse processing, respectively (E. Zaidel, in press). Unfortunately, performance on the battery depends on a good memory, and presupposes intact intelligence and a fairly high educational level, especially in mastery of vocabulary. Consequently, the battery may not be best suited for assessing pragmatic deficits in commissurotomy patients.

1.3. Hemispheric Independence

Comparing the competence of the two disconnected hemispheres on a variety of tasks confirms the broad outlines of the principles of hemispheric specialization gathered from patients with hemispheric damage. The LH is specialized for language, especially speech, phonology, and syntax, whereas the RH is superior on certain visuo-spatial and perceptual tasks, such as recognizing emotions in faces. However, the split-brain experiments also demonstrate that each hemisphere is a separate and independent cognitive system, with its own perception, cognition, memory, and language. The two disconnected hemispheres appear capable of working independently and in parallel, using different competencies and strategies. Researchers working with commissurotomy patients quickly adopt the "split-brain lingo," addressing the two hemispheres separately, and referring to them as different "persons" who are occasionally in conflict. Typical expressions are, "The RH did well today," or "The LH was rather upset when the RH could do the task."

Theoretically, the split brain provides the criterial experiment for laterality research. It operationalizes the concept "degree of hemispheric specialization" by demonstrating independent processing of the same task in each hemisphere. Thus, the disconnection syndrome makes it well-defined and coherent to use such expressions as, "Hemisphere *x* scored *s* on test *z*." From this, it is easy to go on to operationalize the concept, "Hemisphere *x* performed better than hemisphere *y* by amount *d*." In other words, the split brain provides a criterial experiment for the concept of relative hemispheric specialization in the normal brain (Harshman, 1980).

The split-brain paradigm introduces into studies of laterality effects in normal subjects a systematic distinction between (1) tasks that can be performed by either hemisphere "direct access"-fashion, albeit using different strategies and exhibiting different abilities, and (2) tasks that can be performed only by one hemisphere with specialized processing machinery, so that stimuli reaching the other hemisphere (in the normal brain) must be relayed across the corpus callosum prior to processing ("callosal relay" model). "Direct-access" tasks reflect hemispheric independence and relative specialization, whereas "callosal-relay" tasks reflect exclusive hemispheric specialization. Direct-access tasks should show comparable laterality effects in the split and normal brain, but callosal-relay tasks should show much larger laterality effects in the split than in the normal brain. Thus, a laterality effect in a direct-access task reflects relative specialization, whereas in a callosal-relay task, it reflects callosal connectivity as well.

2. Stimulus Modalities

The general logic for studying hemispheric specialization in the split brain is to restrict sensory input and motor response to one hemisphere at a time, and compare latency or accuracy in the two conditions. In the case of visual and somesthetic input, predominantly contralateral innervation guarantees that LVF and left-hand information will reach the RH, whereas RVF and right-hand input will reach the LH. In the case of auditory stimuli, contralateral input can be assumed only when two acoustically similar, but not identical, stimuli reach both ears simultaneously (dichotic listening, *see below*) For motor responses, it is assumed that each hemisphere has better control of the contralateral hand, especially at distal joints, but in the chronic disconnection syndrome, both hemispheres develop ipsilateral motor control sufficient for simple actions, such as binary choices. Consequently, most experiments rely on complete or partial lateralization at the input side, although, theoretically, either stimulus or response lateralization should suffice. It is also usually assumed that speech responses originate in the LH.

2.1. Visual Testing

Most experiments rely on lateralized visual stimuli, because it is relatively easy to restrict stimuli to one hemifield, and because this permits presentation of more complex and naturalistic stimuli.

2.1.1. Half-Field Tachistoscopy

The methodology of using brief, lateralized, visual presentations for studying hemispheric specialization in split-brain patients is essentially identical to that used with normal subjects. Lateralized stimuli are presented for less than 150 ms, in order to prevent the confounding effects of involuntary saccadic eye movements towards the stimuli. The latency of such saccades is about 180 ms. One common difference from testing normal subjects is the use of bimanual response buttons, so that, on a given trial, a random presentation of a stimulus to the left or right hemifield is paired with a response by the left or right hand, respectively.

2.1.1.1. STIMULI. The usual methodological concerns about using hemifield presentations recur here. The left hemifields of both eyes project to the right hemiretinae and from there to the right hemisphere, and the right hemifields project to the left hemiretinae and to the left hemisphere. Fibers from the nasal hemiretinae cross at the optic chiasm. The crossed fibers have somewhat stronger anatomical projections than the ipsilateral pathways, and this may confer on the former some advantage for simple psychophysical functions (Davidoff, 1982). This suggests the use of binocular presentations, and possibly the exclusion of subjects with a strong eye dominance, to counteract possible asymmetrical confounds. However, neither factor (dominance of crossed or nasal projections, or eye dominance) has ever been shown to affect laterality effects for higher functions with monocular presentations.

Animal research suggests that there is some bihemispheric anatomical representation around the vertical meridian, up to 5° in cats and 1° in monkeys. However, clinical and experimental studies in humans, including split-brain patients, suggest no overlap to within several minutes of arc. At any rate, the standard procedure of presenting stimuli with their centermost edges at least 1° away from fixation is safe.

When testing commissurotomy patients, it is customary to alternate visual fields in a pseudorandom order to ensure central fixation and avoid the possible set effects of blocked trials. Even so, it is theoretically possible to circumvent proper lateralization by fixating laterally (e.g., to the left) so that both left-sided and right-sided stimuli actually reach the same (e.g., the left) hemisphere. This is easy to check by EOG measurement of eye movements, by videotaping the eyes, or by direct inspection from behind and through a small hole in the projection screen. In our experience, such precautions are unnecessary with experienced split-brain patients who fixate properly on a vast majority of trials. Similarly, changing eye accommodation to focus on a plane in front of or behind the screen can change lateralization, but this will involve some loss of acuity, and there is no evidence that such accommodation ever happens. More serious is the possibility of divergent focus in the two eyes as a result of dyplopia or strabismus. For the above reasons, we prefer to test the patients in monocular vision with an eye patch over the nondominant (usually left) eye.

2.1.1.2. Responses The standard procedure is to require the patient to respond with the left hand to LVF stimuli and with the right hand to RVF stimuli. This is the patient's natural tendency; it requires little explanation or training, and results in few crossed responses. However, binary-choice reaction-time experiments, with blocked unimanual responses, reveal no interaction between response hand and visual field of presentation (Zaidel E., 1983). This is because of effective ipsilateral motor control in both hemispheres for simple manual choices. Thus, unimanual responses cannot be assumed to reflect decisions in the contralateral disconnected hemispheres. Curiously, a significant VF by hand interaction does occur in these patients in simple reaction time (RT) to light flashes. The crossed-uncrossed difference in complete commissurotomy patients ranges from 30–90 ms, as compared to a normal mean of 2–3 ms (Clarke and Zaidel, 1989). The discrepancy may be caused by differences in subcallosal transfer of certain signals, and/or ipsilateral motor control. We also find that crossing the hands has no effect on either simple or choice RT, implying the absence of spatial compatibility effects.

In contrast with unimanual responses, vocal output is generally assumed to reflect LH processing. Although this issue has incurred some recent controversy *(see below)*, we believe that in the absence of early, massive damage to language areas in the LH, speech does not develop in the disconnected RH. Conceivably, failure of verbalization of stimuli that are restricted to the LVF or to the left hand might, in some cases, reflect access to part of the LH that is intrahemispherically disconnected from speech centers in the same side. This was never tested experimentally although, in general, it is quite unlikely.

2.1.1.3. CHIMERIC PRESENTATIONS. One variant of hemifield tachistoscopy employed successfully with split-brain patients, and subsequently extended to normal subjects, is the use of chimeric stimulus figures divided down the vertical midline, so that each hemifield receives one-half of a different picture (Levy et al., 1972). Under these conditions, each hemisphere seems to complete its half of the chimera. Levy et al. required patients to respond vocally or by pointing unimanually to multiple choice arrays exposed in free vision. They found different patterns of hemispheric dominance, depending on the nature of the stimulus and the task. This competitive paradigm allowed them to relate hemispheric processing styles to patterns of interhemispheric control. This paradigm had been adapted for testing normal subjects by placing a narrow vertical strip along the edge of the two half-chimeras, so that the stimuli are not recognized as chimeric (*see* Bradshaw and Nettleton, 1983, for a review).

2.1.2. Alternative Media for Stimulus Presentation

In our laboratories, we have recently shifted from using box and projection tachistoscopes to a computerized system for lateralized CRT presentations. We use the same system for testing normal subjects and split-brain patients. One important advantage of the computerized system is the ability to easily randomize and counterbalance the order of stimulus presentations. The system has facilities for designing alphabetic and graphic stimuli, specifying the experimental design, running the experiment on-line, gathering data, and performing statistical analyses. Preliminary experiments with both normal subjects and split-brain patients suggest that light computer fonts on a dark background exhibit smaller laterality effects than similar fonts on slides rear-projected onto a screen by a projection tachistoscope. On the other hand, reverse video presentations, with dark letters on a light background, exhibit the same or greater laterality effects than slides. The reasons for this are not clear and are now under study. They may include the effects of persistence of CRT phosphor, brightness, contrast, and spatial frequency.

2.1.3. Techniques for Hemispheric Scanning of Complex Arrays

2.1.3.1. THE Z-LENS. In 1970, E. Zaidel developed a contact-lens based system that allowed free ocular scanning of complex visual arrays by one hemisphere without restriction in time (E. Zaidel, 1975). The system is a variation of the contact-lens technique for stabilizing retinal images. Here, however, it is not the stimulus image itself, but rather a half-field occluding screen, that is stabilized on the retina. The system has three components:

1. The stimulus board in the subject's lap in a dental chair
2. An optical system, with photographic lenses and a mirror, for projecting a reduced image of the stimulus board close to the subject's eye and
3. The contact lens system, which carries a collimator for focusing on the stimulus image near the eye and for supporting the half-field occluder.

The subject sits in a dental chair with the stimulus board in his or her lap and with the left (nondominant) eye patched (Fig. 1). On the dominant eye, the subject wears a contact lens with a short-

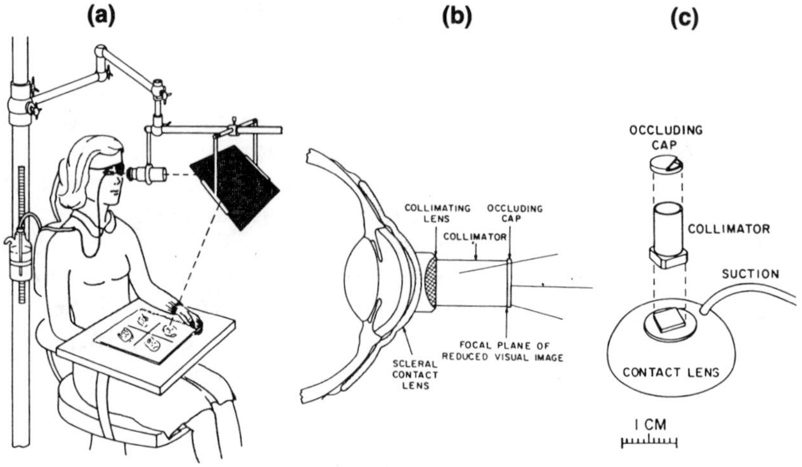

Fig. 1. The contact lens system for the scanning of complex visual stimuli by one hemisphere at a time. (a) The experimental setup. (b) A cross-section view of the eye, contact lens, and collimator. (c) The contact lens, collimator, and cap for occluding part of the visual field (E. Zaidel, 1975).

focus collimator mounted on the corneal region. The image of the stimulus board is reflected by a front surface mirror, inverted by a dove prism, reduced by a photographic objective, and projected very close to the eye, at the focal plane of the collimator. On the same plane, at the endpoint of the collimator, there is mounted a screen that occludes one-half of the visual field. The reduced aerial image of the stimulus board is stationary, and its virtual image when viewed through the collimator appears at essentially normal size and distance, though displaced along the visual axis of the subject. As the subject moves his or her eyes, the contact lens follows along faithfully, and with it, the collimator and the half-field screen. Thus, the subject can continuously scan the stimulus board and his or her own hand on it, but at each point only the same visual half-field is stimulated. Alternatively, the mirror can be replaced by a rear-projection screen for slides or films, and the dove prism is then rotated to correct the final image.

The lenses used here are individually molded, triple-curvature scleral (haptic) lenses, flush-fitting at the sclera, with minimal clearance at both the cornea (to allow back-surface optical correction) and the limbus (where pain receptors abound.) The lens has a ring of contact on the corneal surface just above the margin of the limbus. This provides superior stability, similar to that provided by lenses with close fit at the limbus and with complete corneal contact, without, however, sacrificing comfort.

Subjects wear their lenses for no more than 30 min each session, and no more than two sessions a week, in order to avoid corneal damage from the tight fit. Wearing the lens during testing incurs no discomfort, and it is not necessary to apply a local anesthetic. It is important to note that even very small air bubbles behind the lens can seriously raise the hydrostatic pressure and must be eliminated by reinserting the lens.

Before insertion, the lens is filled with a buffer solution that has virtually the same refractive index as the aqueous humor, the cornea, and the material of the contact lens, so that they act together as a single optical medium. To enhance contact lens fit and further reduce slippage, especially with large eye movement, suction is applied to the buffer solution between the lens and the cornea with a simple manometer. Optimal manometer pressure during testing was found to be –23 cm of water.

The collimator (Fig. 1) incorporates a lightweight (50 mg with a 5-mm dia), short-focus (10 mm) glass lens with acceptable aberrations. The collimator consists of a light (about 50 mg) aluminum tube, approximately 5 mm in diameter and 12 mm in length, and with walls 0.175 mm thick. The square base of the collimator fits a machined, polished step on the corneal region of the lens, ensuring consistent orientation. The collimator can be mounted on or removed from the contact lens while the lens is in place; it is not glued permanently to the contact lens, in order to avoid vapor formation in the clearance between the contact lens and the small collimator lens as a result of temperature differences. A number of aluminum caps were constructed to occlude the visual field at different longitudinal meridians. The total weight of the contact lens and collimator assembly is 800–900 mg.

During unilateral testing, one-half of the visual field is occluded, as a rule about 1.5° past the vertical meridian, the position of which is determined empirically. Because scanning eye-movements can cause lens slippage of up to 1°, the effective occlusion is approximately 0.5–1.5° past the center of the fovea. The partial foveal occlusion of the stimulated visual half-field counteracts lens slippage as well as possible information leakage to the wrong hemisphere owing to bilateral cortical representation of the retina about the vertical midline.

The clear advantage of the contact lens system over the tachistoscope is that it allows prolonged scanning of natural stimuli. In this manner, the system also circumvents any memory component that may be present in brief presentations. Once constructed, the system is also simple to use and is well-tolerated by the patients. In our laboratories, the lens was found to be far superior to tachistoscopy in eliciting RH competence and minimizing LH dominance and interference. The disadvantages of the system are:

1. The lenses need to be individually fitted and are not transferable to other patients
2. Some brain-damaged patients and children may not tolerate the lenses well
3. The system is difficult to modify and somewhat cumbersome to adapt to different testing conditions
4. The lens should not be worn for more than 45 min at a time
5. Small head movements may cause loss of focus
6. The collimator introduces some optical aberrations
7. The method does not lend itself to speeded tasks and RT measures and
8. The system is monocular.

2.1.3.2. OTHER LENS SYSTEMS. The individually molded, flush-fitting scleral lens with limbal clearance and buffer-solution manometer suction used by Zaidel can, in principle, be replaced by universal lenses, but these are difficult to align, often incur pain, and vary widely in amount of slippage caused by eye movements (E. Zaidel, 1973).

Another approach adopted by some is to use standard corneal lenses, painted black except for a small slit on the nasal or temporal side (Dimond et al., 1975). This technique relies on the eccentricity of the slit, and on the relatively small distance between eye and stimulus to avoid refraction into the wrong hemifield. However, the system does not allow for good acuity together with precise control of the extent of visual-field occlusion, because of excessive lens slippage. Bradshaw (this volume) tried the technique and found it difficult, uncomfortable, and unreliable. Sivak et al. (1985) attempted to reduce the slippage problem by using a longer, soft lens that had a flat edge and weight on the bottom to prevent rotation. However, slippage cannot be completely controlled, and problems of alignment and acuity most probably persist. Finally, Francks et al. (1985) occluded part of close-fitting goggles, but this allows eye movements and requires opaquing an area large enough to account for the most extreme lateral eye movements and the furthest distance between goggle and stimulus. The lateral limits technique, described next, uses a similar concept.

2.1.3.3. LATERAL LIMITS METHOD. This technique was developed by Myers and Sperry (1982) and replicated by Trope et al. (1988). With this method, which requires no attachments to the eye, stimuli can be presented to either visual hemifield at the corresponding lateral limits of horizontal eye rotation, where further eye movements cannot be used to transfer the stimuli into the view of the unintended hemisphere. A biteboard clamped to the edge of a table is used to hold the head of the subject in a fixed position, and the visual midlines at the limits of lateral eye rotation are determined with monocular vision, using the blindspot of each eye as a reference. Once these limits have been determined, no eyecover is needed, and lateralization to the right hemisphere can be achieved by having the subject look to the extreme left, while stimuli or response arrays are presented to the left hemifield just beyond the left lateral limit of the center of gaze (and vice versa for input to the left hemisphere). Movable panels, placed in front of the stimuli or response arrays, are used to control the timing of presentation.

The technique is monocular, with the temporal side of the visual field of the left eye feeding into the RH, and the temporal side of the RVF of the right eye feeding into the LH. The technique may be uncomfortable, and does not allow normal exploratory eye movements, because the stimulus image lies fixed beyond the subject's most lateral gaze. The eye movements themselves may introduce contralateral hemispheric bias, and headturns can also affect hemispace asymmetries (Bradshaw, this volume). In conclusion, none of the alternatives to the Z-lens provide both true hemispheric scanning and freedom from slippage; the Z-lens remains a rewarding method for extensive testing of a few selected patients.

2.1.3.4. UNIVERSAL EYE-TRACKING SYSTEMS. Instead of yoking the hemifield occluder directly to the eye via a contact lens, it is possible to track eye movements noninvasively and use the eye movement's horizontal component to control hemifield occlusion. The critical needs are to separate eye movements from head movements, obtain an accurate measure of eye movements (for example, with an error $< 30^1$ of arc) within a relatively wide visual scanning area (for instance, $\pm 10°$ of arc), and ensure occlusion in real time. Such systems are expensive, but when user-friendly enough, they can be used with commissurotomy patients, hemisphere-damaged patients, normal adults, and children. Three versions of this approach, differing in the method of hemifield occlusion adopted, have been used to a limited extent. All three versions can be used to stabilize an occluder with an arbitrary shape, i.e., to simulate an arbitrary scotoma.

2.1.3.4.1. Small Mechanical Shutter. In 1977, E. Zaidel and Frazer (1977) implemented the breadboard of a universal hemifield occluder based on a monocular generation III SRI Dual Purkinje Image Eyetracker yoked to a motor that drove a small shutter located in the plane of a reduced image of the stimulus. This solution is similar to the Z-lens, except that the occluder is mechanically driven by a motor rather than by a contact lens, so that here the optical system is stationary rather than attached to the eye. This avoids the focusing problems resulting from head (i.e., collimator) movements and permits single adjustment for optical correction with subjects who wear glasses. Again, the experimental arena is in the subject's lap (or, alternatively, a screen for projecting slides or films) to permit monitoring and visual control of hand movements on the stimulus board.

The SRI Tracker uses a collimated infrared light source to create and track reflections from the front surface of the cornea (the virtual first Purkinje image) and from the back surface of the lens (the real fourth Purkinje image). Although the fourth image is much dimmer than the first, the two are almost coincident and lie in the same focusing plane. These two images move similarly during eye translation, and differentially under rotation. The change in separation is used to determine eye rotation, free of artifacts introduced by translation. The first Purkinje image is brought to focus on a stationary photodetector by the receiving optics. Tracking is accomplished by means of centering the first image on the photodetector, using a scanning (servo) system. The fourth image is, in turn, passed onto a second photodetector after reflection from another servo-driven mirror. Since the first image is stabilized in space, the motion of the other image is exactly equal to the relative motion of the two images. From this relative motion, the direction of gaze can be accurately determined.

The early SRI tracking system incorporated a "blink" circuit and a search mode that were activated and caused instantaneous occlusion of the total field when the images were temporarily lost. Head movements within a cubic centimeter were permitted, so that a simple chin and forehead rest provided adequate head stabilization without using a biteboard. There were no attachments to the eye, and the infrared light did not interfere with or harm normal vision. The machine could track with an accuracy of better than 20^1 of arc in a visual field of $\pm 10°$ with a delay of 1–2 ms.

The occluding mask moved mechanically in a reduced image plane of the visual stimulus produced by a short focus lens (photographic objective), such as a wide-angle camera lens with f = 35 mm. Then a 40-cm stimulus 100 cm away (20° field) yields a real reduced image 1.45 cm in length, and this defines the range of movement of the mask. A second and comparable lens (Fig. 2) then reconstructed a virtual image at normal size and distance.

An ideal 1-ms response delay following a 10° saccade means about 15^1 of arc delay in shutter movement after the saccade has terminated. To prevent information leakage to the wrong visual half-field, the lightweight occluder needs only to be extended approximately .2 mm past the actual vertical midline. In control tests, we obtained an occluder response of less than 5 ms to a 10° sweep square wave simulating a saccade.

Linearization of the eye-movement signal was accomplished by calibrating a hand-wired electronic potentiometer board, adjusted on a 10×10 grid along the horizontal and vertical meridians for each subject.

Concurrently with half-field occlusion, a continuous record of the subject's eye movements was charted on an X-Y plotter and stored in videotape in the form of a fixation mark superimposed on each video frame of the picture of the stimulus board for further analysis. In this way, possible hemispheric differences in temporal and spatial patterns of ocular scanning and visual search can be probed.

2.1.3.4.2. CRT Occluders. McConkie and Rayner (1976) used a Biometrics Nacro-systems Model 200 eye-movement tracker

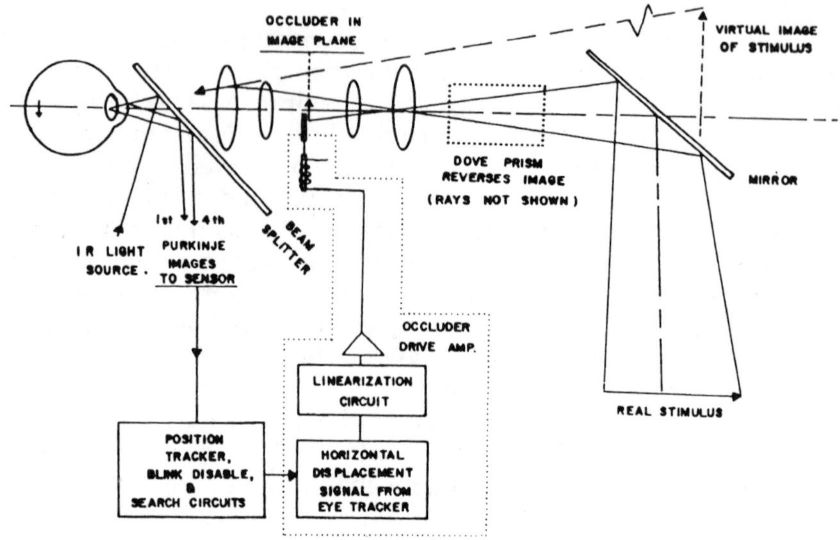

Fig. 2. A schematic diagram of a universal hemifield occluder, using the dual Purkinje image eye-tracker and a small mechanical shutter (Zaidel and Frazer, 1977).

connected to a computer. The subject's eye movements controlled the position of a "window" on the computer CRT as follows: a passage of mutilated text was initially presented on the CRT, with every letter from the original English text replaced by an X or a visual mask consisting of an interlaced square wave grating; whenever the reader fixated, a region around the fixation changed into readable text. By varying the size of the window to the left and right of fixation, it was possible to determine the extent to which the perceptual span in reading is asymmetric. Results showed a perceptual span to the right.

Pollatsek et al. (1981) used a newer version of the system to study the asymmetry of the perceptual span in Hebrew: Eye movements were recorded with a generation III dual Purkinje eye tracker that was interfaced with a Hewlett-Packard 2100A computer. The display was a Hewlett Packard 1300A CRT with a P31 phosphor that decays to 1% of maximum brightness in .25 ms. The tracker had a frequency response of 300 Hz, and its resolution was 10^1 min of arc; its output is reported to be linear over the 14° display. The signal from the eye tracker was sampled every millisecond by the computer through an A-D converter. Over each 4 ms, the horizontal voltage level was compared to the prior 4 ms, and as a result of these values, the computer determined whether the eye was in a saccade for fixation. Calculations and photoelectric testing indicated that the display change was accomplished within 2–10 ms after the termination of the saccade. This value includes the time for the computer to determine the new location of the eye, the lag in the signal from the eye tracker to the computer (about 1 ms), and the time to output the mask to the CRT. The variability in completing the display change was the result of the fact that larger masks took longer to output than smaller masks. Thus, with the smallest mask, the display change was often completed almost instantaneously (2–3 ms), whereas with the largest masks, it is possible that the change took up to 10 ms. The phenomenological experience of all of the subjects was that the window of mask moved in perfect synchrony with the eye. The computer display change could occur within 2–10 ms after the termination of a saccade. Subjects' heads were fixed to a bite bar. This time, the perceptual span was asymmetric to the left.

Nettleton et al. (1983) used a Gulf & Western Applied Science Laboratories Eye-Trac Model 200 research eye movement monitor (formerly Biometrics 200) to measure horizontal gaze relative to head by the differential reflectivity of the iris and the sclera. The system is said to be capable of measuring horizontal eye movements over a range of approximately ± 20°, with a resolution of 1°. The resolution can be improved to a few minutes of arc with a bite bar. The response time of the system is reported to be less than 9 ms. Drift is less than .2° of arc. At any point on the screen, a hemifield mask will be repositioned on each new raster scan in response to the output of the control system once every 20 ms.

It is claimed that this 20-msec delay is not a problem, because the points decay to 10% in 50 μs, but 10% may well be visible and, in any case, may have unknown subliminal effects, perhaps asymmetric. The inherent limitations in image decay and regeneration on faster CRTs can be avoided with point xyz displays possessing fast phosphors, although at a considerable cost of memory for immediate access of images. Head movements were recorded by a transducer, whose output was integrated with the output of the eye movement monitor. The display was a standard video monitor that can be used to present computer-generated stimuli, scenes fed through a video camera, or stimuli prerecorded on videotape.

2.1.3.4.3. SRI Stimulus Deflector (Scotoma Simulator). SRI's third-generation Dual Purkinje Image tracker proved too user-unfriendly for some of our patients, and the mechanical shutter lagged behind the eye by several ms with large saccades. Instead, we have recently installed a fifth-generation SRI tracker (Crane and Steele, 1985) together with an electronic/optical system for stabilizing an arbitrary scotoma on the retina (Crane and Kelly, 1983). In our case, the scotoma is of one hemifield, as in hemianopsia. The servo-controlled stimulus deflector has a much wider bandwidth than the mechanical shutter. The fifth-generation tracker is more accurate and easier to use, for both subject and examiner, than its third-generation predecessor. The tracker platform itself is auto-staged to the subject's head movements, so that tracking is lost less frequently.

One limitation of a computerized display system is the fact that only CRT-generated images can be used, and their ability to realistically simulate complex real-world (i.e., three-dimensional) images is limited. Much existing material for testing perceptual cognitive style, nonverbal intelligence, and other attributes that could be used in laterality research is difficult to implement on CRT graphics. By contrast, the real-world viewing ability of the artificial scotoma system allows use of virtually all visual materials currently available, without the burden of transcription to a digital equivalent.

This artificial scotoma simulator (or stimulus deflector) consists of two high-speed, servo-controlled, deflector mirror systems that rotate in response to signals from the eye-tracker, one about a horizontal axis and the other about a vertical axis. These mirrors serve to stabilize a scotoma (in our case a blind half-field) on the retina, while the target itself passes through the system twice and, thus, remains unaffected.

The stimulus platform is above the subject's lap, and optical correction is easy to adjust by focusing a photographic lens in the optical path. A simple, objective procedure for calibrating the system for a brain-damaged patient remains to be implemented.

A significant amount of interfacing will be required to take advantage of the eye-tracker and artifical scotoma generator in our present computer installation. This will principally involve modifying and extending the existing software to allow digitization of the horizontal and vertical eye position outputs from the eye-tracker, and utilizing this information on-line in performing experiments. Three major modifications or extensions must be made to the existing software: Routines must be added to allow the eye-tracker system to be calibrated and linearized, and to verify that the specified half of the image is occluded; the display software must be modified to allow correction for dynamic errors in the eye-tracker/artificial scotoma system, and the display software must be modified to support the nontachistoscopic presentation of images made possible by the availability of the artificial scotoma. Additionally, software must be added to enable computer control of the artificial scotoma generator in real time, but the universal occluder can be used as a stand-alone system for a variety of experiments without computer control.

2.2. Auditory Testing: Dichotic Listening

2.2.1. The Right-Ear Advantage

When different nonsense consonant–vowel (CV) syllables from the set ba, da, ga, pa, ta, ka are presented simultaneously to both ears (dichotic listening), normal subjects report verbally the right-ear stimuli more accurately than the left-ear stimuli. This small but significant and reliable "right-ear advantage" (REA) is interpreted to reflect LH specialization for phonetic perception. Commissurotomy patients show a much larger REA than do normal subjects, and they allow an incisive analysis of the mechanisms that produce it. Because the REA varies with stimuli and procedure, it is necessary to ascertain the mechanisms for a specific test with commissurotomy patients before reaching a definitive conclusion on the meaning of the REA on that test in normal subjects. Dichotic listening to nonsense CVs is a particularly effective way of establishing auditory disconnection (often attributed to an interrupted callosal isthmus) because it is impossible to simulate *(see below)*.

The auditory system represents both ears in both hemispheres, with crossing fibers at the level of the brain stem (superior olive), the midbrain, and the corpus callosum (probably at the isthmus). Nonetheless, the contralateral ear-hemisphere projections dominate the ipsilateral fibers, so that under conditions of dichotic listening, the ipsilateral projections are suppressed. Then the right-ear stimuli project directly to the LH, whereas left-ear stimuli project directly to the RH and need to be relayed through the corpus callosum prior to processing in the LH.

Dichotic listening experiments have commonly manipulated

1. The acoustic structure of the stimulus pair
2. The meaningfulness of the stimuli and
3. The memory load, i.e., the number of sequential stimulus pairs to be reported in each trial.

Set, attentional, and order-of-report variables are known to affect the REA in normal subjects. The most common response procedure in dichotic listening is verbal report. This is inappropriate for tapping the disconnected RH. Variants include monitoring the ears for target stimuli, reporting the laterality of predetermined targets, and indicating by pointing whether or not a lateralized picture that follows the dichotic pair matches one of the auditory stimuli.

2.2.2. The Three Assumptions

Consider a dichotic tape with linguistic stimuli. More accurate perception of right-ear stimuli is commonly considered to be evidence of LH specialization for language. Three independent assumptions are made in this interpretation (E. Zaidel, 1983). First, it is assumed that the LH is specialized for processing the input signal. Second, it is supposed that the ipsilateral signal from the left ear to the LH is suppressed, perhaps at a subcortical level. Berlin (1977) suggests that ipsilateral suppression occurs at the medial geniculate bodies. Third, it is assumed that stimuli presented to the left ear will first reach the RH, and then cross the corpus callosum to be processed in the LH. This left-ear signal then competes or interferes with, but does not dominate, the direct contralateral right-ear signal, resulting in the observed REA.

Although most experiments interpret the observed ear advantage as evidence for hemispheric specialization in the perception of the auditory stimuli, many of the studies include other task components, such as verbal responses or memory requirements, that could separately contribute to the laterality effect. Most importantly, individual differences in the REA could reflect not only differences in hemispheric specialization, but also differences in callosal connectivity, as well as brain stem asymmetries across individuals. Similarly, the assumption of ipsilateral suppression is usually made without any direct evidence. Although some partial, early supporting animal models exist (e.g., Rosenzweig, 1951), there is no definite information on the mechanism or anatomical locus of ipsilateral suppression. In particular, it is still generally unknown whether ipsilateral suppression occurs subcortically or whether it occurs in the cortex or shows cortical influences. The split brain can provide direct evidence for ipsilateral suppression for particular dichotic stimuli.

2.2.3. Probing the Disconnected Right Hemisphere

The verbal report procedure is inadequate for probing RH processing. Lateralization of the response by using left-hand pointing is inadequate, too, because the LH can control left-hand responses, presumably through ipsilateral efferent manual projections. Instead, E. Zaidel (e.g., 1983) used a lateralized visual probe technique. In the first experiment, the stimuli were dichotic pairs of natural tokens of the stop CVs ba, da, ga, pa, ta, and ka prepared with a computerized system at Haskins Laboratory in New Haven. Perception of the dichotic pairs was assessed separately, by verbal report and by lateralized visual probes. In the visual probe conditon, each dichotic pair was followed immediately by a triplet of letters (from the set B, D, G, P, T, and K, corresponding to the six dichotic syllables) representing the left-ear syllable, the right-ear syllable, and a syllable differing from both in one or two phonetic features (voicing and place of articulation). The triplet was flashed quickly to the left or right visual hemifield. The subjects were required to point to the letter representing the sound they were most sure of having heard in either ear. These tests were administered to normal subjects, hemispherectomy patients, and commissurotomy patients.

The results showed a small, but reliable, REA in normal subjects, both in verbal report and in either visual half-field with visual probes. Commissurotomy patients showed a massive REA, but not as complete as that of a case of right hemispherectomy. Monotic presentation resulted in good and equal verbal report (LH) from either ear. The RH could not perceive signals from either ear in either the dichotic or the monotic condition. Thus, all three assumptions were verified. In particular, given the verification of exclusive LH specialization and of ipsilateral left-ear suppression, the large difference between the REA in normals and in split-brain patients verified the assumption of callosal interference and demonstrated that this is a callosal relay task. Furthermore, there was some evidence for different amounts of left-ear suppression, depending upon the task and phonetic feature differences between the two ears, suggesting that ipsilateral suppression was affected by cortical processes, particularly by hemispheric specialization itself.

2.2.4. Testing the Significance of the Left-Ear Score

Clarke described the following procedure for assessing the significance of the left-ear score relative to chance when probing the right hemisphere and obtaining one response on each trial (Clarke et al., 1989). First, given a stimulus set of six CV syllables and considering the left ear as an independent channel, a left-ear report of 1/6, or 16.7%, is at chance. However, when there is a reciprocal relation between the left- and right-ear score, the left-ear accuracy may be artifically low. Therefore, even when left-ear accuracy is at or below chance, Clarke applies the following test: Consider only those trials in which a left-ear item or an error (corresponding to neither ear) occurs. On a given trial, the left-ear item can be one of five unique stimuli, after excluding the right-ear item, so that chance performance in the left ear is 1/5, or 20%. The difference of the left-ear score from chance can be tested using the normal approximation to the binomial guessing distribution:

$$z = (x - c)/(npq)^{1/2} \quad (1)$$

where x is the left-ear score, n is the number of items in the test, p is the guessing probability, $q = 1 - p$, and c is the chance correct score. Then

$$z = (x - n/5)/(.16n)^{1/2} \quad (2)$$

and x can be determined for $z = 1.96$ at the .05 level of confidence or $z = 2.57$ at the .01 level of confidence.

2.2.5. Manipulating Meaningfulness, Delay, and Attention

A more recent study in collaboration with B. Kashdan has extended the lateralized probe technique in several important ways:

1. Only one lateralized probe followed each dichotic pair
2. The meaningfulness of the CV syllables was manipulated
3. The delay between the auditory pair and the probe was varied and
4. Ear attention was manipulated (Kashdan, 1979; E. Zaidel, 1983).

The main questions were whether and how the common variables of meaningfulness and attention separately affected hemispheric specialization and ipsilateral suppression in contributing to a possible difference in overall laterality effect (REA).

2.2.5.1. Meaningfulness. A dichotic tape with pairs of syllables from the set Bee, Dee, Gee, Pee, Tee, and Kee was produced at Haskins Labs from natural tokens. These syllables are phonetically similar to the usual CV's Ba, Da, Ga, Pa, Ta, and Ka, but each can refer to a letter (e.g., B) or an object (e.g., the insect bee). Each dichotic pair was followed by a picture flashed briefly and randomly either to the left or to the right of a central fixation dot. The subject then pointed with the hand ipsilateral to the stimulated half-field to the word "yes" or "no," in order to indicate whether the picture did or did not match the sound heard in either ear. In the "letter" condition, the flash consisted of an uppercase letter (B, . . . ,K), and in the "picture" condition, the flash contained a simple line drawing (a bee, a girl named Dee, a boy named Guy, a pea pod, a tea cup, and a key). This cross-modal lateralized task allowed each disconnected hemisphere to respond separately. Monaural and binaural control conditions were also administered.

In addition, the delay between the dichotic pair and the lateralized flash was varied from 0–.25 s and .5–1 s. Attention instruction varied between attending to both ears in the usual manner, attending to one ear for the whole test, or attending to the randomly selected ear receiving a brief beep 1 s before the dichotic pair.

Right visual half-field (LH) performance of commissurotomy patients in the zero delay and no attention condition showed a large REA, which was variable with performance level and higher for consonants than for picture probes. LVF (RH) score was consistently above chance in only one patient (LB), and occasionally in another (NG). Thus, LH specialization for the task was verified.

An analysis of variance was performed on the five patients' data with visual half-field (L, R), ear (L, R), and probe type (consonant, picture) as independent within-subject variables and with d', a bias-free measure of sensitivity, as a dependent variable. From the accuracy data for each subject in all conditions d' was generated by pairing the probability of hits with the probability of false alarms. The ANOVA disclosed a significant REA, confirming hemispheric specialization and a significant field X ear interaction. For the left ear in the LVF, d' was significantly above zero, but d' for the right ear in the LVF was not. Similarly, d' for the left ear in the RVF was essentially zero, confirming ipsilateral suppression. No significant effects or interactions resulting from probe type occurred.

Restricting our attention to LB and NG, the only commissurotomy patients who showed some (albeit minimal) RH competence, we found a (nonsignificant) VF × probe interaction, with picture probes yielding greater sensitivity in the LVF, and consonants yielding greater sensitivity in the RVF. Only consonant probes in the LVF yielded sensitivities less than 1.

The right ear (RE) score in the RVF (LH) is unchanged whether the signal is dichotic or monotic, that is, regardless of whether there is a competing signal in the left ear (LE). This also verifies the ipsilateral suppression of the LE in the LH. Further, with monaural presentations of one channel to only one ear, the LE signal is reported somewhat less accurately than the RE in the LH. Thus, the ipsilateral LE-to-LH channel is somewhat weaker than the crossed RE-to-LH channel, even without dichotic competition. This laterality effect disappeared, and somewhat lower scores for either ear were obtained, with binaural presentations of the same signals to both ears. Therefore, the ipsilateral signals would seem to have some functional significance, even when they simply duplicate the contralateral ones.

A similar subtle asymmetry was observed in initial training, with one hand pointing to the picture of the stimulus among six exposed in free vision. Here, monaural LE signals yielded slightly higher initial error rates with right-hand pointing; RE signals first showed more errors with left-hand pointing, and binaural signals showed more initial left-hand errors, thus demonstrating LH control. However, either hand, pointing to one of the six choice pictures in free vision in the dichotic condition, shows the same massive REA. Thus, hand pointing is not a reliable index of contralateral hemispheric control.

2.2.5.2. Delay. An ANOVA applied to the delay data, where trials were partitioned into delay (0.5 or 1 s) and no-delay conditions, revealed the usual significant VF, ear, and ear X VF effects, as well as a significant VF X ear X delay interaction. Although delay had no effect on the LH, it affected the RH in complex ways, both interacting with probe type and showing an effect of the length of the delay.

LB's LH showed a massive REA at all delays and, equally, for letter and picture probes. By contrast, at zero delay, the RH showed a significant LEA for pictures, but a nonsignificant REA for letters. A simple mathematical model suggested RH response control for the pictures but LH cross-cueing and control (with poor ipsilateral visual transfer LVF–LH) for the pictures. Thus, the degree of LE suppression in the RH for the auditory dichotic pairs varied as a function of stimulus meaningfulness (letter vs picture).

When the visual probes followed the auditory dichotic pair by 0.5s, the RH controlled performance, to produce an LEA for both pictures and letters. The LEA was larger for pictures. With a 1-s delay, the RH showed a massive LEA for pictures, but a reversal to a nonsignificant REA, signaling LH cross-cueing and control, for letters. Thus, the hypothesis of uniform subcortical ipsilateral suppression in dichotic listening is not supported by this data. Rather, ipsilateral suppression is seen to depend upon a variety of cognitive variables, and particularly on hemispheric specialization. Whether the lability of ipsilateral suppression is associated with poor competence in either hemisphere, or only in the RH, remains to be found.

2.2.5.3. Attention. The effects of attention are even more complex. In LB, instructions to attend to one ear throughout the test had the result of reducing the laterality effect in the hemisphere contralateral to the unattended ear, without changing the laterality effect in the hemisphere contralateral to the attended ear. In other words, in each hemisphere, attention to the contralateral ear had little influence on the laterality effect, whereas attention to the ipsilateral ear resulted in a substantial change. This change was especially strong and unpredictable in the RH. The attention set can affect both the contralateral and ipsilateral ear signals in the "unattended hemisphere." In contrast, with delay, attention affected both hemispheres and resulted in substantial changes in laterality effects, especially in the blocked (set) condition.

However, when attention was signaled by random beeps to one ear before the dichotic pairs, LE beeps decreased the laterality effect in both hemispheres but primarily in the LH, whereas RE beeps increased the laterality effect in the RH without affecting that in the LH. It seemed that the effect of attention here was mediated by the LH, either to decrease its REA or to decrease its interference with the LEA in the RH.

Parallel experiments with normal subjects showed no effect of probe lateralization, of probe type, or of attention (E. Zaidel, 1983). The failure of the REA in normal subjects to be affected by probe type (that is, stimulus meaningfulness) parallels the pattern observed in the disconnected LH, but not in the RH. This again confirms exclusive LH specialization and callosal relay to the LH on this task. The effects of attention were seen to be mixed. It can counteract ipsilateral suppression while reducing the contralateral signal. Attention seems to have a much larger effect on the disconnected LH than on the intact LH, suggesting that its effect on ipsilateral suppression can be partly mitigated by the commissures.

2.2.5.4. LABILITY OF ATTENTION In a subsequent experiment (Clarke et al., 1989), four commissurotomy and two hemispherectomy patients listened to the nonsense CV syllable Bee . . . Kee tape, and attention was manipulated in blocks. Responses were by unimanual pointing to a response sheet containing the six consonants B, D, G, P, T, and K, positioned at the patient's midline. This is in contrast to the previous experiment, which included lateralized visual probes and required simpler yes/no recognition. For each set of trials, the patient was instructed to report only the left- or right-ear stimuli. In addition, an arrow positioned centrally above the response sheet pointed either to the left or right, and the experimenter tapped the appropriate shoulder of the patient every five trials. Two commissurotomy patients and the two hemispherectomy patients showed no effect of attention. Two commissurotomy patients, LB and NG, did report more right-ear items with right-ear attention and more left-ear items with left-ear attention. Moreover, in these two patients, attention improved left-ear scores above chance (relative to errors). Thus, attention can affect ipsilateral suppression, but does not do so reliably and uniformly across patients.

2.2.6. Summary

Other experiments manipulated the interaural lag and the intensity difference between the signals to the two ears, and failed to show systematic hemispheric effects in commissurotomy patients (Cullen, 1975). Together, the fragility of changes in laterality effects in the disconnected brain as a function of stimulus meaningfulness, delay between dichotic pair and probe, attention, lag between the two ear signals, and intensity differential between the two ears, all show that the effects of those variables on the REA in normal subjects, when they occur, do not have a simple interpretation in terms of hemispheric competence. Rather, they may have more to do with callosal connectivity. Until such effects become less labile and more interpretable, any effect observed in normal subjects needs to be confirmed using the identical paradigm with commissurotomy patients.

2.3. Somesthetic Testing

2.3.1. Clinical Tests

Hemisphere disconnection can be demonstrated with respect to somesthesis (including touch, pressure, and proprioception) in a variety of ways.

2.3.1.1. CROSS-RETRIEVAL OF SMALL TEST OBJECTS. Unseen objects in the right hand are handled, named, and described in a normal fashion. However, attempts to name or describe the same objects held out of sight in the left hand consistently fail. In spite of the patient's inability to name an unseen object in his or her left hand, identification of the object by the right hemisphere is evident from appropriate, adroit manipulation of the item, and retrieval of the same object with the left hand from among a collection of other objects screened from sight. Split-brain patients routinely have excellent same-hand retrieval (with either hand). What distinguishes the split-brain patients from normal subjects is their inability to retrieve with one hand objects felt with the other.

2.3.1.2. CROSS-REPLICATION OF HAND POSTURES. Specific postures impressed on one (unseen) hand by the examiner cannot be mimicked with the opposite hand. For example, one can place the tip of the thumb against the tip of the ring finger and have the other three fingers fully extended. The split-brain patient cannot mimic with the other hand a posture thus impressed on the first hand. This procedure should be repeated with various postures and in both directions.

2.3.1.3. CROSS-LOCALIZATION OF FINGER TIPS. The split-brain patient has a partial loss of the ability to name exact points stimulated on the left side of the body. This defect is least apparent, if at all, on the face, and it is most apparent on the finger tips. This is not a deficit dependent upon language, since it can be carried out by nonverbal means either from right hand to left hand or from left hand to right hand. An easy way to demonstrate the defect is to have the subject's hands extended, palms up (with vision excluded). One touches the tip of one of the four fingers with the point of a pencil, asking the patient to then touch the same point with the tip of the thumb of the same hand. Split-brain patients do this at a normal level (above 90%) with either hand. One then changes the task, so that the finger tip is to be indicated by touching the *corresponding* finger tip of the other hand with the thumb of that (other) hand. Sometimes the procedure should be demonstrated with the patient's hands in full vision until the patient understands what is required. This cross-localization cannot be done by the split-brain patient at a level much better than chance (25%). Normal adults almost always do better than 90%. The same test can be refined by utilizing the volar surfaces of each of the three phalanges. Another refinement is to use a calibrated aesthesiometer (Volpe et al., 1979).

The effectiveness of these simple clinical procedures depends on adequate precautions against cross-cueing and ipsilateral transfer of identifying features. This can be easily accomplished by including enough different objects, and so on, without prior exposure to them, so that one or two simple features will not suffice for identification.

2.3.2. Use of Somesthetic Input in Experimental Tests

Somesthetic input has been used to demonstrate disconnection or to ensure lateralized input or output in many experiments with other modalities or purposes, but, to date, there has never been a systematic comparison in commissurotomy patients of laterality effects in somesthesis with effects in other modalities. Nonetheless, some hints exist.

First, when the tactile component of a task is incidental to its higher-order processing demands, then it is easy to show that the laterality effects obtained hold across different stimulus modalities. Thus, D. Zaidel and Sperry (1973) demonstrated a trend toward RH superiority in a modified cross-modal version of Raven's Colored Progressive Matrices, in which the problem (incomplete patterns) was exposed in free vision, but the alternative answers (the missing parts) were palpated unimanually out of view. The standard visual form of the same test was then readministered unilaterally, using the contact lens technique for hemispheric ocular scanning, and confirmed the earlier result (E. Zaidel, et al., 1981). Some bilateral increase in performance occurred on the visual relative to the tactile form of the test, but the disconnected RHs remained slightly superior.

Second, if the tactile component of a task is integral to it (e.g., when tactile presentations of complex stimuli require temporally sequential samplings that need to be integrated in time), then changing the modality of the stimuli may change the processing demands and, consequently, also the laterality effects. This is just what we found when we compared unilateral scores on the standard version of the Visual Sequential Memory Subtest of the Illinois Test of Psychologistic Abilities to scores on a tactile adaptation of it (E. Zaidel, 1973, 1978). This test required the subject to rearrange, from memory, a set of "nonmeaningful" (though probably verbalizable) geometrical figures in the order in which they had been presented before being scrambled. The LHs of the two commissurotomy patients were superior on a lateralized presentation of the visual form of the test. However, this significant left–right difference vanished when a lateralized tactile version of the test was administered to the same patients by using raised zinc models of the same patterns. The observed LH advantage on the visual version may be attributable to LH dominance in visually guided behavior, or to LH use of verbal encoding in order to benefit from rehearsal in a superior short-term-verbal memory. In any case, the modality change must have affected the solution strategy, thus erasing the original dominance pattern.

Third, when the purpose of the task was to study the tactile recognition of two-dimensional geometric shapes, e.g., using Benton's Stereognosis test, the tested commissurotomy patients showed a bilateral deficit, no consistent hemispheric superiority, and greater deficit in the hand contralateral to the hemisphere with predominant extracallosal damage (E. Zaidel, 1978, 1989a). Two patients showed a significant right-hand advantage, and two showed a significant left-hand advantage. The stereognostic deficit, (which involved shape recognition, rather than apprehension of meaning), occurred in the absence of primary somesthetic impairment of constructional apraxia, but was not correlated with supramodal hemispheric specialization effects. A selective stereognostic deficit of the complete commissurotomy patients relative to the partial commissurotomy patients suggests that disconnection contributed to the disability. The contribution of hemispheric lesions to the disability appeared partly asymmetrical: patients with predominantly LH lesions tended to have a much more severe impairment in the contralateral hand, whereas patients with predominant damage to the RH seemed moderately impaired bilaterally (cf Carmon and Benton, 1969).

Fourth, somesthetic input has greater ipsilateral projection than visual input. This appeared in two commissurotomy patients, tested in various pairings of unimanual stimulus exploration and hemifield viewing of the stimulus with hemifield scanning of the choice card using the contact lens system (E. Zaidel, 1989a). Patients LB and NG both named figures palpated by the left hand better than they named the same figures exposed in the LVF. This contrasts with better naming of pictures in the RVF than of the actual objects felt in the right hand. Perhaps, following complete cerebral commissurotomy, continuous and bilateral eye movements provide adequate interhemispheric integration in the visual sphere, whereas somesthetic cross-integration relies heavily on functional reorganization of ipsilateral efferent/afferent tactile–kinesthetic control. The fact that naming of left-hand stimuli is better than of LVF stimuli, even while the converse is true for the right hand and RVF, is consistent with ipsilateral control, rather than with RH speech. Furthermore, correct naming of left-hand stimuli deteriorates rapidly when the choice set increases and when it is not known to the patient in advance.

Fifth, the ipsilateral projections appear to be asymmetric; exposing the choice card in the RVF while exploring the stimulus with the left hand results in a better performance than exposing the choice card in the LVF while exploring the stimulus with the right hand. This can be interpreted to mean that LH control over the left hand is stronger than RH control over the right hand. Unlike the motor system, ipsilateral somesthetic and kinesthetic manual afference does not appear to be stronger for proximal than distal extremities.

A critical feature of sophisticated testing is the occasional random request for a verbal reply with left-hand or LVF presentations. Incorrect replies ensure that information has not leaked into the LH. Failure to incorporate and properly interpret this maneuver is a notorious oversight.

In conclusion, experiments designed to assess hemispheric specialization in the disconnected hemispheres for a particular task by employing unimanual stimulus or response exploration should be interpreted with some caution. Disconnection itself seems to produce some bilateral somatosensory deficit, and extracallosal damage appears especially detrimental to somesthetic function.

Moreover, by lateralizing different components of the task, the experimenter is likely to change the solution strategies adopted by both hemispheres and activate complex, variable patterns of interhemispheric interaction. Even the mere exposure of the multiple-choice card of Benton's Stereognosis test in bilateral vision, rather than the hemifield homolateral to the exploring and responding hand, improves performance. Similarly, manual stimulus exploration by one hand and pointing response selection with the other can help the RH and suppress the LH (E. Zaidel, 1989a). More generally, our results do not support the hypotheses that the RH is superior in manipulo-spatial tasks in general (Le Doux et al., 1977; but *see* Bogen and Bogen, 1983) or that the RH is superior for tactile perception and the LH for visually guided stimulus exploration. Rather, it may be that, in the normal brain, the right hand plays a special role in sequentially constructing an image from successive tactile impressions, whereas the RH is instrumental in refining the resulting integrated image and maintaining it in memory.

2.4. Motor Skills and Apraxia Testing

One would think that severing the connection between the two hemispheres should result in a wide spectrum of problems for motor integration. However, patients with complete commissurotomy have been observed to retain preoperative motor skills requiring bimanual coordination, such as tying shoe laces, cooking, shuffling cards, and even swimming or bicycling. This implies that well-rehearsed motor skills are regulated by brain centers (cerebellar?) not directly affected by the commissurotomy.

On the other hand, when such skills as fastening a row of buttons were timed for speed, complete commissurotomy patients were found to be appreciably slower than normal control subjects (D. Zaidel and Sperry, 1977). Presence of general brain damage is one possible explanation for the observed reduced speed. However, when a new bimanual motor skill that requires continuous mutual monitoring between the two hands was attempted by both partial and complete commissurotomy patients, marked impairment in ability to learn was observed (Preilowski, 1972). These two examples demonstrate the important role that the forebrain commissures play in motor coordination.

In order to obtain a general assessment of the long-term effects of cerebral commissurotomy on motor skills and coordination, standardized motor tests measuring new skills as well as preoperatively well-rehearsed motor skills may be used. Unfortunately, most of them require speeded performance. Yet, they can be invaluable in providing clues to patterns of existing deficits. It is also crucial to ensure that the levels of perception and memory required for performance are minimal. Otherwise, failures could be attributed to these factors. In addition, it is essential that scores on standardized tests be supplemented by performance on conventional clinical tests for apraxia. It goes without saying that presence of dyspraxia would invalidate measures of motor skills. Together, both types of tests may provide a well-rounded picture of motor performance competency. A description of three illustrative nonapraxia tests is provided below. A more detailed and complete description of both types of tests is available in D. Zaidel and Sperry's report (1977).

2.4.1. *Crawford Small Parts Dexterity Tests* (Crawford and Crawford, 1956)

In the first part, a pair of tweezers is used with the right hand to transfer small pins from a bin into close-fitting holes and to put collars over each protruding pin. In the second part, 36 small screws are lifted one by one using both hands and threaded into holes with a screwdriver. Score in each part is the time required to complete the task.

2.4.2. *Purdue Pegboard (PPB)* (Tiffin, 1968)

Thirty seconds are allowed for transferring as many pins as possible from bins into separate holes with the right hand alone, left hand alone, and both simultaneously. Another task involves the same transfer, with the additional assembly on the inserted pins of washers and collars, using both hands alternately, allowing 60 s. Score is the total achieved for three repeated trials.

2.4.3. *Pursuit Rotor* (Heap and Wyke, 1972)

Employing a standard rotary pursuit apparatus, subjects using a metal-tipped stylus attempt to keep contact with a penny-sized metal disc rotating at 60 rpm. The total contact time is automatically measured. Left hand always follows the right, after an interval of 2 min. Score is the average contact time for ten trials.

When assessment time is limited, the best method for determining presence or absence of motor impairment caused by forebrain commissurotomy may be the use of a test requiring asynchronous bimanual motion. Of all the bimanual tests used in the D. Zaidel and Sperry study to assess the effects of cerebral commissurotomy, this type proved the most sensitive. Earlier, Preilowski (1972) reported the deficient performance of these patients in a bilateral crank-turning task requiring interdependent, asynchronous bimanual control. Pantograph tracing of a star is an example of a test in this category that is easily obtainable (Wyke, 1971). This test requires the manipulation of a standard pantograph with both hands to trace a line inside a double-line star. The subject is required to avoid going outside the printed lines. Another variation involves tracing a line, also inside a double-line star, by manipulating the two knobs of an Etch-a-Sketch apparatus.

Since eye–hand coordination is essential in any motor testing, the best experimental conditions are afforded in free vision, where the input is available to both hemispheres continuously. The resulting performance provides information about the role of the forebrain commissures in the motor execution, rather than about the specific hemispheric contribution to the task at hand. Under such conditions, tasks on which manual asymmetry is nevertheless observed become particularly important for understanding the hemispheric contribution. For example, the clinical apraxia tasks administered in free vision to complete commissurotomy patients revealed in some of them a left-sided ideomotor apraxia, a right-hand dyscopia, and a left-hand dysgraphia (D. Zaidel and Sperry, 1977).

2.4.4. Apraxia Tests

In the apraxia tests, subjects are asked to perform different gestures to spoken commands. *Ideomotor apraxia:* Make the sign of the cross, salute, wave goodbye, threaten somebody with your hand, show that you are hungry, thumb your nose, snap your fingers, and so on. *Ideational apraxia:* The following objects are picked by the subject, who demonstrates their use: hammer, toothbrush, scissors, revolver, eraser, lock and key, match and match box. *Nonrepresentational movements:* Place hand under chin, place hand in front of nose, touch index finger to ear, put hand behind head, touch thumb to forehead. *Facial praxis:* Blow out match, sip on straw, protrude tongue, cough, sniff flower, close eyes, lick upper lip, puff out cheeks, whistle, wrinkle nose. *"Intransitive" movements:* Salute, scratch head, throw kiss, indicate full stomach, beckon, hitchhike, make fist. *"Transitive" movements:* Brush teeth, shave, comb hair, drink with spoon, file fingernails. *Leg praxis:* Stamp out cigaret, press gas pedal, tap foot, kick ball, slide foot in slipper. *Whole body:* Walk backwards; stand like boxer, stand like a golfer; stand like a batter; shovel dirt; jump; squat; bow; shrug. *Bilateral hand movements:* Play piano, clap, circle hands in air, pray, jump rope.

In all of the above unimanual tests, the patients are asked to perform the entire series, first with the left hand and then with the right. All initial instructions are given without demonstration, either by pantomime or through pictures. For every item, a response is "correct" if it is immediate or preceded by slight hesitation, and "incorrect" if protracted and irrelevant.

To test for dyscopia (constructional apraxia) patients are asked to copy seven geometric figures, first with the right hand and then with the left: square, triangle, hexagon, cube, diamond, cross, simple nonsense figure. Dysgraphia is tested by having the patient write to dictation, first with the left hand and later with the right: e.g., mother's first name, "Today is Friday," "baseball," "car."

3. Methodological Issues

3.1. Statistics and Metrics

3.1.1. Laterality Index

Whereas a behavioral laterality effect in a normal subject may incorporate both hemispheric specialization and callosal connectivity components, the laterality effect in a commissurotomy patient is essentially a measure of hemispheric specialization. In the case of a pure direct-access task, the laterality effect in the normal subject is the same as the one in the split brain, and both are measures of hemispheric specialization. The actual laterality indices used in split-brain research are the same ones used in experiments with normal subjects (E. Zaidel, 1979a, 1980; Bryden and Sprott, 1981; Harshman and Lundy, 1989). ANOVA with LVF and RVF as a within-subject experimental variable, and difference or ratio measures, such as LVF–RVF or LVF/RVF, are poor laterality indices. A relative ratio such as Marshall's f [(LVF–RVF)/(LVF+RVF) or (LVF–RVF)/([1-LVF]+[1-RVF]), depending on whether total accuracy is low or high, respectively] or Bryden and Sprott's Lambda is preferable, and may be less dependent on total accuracy.

In commissurotomy patients, as in normal subjects, accuracy and latency measures are not equivalent, since they show complex interactions with visual field, suggesting different resource assignments in latency–accuracy tradeoff. The disconnected RH is also more variable and more labile than the disconnected LH in the laboratory setting.

The signal detection model can be adapted to testing commissurotomy patients by assuming hemispheric independence and computing sensitivity (d') and bias (β) for each side. However, it should be remembered that the assumptions of the signal detection model may not apply to the process under investigation. Statistical assumptions about normality and equality of variances may not be satisfied, or the data corresponding to different criterion levels may not fall on a straight ROC line in a double-probability plot. However, even if the ROC curve is "well behaved," the signal detection model may be inappropriate for a more fundamental reason. For example, the signal detection model may be inappropriate for lexical decision tasks, since the model assumes a discrimination between two populations, a signal (such as words) and a noise (such as nonwords) according to some criterion. On the contrary, it may be that in the right hemisphere (or in each hemisphere) words and nonwords are decided by two separate and parallel processes. An alternative to the use of signal detection is to use the classical proportion of trials correct, adjusted for guessing and response bias (Woodworth and Schlosberg, 1954):

$$tc = (\text{proportion hits} - \text{proportion false alarms})/(1 - \text{proportion false alarms}) \quad (3)$$

This measure, *tc*, can be computed for both words and nonwords in a lexical decision task. Apparently, natural analogies between laterality indices, such as Marshall's f, and signal detection's d' can be misleading (*see* detailed discussion in E. Zaidel, 1979a). Similarly, some analogies between signal detection criterion levels and certain experimental variables (such as shared phonetic features in dichotic CV pairs) are at best partial, and make strong and usually unjustified assumptions (*see* examples in E. Zaidel, 1979a). Actual application of signal detection to lexical decision in split-brain patients reveals consistently larger d' in the disconnected LH than in the disconnected RH. βs vary widely between the two hemispheres in individual patients, but they show no systematic patterns across patients.

3.1.2. *Single-Case Statistics*

Commissurotomy patients can be analyzed individually with statistics appropriate for single-case designs. Latency differences between the two disconnected hemispheres can be tested by t-tests for correlated means with items as the random variable.

3.1.3. *Theoretical Metrics of Hemispheric Competence*

Theoretically motivated hemispheric comparisons across commissurotomy patients fall into two classes: (1) qualitative analyses, in terms of hemispheric dissociation along some information processing component or stage, and (2) quantitative comparisons, in terms of existing metrics of theoretical relevance. In the past, for quantitative comparisons we have used (1) equivalent mental-age norms on age-standardized developmental tests, and (2) percentile ranks relative to aphasics or hemisphere-damaged patients (E. Zaidel, 1985a). By expressing the ability of a given hemisphere on some task in terms of a normal child who obtained the same score, we can learn about the cognitive developmental stage of that hemisphere. Such data are relevant to the issue of the ontogenesis of hemisphere specialization (E. Zaidel, 1978). Similarly, by expressing hemispheric ability in terms of a percentile rank relative to some aphasic population, we can learn about the role of that hemisphere in accounting and/or compensating for linguistic deficit. Each metric also permits a direct comparison of competence on a test across hemispheres and patients.

3.2. Special Problems of Testing the Disconnected Right Hemisphere

3.2.1. *"Passivity"*

The disconnected RHs appear to be uniquely passive during experimental sessions. They rarely generate spontaneous behavior, and seem to have a limited competence for constructive actions during formal testing. This is true not only for speech and writing, but also for drawing and building. Most of the time, however, the disconnected RHs are capable of simple actions in response to very specific task demands, such as pointing to a multiple choice array or arranging a sequence of pictures. Because of this passive behavior, it is difficult to penetrate the mental life of the disconnected RH, and the experimenter is deprived of the opportunity to observe patterns of natural behavior. Instead, the experimenter has to rely on hypothesis-verification experiments, where competing conjectures have to be formulated and operationalized in each test. There is always the danger that the operationalization is unnatural, so that the disconnected RH fails even though it is competent to handle the relevant constructions. Conversely, it is also possible that the RH can make the distinction at issue for reasons other than those hypothesized by the experimenter. Consequently, there is the ever-present danger of either underestimating or overestimating RH competence.

A recurring issue in documenting perceptual, cognitive, linguistic, or mnestic incompetence in the disconnected RH is whether it has understood the task. This problem is not unique to the RH and applies equally to children and brain-damaged patients. When testing commissurotomy patients, every attempt is made to explain and illustrate the task redundantly, both verbally and nonverbally, whenever appropriate. In fact, minimal instructions usually suffice for appropriate test behavior, suggesting that the disconnected RH has very effective auditory language comprehension in context. Nonetheless, when attempting to document incompetence, the goal is to construct control tests that are identical to the experimental task in all but the relevant dimension, and show that the disconnected RH can perform them.

3.2.2. The Multiple-Choice Paradigm

The task most commonly employed in testing the disconnected RH involves matching the target stimulus with one of 3–6 pictures or tactile displays in a multiple-choice array. The pictures can be lateralized to the LVF using the contact lens system, which permits free ocular scanning of the array. The stimulus itself need not be visual. For example, we have used this paradigm for extensive assessment of the auditory vocabulary of the disconnected RH. Both hemispheres hear the target word, but only the RH sees the multiple-choice array and only it can control the left hand to point to the correct picture (E. Zaidel, 1976).

The multiple-choice paradigm clearly belongs in the hypothesis verification, rather than hypothesis generating, category, yet it is remarkably rich. The association between the stimulus and the target picture can be arbitrarily complex, and the decoys can be chosen to make just the contrasts of theoretical interest. For example, in a word recognition experiment, the foils can be semantic, auditory, and visual errors. On the other hand, the multiple-choice paradigm is rather demanding cognitively. It is metalinguistic, since it requires the evaluation of proposed solutions (the multiple choices) and the rejection of incorrect ones. Elsewhere, we have argued that linguistic monitoring operations, such as error detection in reading, are modular relative to the linguistic operations proper (E. Zaidel, 1987). In that case, competence in the former can be independent of competence in the latter. Moreover, the multiple-choice paradigm presupposes prerequisite cognitive skills, including sequential sampling of alternatives and rehearsal in short-term memory (cf E. Zaidel and Peters, 1981), as well as the ability to operate in a context-free environment. Each of these skills may be asymmetrically represented in the two disconnected hemispheres.

3.2.3. Left-Hemisphere Dominance over Motor Pathways

The failure of the disconnected RH to speak, construct, or generate spontaneous behavior during laboratory testing may reflect LH dominance over the motor pathways designed to preserve unified behavior and thus, perhaps, the integrity of the self. On the one hand, the RH does have adequate access to the articulators and to motor programs for activating the left hand. Excellent RH control of articulation is commonly apparent in aphasics or following dominant hemispherectomy, and RH control of left-hand praxis is easy to demonstrate with nonverbal imitation in the split brain. Thus, failure of the disconnected RH to perform these functions may reflect active inhibition or interference by the LH, made possible by subcortical integrative mechanisms. In turn, proper motor responses during testing may reflect LH cooperation in praxis control or, alternatively, just temporary release of RH praxis. On the other hand, we have observed occasional apraxic left-hand behavior during testing in patient NG. On one occasion, her hand pointed randomly instead of to response pictures; on another, she could not grasp a pawn and place it appropriately in Piaget's Landscape test (E. Zaidel, 1978). This could mean that LH participation is sporadic, and provides active support of left-hand control or RH programming of praxis. Thus, LH participation in RH praxis may be supportive, disruptive, or absent.

LH dominance is also frequently apparent when the inferior LH controls behavior on a task exposed to both hemifields (E. Zaidel, 1978). This may result from a suboptimal evaluation by a control mechanism misled by the nature of the task or stimuli (Levy and Trevarthen, 1976). Most dramatic is the persistent verbal denial by the subject of previously demonstrable RH competence (E. Zaidel, 1978). Such denial is paradoxical not only because of the patients' longstanding experience of disconnection, but also because of effective noncallosal exchange of partial complex information between the disconnected hemispheres in the chronic syndrome.

3.2.4. Set and Superstition

Researchers who have worked intensively with commissurotomy patients over a long period of time have developed testing rituals designed to optimize performance by the disconnected RH. This is because LH interference often obscures RH competence. First, we have observed that even in cases where the LH is inferior it is more likely than the RH to assume control over behavior when both hemispheres have access to the input. Second, the LH seems to possess better functional use than the RH of the sensory–motor, visual, and tactile–kinesthetic ipsilateral projection systems. Third, many data converge to demonstrate a stronger resiliency of LH performance level in the face of cognitive perturbations. Put conversely, observed RH superiority can often be reversed by small changes in the conditions of the task, such as response delay, solution strategy, ambiguity of the possible answers, and input modality. Fourth, and foremost for a theory of consciousness, is the persisting and active neglect and denial of RH experiences by the LH (E. Zaidel, 1978).

Methods to overcome LH dominance include:

1. Inducing an "RH mood" before testing by listening to music and/or minimizing talking
2. Nonverbal demonstrations of the task whenever appropriate
3. Testing the RH first to prevent the LH from becoming familiar with the problem and dominating the responses during RH presentations

4. Frequent praise of correct RH performance
5. Intermittent exhortations, such as, "Let your hand go," "Don't force it," "Don't try to understand or verbalize what you do" and
6. "Encouraging" the RH by allowing it a second chance if wrong

Such methods may encourage the patients to "cooperate" with the examiner and show RH superiority, but the need to break LH sets is documented by showing that commissurotomy patients have alternating runs of correct and of incorrect responses much longer than would be expected by chance. Changes in LH dominance during the test may also explain the greater variability in performance in the disconnected RH observed in test–retest comparisons (E. Zaidel, 1979b).

Since the disconnection syndrome entails short-term memory loss, it is important to design hemispheric testing paradigms that do not depend on memory load. In testing the disconnected RH in particular, we found that performance on complex tasks tends to deteriorate when the choice set is large, e.g., with more than three or four multiple choices. A case in point is a cross-modal version of Raven's Colored Progressive Matrices, administered by D. Zaidel and Sperry (1973). The problem was exposed in free vision, but the possible answers were converted to etched, raised zinc patterns palpated in turn by one hand out of view. The disconnected RH performed disproportionately better with three choices than with the standard six.

3.3. Counterfeit Disconnection

For a glory-seeking or psychotic person who likes center stage, being a commissurotomy patient can be a satisfying full-time occupation. As a professional subject he or she can enjoy travel to exotic locations and be rewarded both socially and financially. How can we tell whether such a person has the real disconnection syndrome or, instead, is familiar with its features and can play the role well? The simplest answer is to obtain an MRI and inspect it for a full section in a midsagittal view. However, it is difficult to assess the status of the anterior commissure by MRI (Bogen et al., 1988), and failure to disconnect this structure might leave the patient free of some disconnection symptoms (Hamilton, 1982). Also, disconnection may be the result of secondary lesions to structures that project fibers to the corpus callosum and, thus, not show up on a midsagittal view.

3.3.1. Supranormal Effects

Some laterality effects occur only following callosal disconnection and cannot be simulated by a person with intact neocommissures. A good example is the REA in dichotic listening to nonsense CVs. When the two channels are carefully aligned (and the fundamental frequency of both syllables is the same), the pair fuses, and commissurotomized subjects tend to hear one sound, much more often the one in the right ear. Indeed, attending to the left ear has little or no effect on the REA (E. Zaidel, 1983). Consequently, the counterfeit subject cannot simply report the right ear accurately and feign chance performance with left-ear stimuli, since he or she often will not be able to tell from which ear the sound came. Moreover, for nonsense CV pairs, there is a reciprocal relationship between right-ear and left-ear scores (Berlin and McNeil, 1976), so that suppression of the left ear signal in the disconnected LH results in an above normal right-ear score. This cannot be faked.

3.3.2. Involuntary Effects

As far as we know, measures of cerebral activation cannot be simulated, and standard monitoring techniques reveal effects that are unique to commissurotomy patients with the full-fledged disconnection syndrome. One example is the Event Related Potential (ERP) to linguistic semantic anomalies. Kutas et al. (1988) presented auditory sentences followed by a visual word that varied in "cloze" probability. Words with a low cloze score seem unexpected, incongruous, and anomalous (e.g., "Every Saturday morning he mows the *chair*".) Kutas et al. found an enhanced central-parietal negativity (N400) that correlated highly with the cloze index of semantic anomaly and showed a larger amplitude over the normal RH than over the LH. By contrast, commissurotomy patients with no RH speech (LB, NG, JW) who received two completing words to the two hemifields simultaneously show the N400 in response to anomaly in the RVF, but not in the LVF, and only over the LH. The disconnected RHs failed to exhibit the N400 even though they could detect the anomaly behaviorally. It is unlikely that normal subjects can "train themselves" to elicit no N400 to LVF stimuli over the RH.

3.3.3. Standard Effects

The counterfeit patient would need to be exceptionally up-to-date and incredibly well-practiced to be able to simulate all the known laterality effects. Since many of the effects are automatic, it may be impossible to counteract them even with much practice. For example, simple manual reaction times to light flashes show a crossed–uncrossed difference (CUD) of from 30–90 ms in commissurotomy patients, as opposed to CUDs from 1–6 ms in normal subjects (Clarke and E. Zaidel, 1989). Can normal subjects train themselves to respond with a 30-ms delay in a crossed hemifield–hand condition?

3.3.4. Possible Effects

It is easy to produce experiments that show performance by commissurotomy patients that is better than, or different from, normal subjects. Dual-task interference paradigms (sharing tasks between the two hemispheres) should show greater interference and, thus, greater performance decrement in the normal than in the split brain. Tasks that are stimulus-specific, e.g., priming of lexical decision with a specific set of semantically related words presented once, cannot be anticipated and, thus, cannot be faked with practice.

3.3.5. Pseudodisconnection

No single sign is a sufficient index for disconnection. Some interhemispheric disconnection effects may be attributable to intrahemispheric disconnection in either hemisphere. For example, it is theoretically conceivable that LVF stimuli are available to LH processes that are disconnected from language centers in the same side. Cases of implicit knowledge or memory may be good examples. Thus, failure to name LVF stimuli may be insufficient to establish interhemispheric disconnection. (In this case, a possible way to demonstrate failure of disconnection may be to show that RVF stimuli cannot be named either.) Similarly, left-ear suppression can occur with LH lesions (paradoxical ear extinction, e.g., Damasio and Damasio, 1979), although this is usually interpreted as auditory disconnection.

3.4. Right-Hemisphere Speech or Noncallosal Interhemispheric Transfer?

Occasionally, split-brain patients can name stimuli shown in the left hemifield (LVF) or palpated with the left hand (Lh) (Butler and Norsell, 1968; Levy et al., 1971; Teng and Sperry, 1973; Johnson, 1984; Myers, 1984). This could be a result of:

1. Improper lateralization of the stimuli
2. Ipsilateral projection of sensory information from the LVF or Lh to the left hemisphere where verbalization occurs
3. Subcortical transfer of cognitive information sufficient to identify the stimulus to the LH following recognition by the RH
4. Cross-cueing from the RH to the LH, using shared perceptual space (e.g., the RH may fixate on a related item in the room, thus identifying it to the LH, or it may trace the shape of the object in question with the head so the movement can be "read off" by the LH) (Bogen, 1987) or
5. RH speech (e.g., the patient P.S. of Gazzaniga et al., 1979).

Only when all other alternatives are ruled out can RH speech be considered seriously, given the weight of evidence so far. For example, it was never investigated whether LVF or Lh stimuli could be named at the same time that nonverbal Rh identification of these stimuli failed. To date, there is no compelling evidence for RH speech in any of the patients in the California series (Myers, 1984).

What appears to be RH speech may reflect improper stimulus lateralization to the RH. Improper lateralization with tachistoscopic presentations can occur not only by failure to fixate on the central mark (e.g., deviating to the left so that both lateralized stimuli fall in the RVF) or by saccades to the stimuli when the presentation is too long, but also by fixating on a point behind or in front of the plane of the image, or, with binocular presentations, by divergent fixations of the two eyes. Improper lateralization with the contact lens or the lateral limits method can be the result of faulty calibration.

Verbalization of left-field stimuli resulting from ipsilateral sensory projection is limited to simple sensory features and to items uniquely identified by them, such as curved vs straight contours or sharp vs dull edges (Trevarthen and Sperry, 1973). Effective cross-cueing depends critically on the size of the stimulus set and on prior exposure to it, so that simple cues suffice to eliminate alternatives. LB often uses a verbal cross-cueing strategy in which the LH seems to guess in turn each letter making up the name of the stimulus by going subvocally through the alphabet, with the RH apparently signaling when the correct one is reached (D. W. Zaidel, 1988). Subcortical transfer appears effective for semantic features abstracted from the meaning of the stimulus without necessarily identifying it uniquely, and thus, without making naming generally possible. These features include affective and connotative information (happy, sad, pleasant, and so forth) (E. Zaidel, 1976; Sperry et al., 1979), associative (sensory and semantic) (Myers, 1984; Myers and Sperry, 1985), categorical ("animals that go in the water," one picture shown to each field simultaneously), functional ("shoe–sock"), or abstract (communication: envelope–telephone) relations (Cronin-Golomb, 1986). Sergent (1987) showed that commissurotomy patients could integrate bilateral dot patterns to decide whether their sum was odd or even. However, it is not clear whether the information transferred involved numerosity, parity, or more concrete sensory information. Visual images do not seem to transfer subcortically. In any case, naming of left-field stimuli in the absence of cross-field matching is not sufficient evidence for RH speech. For example, cross-matching may fail because of a tendency to neglect one hemifield with bilateral presentations.

D. W. Zaidel (1988) studied correct verbalizations and presented elegant examples of writing of the names of pictures restricted tachistoscopically to the LVF of complete commissurotomy patient LB. She concluded that the verbalizations did not represent RH speech, but that his RH could often write in cursive with the left hand the names of simple line drawings, without his being able to name them. Thus, the disconnected RH has some writing, but little or no speech (cf also Levy et al., 1971 for examples of writing the names of objects palpated with the left hand).

3.5. Issues of Interpretation and Generalizability

3.5.1. Generalizing to Other Commissurotomy Patients

Do results with selected commissurotomy patients generalize to other such populations? The answer depends critically on the neurological histories of each population. In brief, there is little, if any, evidence in LB, NG, or any of the other commissurotomy patients of the California series for the kind of LH lesion that would result in RH takeover of language functions. Moreover, no patient has ever had a history of linguistic deficits that could be said to recover through reorganization. Indeed, the disconnected LHs of these patients now reveal only selective pragmatic and paralinguistic deficits, including receptive prosody, pictorial metaphor, and discourse memory, that parallel those observed following right-hemisphere damage, and presumably reflect loss of normal interhemisphere cooperation (E. Zaidel, et al., in preparation). Moreover, the California patients have diverse neurological histories, including age of onset, extent and location of lesion, as well as age at surgery, and yet they fall into a similar behavioral pattern. In fact, of the six complete commissurotomy patients that we have studied intensively, four are thought to have predominantly RH extracallosal damage (NG, LB, RY, and NW), and only two have predominantly LH damage (CC, AA) (Campbell et al., 1981). Numerous somesthetic (Milner and Taylor, 1972), visual (E. Zaidel, 1978), and auditory (E. Zaidel, 1983) laterality tasks show hemispheric patterns that are consistent across patients and cannot be explained by side of extracallosal damage.

The California patients are largely free of the severe deficits that commonly follow focal brain damage. All of these patients have, by now, shown evidence of RH language aspects. Moreover, they show evidence of the same upper and lower limits on RH language that had been demonstrated in more detail for patients LB and NG. The data come from hemifield tachistoscopic and dichotic listening experiments. Thus, all patients, including RY, whose epilepsy is attributable to a car accident at the age of 13 (Bogen, 1969), show the ability to perform lexical decision between concrete nouns and orthographically regular nonwords in their disconnected RHs (E. Zaidel, 1989b; cf also Hamilton et al., 1986), yet they are generally unable to perceive dichotic nonsense CV syllables in the same RHs. Good lexical decision and semantic facilitation in the disconnected RH verifies the existence of a rich lexical semantic system. However, poor dichotic perception of nonsense CV syllables in the same RH verifies the absence of a phonetic apparatus in the RH's language repertoire. Good lexical semantics and poor phonetics are precisely the upper and lower limits observed in NG and LB using more extensive tests and more elaborate techniques.

By contrast, many patients in other series have massive extracallosal damage, often with hemispheric atrophy, making independent hemispheric testing impossible. Others have severe intellectual deficits that make any testing unrewarding (E. Zaidel, in press). In still others, the damage has resulted in speech development in the RH (Sidtis et al., 1981).

3.5.2. Generalizing to Hemisphere-Damaged Patients

In general, the pattern of complementary hemispheric specialization observed in commissurotomy patients confirms the data from hemisphere-damaged patients: The LH is specialized for language, especially speech and syntax, whereas the RH is specialized for visuo-spatial processes. Yet, there are some discrepancies: The disconnected hemispheres are generally free of the dramatic deficits that sometimes follow hemispheric damage. For example, posterior RH lesions can result in severe contralateral neglect and denial syndromes or in prosopagnosia, whereas the disconnected LH has never shown evidence of neglect or prosopagnosia (E. Zaidel, 1975; E. Zaidel, et al., 1981; Plourde and Sperry, 1984). Similarly, localized LH damage can result in word deafness or word blindness, whereas the disconnected RH has a rich auditory lexicon and a substantial reading vocabulary (E. Zaidel, in press). It would seem that certain severe cognitive deficits following hemispheric damage reflect pathological inhibition of residual competence in the healthy hemisphere.

The pattern of language competence in the disconnected RH does resemble that observed in adults with dominant hemispherectomy for late lesions (Burklund, 1972) and in temporal lobe epileptics whose LH is temporarily anesthesized by sodium amobarbital (Rasmussen and Milner, 1977). Also, the ranking, and often the level, of linguistic abilities in the disconnected RH are the ones observed in a large heterogenous aphasic population: auditory comprehension of single words is best and superior to reading; sentences are more difficult to decode than words; and speech and writing are most impaired (E. Zaidel, 1976).

Indeed, evidence for language competence in the disconnected RH has spurred a reexamination of the role of RH in compensating for aphasia because of LH lesions, showing takeover in various language functions (E. Zaidel, in press). Persisting discrepancies, such as pure alexia with posterior LH damage, are gradually succumbing to further analysis. There is now converging evidence from more than half a dozen studies that implicit reading comprehension in pure alexia may be "released" with quick presentations and nonverbal responses that bypass the maladaptive reading control system. (It remains to be shown that such "release" actually activates RH functions.) Impaired reading control in pure alexia contrasts with adaptive release of control of lexical access to the RH in deep dyslexia, presumably when LH access fails (Schweiger et al., 1989).

A similar pattern emerges when comparing pragmatic linguistic deficits in the disconnected RH to those observed following RH damage (Foldi et al., 1983). The disconnected LH is impaired on some, but not all, of the functions lost after RH damage (E. Zaidel, et al., in preparation). Some impaired functions, including prosody, pictorial metaphor, and discourse, reflect genuine RH specialization, others reflect partial RH contributions, and still others reflect the disruptive effects of the lesions. Thus, just as the phenomenology of aphasia obscures some RH language competence, so RH damage seems to underestimate the language capacity of the disconnected LH.

3.5.3. Generalizing to Normal Subjects

The linguistic profile of the disconnected RH seems to underestimate the contribution of the normal RH to language functions. Cerebral blood-flow studies show that both hemispheres are involved in speaking, reading, and listening (Ingvar and Lassen, 1977). Hemifield tachistoscopic and dichotic listening studies of hemispheric specialization in the normal brain also provide evidence for RH involvement (e.g., Silverberg et al., 1979). The absence of a laterality effect in such an experiment is not evidence for bilateral language representation, and the occurrence of a laterality effect need not mean that the inferior hemisphere is not involved.

These ambiguities can be resolved by interpreting laterality effects in the normal brain in terms of direct access and callosal relay models (E. Zaidel, 1983, 1985b; E. Zaidel et al., in press).

We have carried out a series of lateralized lexical decision experiments that manipulated a variety of lexical variables and were administered to commissurotomy patients and to normal subjects. The results show, first, that normal subjects exhibit a pattern of processing dissociation (interaction of VF with the independent lexical variable) that is indicative of "direct access," i.e., independent processing in the two intact hemispheres. Second, the competence observed in the intact RH is far superior to that observed in the disconnected RH. Indeed, we found evidence for RH processing of word concreteness and emotionality (semantic) (Eviatar et al., in press), as well as of length (orthography?) (Eviatar and E. Zaidel, 1989), morphology (Emmorey and E. Zaidel, 1989), and perhaps even phonology (Rayman and E. Zaidel, 1989). Similarly, we have observed effective semantic (E. Zaidel et al., 1988) as well as grammatic (Menn et al., 1989) priming in the intact RH.

4. Conclusion

Each experimental population and its paradigms have their own methodological advantages and disadvantages, and patients with complete cerebral commissurotomy are no exception. They have early brain damage and require subtle and specialized testing skills. However, they also offer an unusual opportunity for comparing the positive competence of each hemisphere with its "sibling," already matched for age, sex, and developmental history. The "final account" of hemispheric specialization and independence is unlikely to come from any single clinical population. Rather, converging evidence is necessary from both patients and normal subjects, using diverse experimental paradigms. It is particularly instructive to develop tests and paradigms that can be applied with little or no modification to commissurotomy patients, to hemisphere-damaged patients, and to normal subjects. When empirical discrepancies between paradigms emerge that cannot be attributed to their inherent limitations, the resolution is likely to constitute a theoretical breakthrough.

Acknowledgments

This work was supported by an NIMH RSDA MH 00179, an NIH award NS 20187, the David H. Murdock Institute for Advanced Brain Studies, and a Biomedical Research Support Grant to UCLA.

References

Berlin C. I. (1977) Hemispheric asymmetry in auditory tasks, in *Lateralization in the Nervous System* (Harnad S., Doty R. W., Goldstein I., Jaynes J., and Krauthamer G., eds.), Academic, New York, pp. 303–323.

Berlin C. and McNeil M. R. (1976) Dichotic listening, in *Contemporary Issues in Experimental Phonetics* (Lass N. J., ed), Academic, New York, pp. 327–291.

Bogen J. E. (1969) The other side of the brain—I. Dysgraphia and dyscopia following cerebral commissurotomy. *Bull. Los Angeles Neurol. Soc.* **34,** 73–105.

Bogen J. E. (1985) The stabilized syndrome of hemisphere disconnection, in *The Dual Brain: Hemispheric Specialization in Humans* (Benson D. F. and Zaidel E., eds), Guilford, New York, pp. 289–304.

Bogen J. E. (1987) Physiological consequences of complete or partial commissural section, in *Surgery of the Third Ventricle* (Apuzzo M. L. J., ed.), Williams and Wilkins, Baltimore, pp. 175–194.

Bogen J. E. and Bogen G. M. (1983) Hemispheric specialization and cerebral duality. *Behav. Brain Sci.* **6,** 517–520.

Bogen J. E., Schultz D. H., and Vogel P. J. (1988) Completeness of callosotomy shown by Magnetic Resonance Imaging in the long term. *Arch. Neurol.* **45,** 1203–1205.

Bradshaw J. L. (in press). Methods for studying human laterality. In *Neuromethods, Vol. 15: Methods in Human Neuropsychology* (Boulton A., Baker G. B., and Hiscock M., eds.), Humana Press, Clifton, N.J.

Bradshaw J. L. and Nettleton N. C. (1983) *Human Cerebral Asymmetry* Prentice Hall, Englewood Cliffs.

Bryden M. P. and Sprott D. A. (1981) Statistical determination of degree of laterality. *Neuropsychologia* **19,** 571–581.

Burklund C. W. (1972) Cerebral hemisphere function in the human: Fact vs. tradition. In *Drugs, Development and Cerebral Function* (Smith W. L., ed.), C. C. Thomas, Springfield, Ill., pp. 8–36.

Butler S. R. and Norsell U. (1968) Vocalization possibly initiated by the minor hemisphere. *Nature* **220,** 793–794.

Campbell A. L., Bogen J. E., and Smith A. (1981) Disorganization and reorganization of cognitive sensorimotor functions in cerebral commissurotomy. *Brain* **104,** 493–511.

Carmon A. and Benton A. L. (1969) Tactile perception of direction and number in patients with unilateral cerebral disease. *Neurology* **19,** 525–532.

Clarke J. M. and Zaidel E. (1989) Simple reaction times to lateralized light flashes: Varieties of interhemispheric communication routes. *Brain* **112,** 849–870.

Clarke J., David A., and Zaidel E. (1989) Effects of attention on ipsilateral suppression in dichotic listening. Manuscript in preparation.

Crane H. D. and Kelly D. H. (1983) Accurate simulation of visual scotomas in normal subjects. *Applied Optics* **22,** 1802–1806.

Crane H. D. and Steele C. M. (1985) Generation V dual-Purkinje image eyetracker. *Applied Optics* **24,** 527–537.

Crawford J. E. and Crawford D. M. (1956) Crawford Small Parts Dexterity Test. The Psychological Corporation, New York.

Cronin-Golomb A. (1986) Subcortical transfer of cognitive information in subjects with complete forebrain commissurotomy. *Cortex* **22,** 499–519.

Cullen J. K., Jr. (1975) Tests of a model for speech information flow. *Dissertation Abstracts International* **36,** 1167A. (University Microfilms No. 75–19, 262).

Damasio H. and Damasio A. (1979) Paradoxical ear extinction in dichotic listening: Possible anatomic significance. *Neurology* **29,** 644–653.

Davidoff J. B. (1982) Studies with nonverbal stimuli, in *Divided visual field studies of cerebral organization* (Beaumont J. G., ed.), Academic, London, pp. 30–55.

Dimond S. J., Bures J., Farrington L. J., and Brouwers E. Y. M. (1975) The use of contact lenses for the lateralization of visual input in man. *Acta Psychologica* **39,** 341–349.

Emmorey K. and Zaidel E. (1989) Morphological complexity and hemispheric specialization. Manuscript in preparation.

Eviatar Z. and Zaidel E. (1989) *The effects of word length and emotionality on hemispheric contribution to lexical decision.* Manuscript submitted for publication.

Eviatar Z., Menn L., and Zaidel E. (in press) *Concreteness: Nouns, Verbs and Hemispheres. Cortex.*

Foldi N. S., Cicone M., and Gardner H. (1983) Pragmatic aspects of communication in brain-damaged patients, in *Language functions and brain organization* (Sigalowitz S. J., ed.), Academic, New York, pp. 51–86.

Francks J. B., Smith S. M., and Ward T. B. (1985) The use of goggles for testing hemispheric asymmetry. *Bulletin of the Psychonomic Society* **23,** 487–488.

Gardner H. and Brownell H. (1986) *Right hemisphere communication battery*. Psychology Service, VAMC, Boston, MA.

Gazzaniga M. S. (1970) *The Bisected Brain* (Appleton-Century-Crofts, New York).

Gazzaniga M. S., Volpe B. T., Smylie C. S., and Wilson D. H. (1979) Plasticity in speech organization following commissurotomy. *Brain* **102,** 805–815.

Hamilton C. R. (1982) Mechanisms of interocular equivalence, in *Advances in the Analysis of Visual Behavior* (Ingle D., Goodale M., and Mansfield R., eds.), Cambridge, MA, MIT, pp. 693–717.

Hamilton C. R., Nargeot F., and Bogen J. E. (1986) Right hemisphere reading. *Society for Neuroscience Abstracts* **12,** 721.

Harshman, R. (1980) *The meaning and measurement of differences in degree of lateralization*. Paper presented in the Symposium on Methodological and Statistical Issues in Neuropsychological Research, chaired by S. A. Berenbaum, Eighth Annual Meeting, INS, San Francisco, January 29–February 2.

Harshman R. A. and Lundy M. E. (1989) Can dichotic listening measure "degree of lateralization"?, in *Handbook of Dichotic Listening: Theory, Methods and Research* (Hugdahl K., ed.), Wiley, New York, pp. 215–282.

Heap M. and Wyke M. (1972) Learning of a unimanual motor skill by patients with brain lesions: an experimental study. *Cortex* **8,** 1–18.

Ingvar D. H. and Lassen N. A. (eds.) (1977) Cerebral function, metabolism and circulation. *Acta Neurol. Scand.* **56,** (Suppl. 64), 1–560.

Johnson L. E. (1984) Vocal responses to left visual field stimuli following forebrain commissurotomy. *Neuropsychologia* **22,** 153–166.

Kashdan B. (1979) *Speech perception and cerebral asymmetry*. Honors undergraduate thesis, Dartmouth College, Department of Psychology.

Kutas M., Hillyard S. A., and Gazzaniga M. S. (1988) Processing of semantic anomaly by right and left hemispheres of commissurotomy patients: Evidence from event-related brain potentials. *Brain* **111,** 583–576.

Le Doux J. E., Wilson D. H., and Gazzaniga M. S. (1977) Manipulo-spatial aspects of cerebral lateralization: Clues to the origin of lateralization. *Neuropsychologia* **15,** 743–750.

Levy J. and Trevarthen C. (1976) Metacontrol of hemispheric function in human split-brain patients. *J. Exp. Psychol. [Hum. Percept.]* **2,** 229–312.

Levy J., Nebes R., and Sperry R. (1971) Expressive language in the surgically separated minor hemisphere. *Cortex* **7,** 49–58.

Levy J., Trevarthen C. B., and Sperry R. W. (1972) Perception of bilateral chimeric figures following hemispheric deconnexion. *Brain* **95,** 61–78.

McConkie G. W. and Rayner K. (1976) Asymmetry of perceptual span in reading. *Bulletin of the Psychonomic Society* **8,** 365–368.

Menn L., Ragman J., and Zaidel E. (1989) *Lateralized noun/verb decision: Part of speech, functor context, and two models of the concreteness effect.* Manuscript submitted for publication.

Milner B. and Taylor L. (1972) Right-hemisphere superiority in tactile pattern-recognition after cerebral commissurotomy: Evidence for nonverbal memory. *Neuropsychologia* **10,** 1–15.

Myers J. J. (1984) Right hemisphere language: Science or fiction? *Am. Psychol.* **39,** 315–320.

Myers J. J. and Sperry R. W. (1982) A simple technique for lateralizing visual input that allows prolonged viewing. *Behavior Research Methods and Instrumentation* **14,** 305–308.

Myers J. and Sperry R. W. (1985) Interhemispheric communication after section of the forebrain commissures. *Cortex* **21,** 249–260.

Nettleton N. C., Wood R. G., Bradshaw J. L., Thomas C. D. L., and Donahoo K. B. (1983) A moving video window or mask yoked to eye movements: A system to permit free ocular scanning within areas of the visual field. *Behavior Research Methods & Instrumentation,* **15,** 487–496.

Plourde G., and Sperry R. W. (1984) Left hemisphere involvement in left spatial neglect from right-sided lesions: A commissurotomy study. *Brain* **107,** 95–106.

Pollatsek A., Bolozky S., Well A. D., and Rayner K. (1981) Asymmetries in the perceptual span for Israeli readers. *Brain Lang.* **14,** 174–180.

Preilowski B. F. B. (1972) Possible contribution of the anterior forebrain commissures to bilateral motor coordination. *Neuropsychologia* **10,** 267–277.

Rasmussen T. and Milner B. (1977) The role of early left-brain injury in determining lateralization of cerebral speech functions, in *Evolution and lateralization of the brain.* (Dimond S. J. & Blizard B. A., eds.) *Ann. N.Y. Acad. Sci.* **299,** 355–369.

Rayman J. and Zaidel E. (1989) Rhyming and the right hemisphere. Manuscript submitted for publication.

Rosenzweig M. R. (1951) Representation of the two ears at the auditory cortex. *Am. J. Psychol.* **167,** 147–158.

Schweiger A., Zaidel E., Field T., and Dobkin B. (1989) Right hemisphere contribution to lexical access in an aphasic with deep dyslexia. *Brain Lang.* **37,** 73–89.

Sergent J. (1987) A new look at the human split brain. *Brain* **110,** 1375–1392.

Sidtis J. J., Volpe B. T., Wilson D. H., Rayport M., and Gazzaniga M. S. (1981) Variability in right hemisphere language functions: Evidence for a continuum of generative capacity. *J. Neurosci.* **1,** 323–331.

Silverberg R., Bentin S., Gaziel T., Obler L., and Albert M. (1979) Shift of visual field preference for English words in native Hebrew speakers. *Brain Lang.* **8,** 184–190.

Sivak B., Sivak J. G., and MacKenzie C. L. (1985) Contact lens design for lateralizing visual input. *Neuropsychologia* **23,** 801–804.

Sperry R. W. (1974) Lateral specialization in the surgically separated hemispheres, in *Neurosciences: Third Study Program* (Schmitt F. O. and Worden F. G., eds.) Cambridge, MA, MIT, pp. 5–19.

Sperry R. W. (1982) Some effects of disconnecting the cerebral hemispheres. *Science* **217,** 1223–1226.

Sperry R. W., Gazzaniga M. S., and Bogen J. E. (1969) Interhemispheric relationships: The neocortical commissures; syndromes of hemisphere disconnection, in *Handbook of Clinical Neurology,* vol. 4 (Vinken P. J. and Bruyn G. W., eds.) Elsevier, Amsterdam, pp. 273–290.

Sperry R. W., Zaidel E., and Zaidel D. (1979) Self recognition and social awareness in the disconnected minor hemisphere. *Neuropsychologia* **17,** 153–166.

Teng E. T. and Sperry R. W. (1973) Interhemispheric interaction during simultaneous bilateral presentation of letters or digits in commissurotomized subjects. *Neuropsychologia* **11,** 131–140.

TenHouten W. D., Hoppe K. D., Bogen J. E., and Walter D. O. (1986) Alexithymia: An experimental study of cerebral commissurotomy patients and normal control subjects. *Am. J. Psychiatry* **143,** 312–316.

Tiffin J. (1968) Purdue Pegboard. Science Research Associates, Chicago, Illinois.

Trevarthan C. and Sperry R. W. (1973) Perceptual unity of the ambient visual field in human commissurotomy patients. *Brain* **96,** 547–570.

Trope I., Rozin P. and Gur R. C. (1988) Validation of the lateral limits technique with a callosotomy patient. *Neuropsychologia* **26,** 673–684.

Volpe, B. T., LeDoux, J. E., Fraser, R. A., and Gazzaniga, M. S. (1979) Spatially oriented movements in the absence of proprioception. *Neurology,* **29,** 1309–1313.

Woodworth R. S. and Schlosberg H. (1954) *Experimental Psychology.* Revised Ed. Holt, New York.

Wyke M. (1971) The effects of brain lesions on the performance of bilateral arm mcvements. *Neuropsychologia* **9,** 33–42.

Zaidel D. W. (1988) Observations on right hemisphere language function, in *Aphasia* (Rose, F. C., Whurr R., and Wyke M. A., eds.), Whurr Publishers, London, pp. 170–187.

Zaidel D. W. (in press) Memory and spatial cognition following commissurotomy, in *Handbook of Neuropsychology*. (Boller F. and Grafman J. eds.), Elsevier, Amsterdam.

Zaidel D. and Sperry R. W. (1973) Performance on the Raven's Colored Progressive Matrices test by subjects with cerebral commissurotomy. *Cortex* **9,** 34–39.

Zaidel D. and Sperry R. W. (1974) Memory impairment after commissurotomy in man. *Brain* **97,** 263–272.

Zaidel D. and Sperry R. W. (1977) Some long-term motor effects of cerebral commissurotomy in man. *Neuropsychologia* **15,** 193–204.

Zaidel E. (1973) Linguistic competence and related functions in the right cerebral hemisphere of man following commissurotomy and hemispherectomy (Doctoral Dissertation, California Institute of Technology). *Dissertation Abstracts International* **34,** 2350B. (University Microfilms No. 73–26,481.)

Zaidel E. (1975) A technique for presenting lateralized visual input with prolonged exposure. *Vision Res.* **15,** 283–289.

Zaidel E. (1976) Auditory vocabulary of the right hemisphere following brain bisection or hemidecortication. *Cortex* **12,** 191–211.

Zaidel E. (1978) Concepts of cerebral dominance in the split brain, in *Cerebral Correlates of Conscious Experience* (Buser P. & Rougeul-Buser A., eds.), Elsevier, Amsterdam, pp. 263–284.

Zaidel E. (1979a) On measuring hemispheric specialization in man, in *Advanced Technobiology* (Rybak B., ed.), Alphen aan den Rijn, Sijthoff and Noordhoff, pp. 365–403.

Zaidel E. (1979b) Long-term stability of hemispheric Token Test scores following brain bisection and hemidecortication. In *Auditory Comprehension: Clinical and Experimental Studies with the Token Test* (Boller F. and Dennis M., eds.), Academic, New York, pp. 135–159.

Zaidel E. (1980) The structuring of language: Clues from hemispheric specialization, in *Signed and Spoken Language: Biological Constraints on Linguistic Form* (Bellugi U. & Studdert-Kennedy M., eds.), Dahlem Konferenzen, Weinheim/Deerfield Beach, FL/Basel, Verlag Chemie, pp. 291–340.

Zaidel E. (1983) Disconnection syndrome as a model for laterality effects in the normal brain, in *Cerebral Hemisphere Asymmetry: Method, Theory and Application* (Hellige J., ed.), Praeger, New York, pp. 95–151.

Zaidel E. (1985a) Right hemisphere language. In *The Dual Brain: Hemispheric Specialization in Humans* (Benson D. F. and Zaidel E., eds.), The UCLA Medical Forum Series. Guilford, New York, pp. 205–231.

Zaidel E. (1985b) Callosal dynamics and right hemisphere language, in *Two Hemispheres—One Brain?* (Lepore F., Ptito M., and Jasper H. H., eds.), Alan R. Liss, New York, pp. 435–459.

Zaidel E. (1987) Hemispheric Monitoring, in *Duality and Unity of the Brain* (Ottoson D. ed.), Macmillan, Hampshire, pp. 247–281.

Zaidel E. (1989a) *Long term stereognosis in the split brain: Hemispheric differences, ipsilateral control, and sensory integration across the midline*, unpublished manuscript. Department of Psychology, University of California, Los Angeles.

Zaidel E. (1989b) *Lexical decision and semantic faciliation in the split brain.* Unpublished manuscript, Department of Psychology, University of California, Los Angeles.

Zaidel E. (in press) Language functions in the two hemispheres following cerebral commissurotomy and hemispherectomy, in *Handbook of Neuropsychology* (Boller F. and Grafman J., eds.), Elsevier, Amsterdam.

Zaidel E. and Frazer R. E. (1977) A universal half-field occluder for laterality research. *Caltech Biology Annual Report* 137–138.

Zaidel E. and Peters A. M. (1981) Phonological encoding and ideographic reading by the disconnected right hemisphere: Two case studies. *Brain Lang.* **14,** 205–234.

Zaidel E., Clarke J., and Suyenobu B. (in press) Hemispheric independence: A paradigm case for cognitive neuroscience, in *Neurobiology of Higher Cognitive Function* (Scheibel A. and Wechsler A., eds.), Guilford, New York.

Zaidel E., Spence S., and Kasher A. (in preparation) Performance of commissurotomy patients and normal subjects on the Right Hemisphere Communication Battery.

Zaidel E., White H., Sakurai E., and Banks W. (1988) Hemispheric locus of lexical congruity effects: Neuropsychological reinterpretation of psycholinguistic results, in *Right Hemisphere Contributions to Lexical Semantics* (Chiarello C., ed.), Springer, New York, pp. 71–88.

Zaidel E., Zaidel D. W., and Sperry R. W. (1981) Left and right intelligence: Case studies of Raven's Progressive Matrices following brain bisection and hemidecortication. *Cortex* **17,** 167–186.

Electrical Stimulation of the Cerebral Cortex in Humans

Catherine A. Mateer, Richard L. Rapport, II, and Don D. Polly

1. History of Cortical Stimulation

After centuries of the theoretical assignment of soul, mind, and bodily functions to various anatomical places, the mid-nineteenth century experienced an explosion of information that allowed accurate cerebral localization to begin. The British philosopher Herbert Spencer anticipated the developments of the next 50 years when he wrote in 1855, "Localization of function is the law of all organization whatever: separation of duty is universally accompanied with separateness of structure: and it would be marvelous were an exception to exist in the cerebral hemispheres" (Haymaker and Schiller, 1970). John Hughlings Jackson, a Spencer student, used clinical observations in patients with epilepsy to begin substantiating theories of cerebral localization, and to define brain regions related to specific functions. Broca, Wernicke, Charcot, and the other great neurologists of the late nineteenth century expanded on these beginnings.

However, the dramatic advances came in 1879, when two young Germans, Eduard Hitzig and Gustav Fritsch, were successful in evoking motor responses from the electrical stimulation of a dog's brain. They concluded that, "Individual psychological functions, and probably all of them, depend for their entrance into matter or for the formation from it upon circumscribed centers of the cerebral cortex" (Clarke and O'Malley, 1968). David Ferrier,

Herman Munk, and Frederich Goltz all expanded these early findings with their own stimulation experiments. These studies, and Ferrier's urging, emboldened Victor Horsley to begin doing surgical operations for the treatment of focal epilepsy at Queen's Square National Hospital in 1886. There followed a quantum leap in the technical sophistication, thinking, and studies of Charles Sherrington and his school (including John Fulton). Better diagnosis and localization were made possible by Berger's invention of the EEG in 1929. Otfried Foerster began to perform regular operations for the management of epilepsy by the early decades of this century, and routinely did stimulation experiments.

The modern era of human cortical stimulation was, however, established by Wilder Penfield at the Montreal Neurological Institute in the 1930s, following his return from studies with Sherrington and Foerster (Penfield and Jasper, 1954; Penfield and Roberts, 1959; Penfield and Perot, 1963). The development of electronic circuitry and reliable pen writing EEG machines suitable for intraoperative corticography led the way for the semiconductor and computer technology of the present era of cortical mapping. Students of the Penfield school have continued to employ stimulation experiments in awake patients, especially for the study of cerebral organization of language. The usual reason for performing these current localization studies in human cortex is the same as Victor Horsley's motivation for the earliest operations—the treatment of focal epilepsy.

As much as 1% of the American population has epilepsy, and as many as 10% of these (or about 200,000 patients) are uncontrolled on anticonvulsant medications. The disorder may be idiopathic or secondary to tumor, arteriovenous malformation, infection, or trauma. Regardless of the cause, some of these patients may be candidates for the surgical treatment of their illness. If the intractable ictus originates focally in a noneloquent part of the brain, and if the patient is motivated to undergo awake craniotomy for the treatment of the illness, then surgical management is an effective option. Regardless of the cause of the disorder, cortical mapping of motor, sensory, and language functions, along with intraoperative identification of the epileptic cortex, increases both efficacy of treatment and safety of the operation. Cortical stimulation mapping has also been effectively utilized to increase the safety of tumor resection and other intracranial neurosurgical procedures.

2. Techniques of Cortical Stimulation

Local anesthesia, sometimes supplemented by reversible intravenous neuroleptic anesthesia in the case of extremely anxious patients, is induced, and a large craniotomy is easily performed. After the dura has been opened, recording electrodes (carbon-ball, silver-ball, wick) are affixed to the skull in a holder and corticography is recorded with the patient fully awake. Subtemporal electrodes are usually included in the array. These are available commercially or may be made in house. Grass Instruments (Quincy, Massachusetts) manufactures electrodes, holder, and harness compatible with most 16-channel EEG machines. The electrodes and leads may be gas-autoclaved and passed off the surgical field to a nonsterile connection box. Reference electrodes are placed on the patient's neck at the time of positioning, which must be carefully done on a well-padded operating table, and the neck is well supported. Sixty-Hz noise is frequently a problem, which requires various operating room devices (X-ray view boxes, electrocautery, EKG, and so on) to be disconnected.

Two modes of stimulation may be used, either constant current or constant voltage. Today constant-current stimuli is the preferred method, since most studies in the last decade have used constant current. This allows more direct comparison between results of various investigators. Using constant-current stimulation, the accepted method is rectangular wave pulses, either monophasic or biphasic. Biphasic is preferred to reduce the possibility of electrode polarization. However, in practice, the short duration of stimulation used in cortical mapping seems to avoid this problem. The usual pulse repetition rate today is 60 Hz, with a pulse duration of 1 ms plus and minus for biphasic stimulation or 2 ms duration for monophasic stimulation, either of which will produce an equal net coulomb flow to the cortex. For the patient's safety, stimulus isolation must be employed.

The bipolar stimulating electrodes, either silver ball or carbon ball, should have an interelectrode spacing of 5 mm. Although the absolute interelectrode space is not critical, the same spacing should be maintained throughout the procedure, since considerable differences in stimulating current threshold may be observed if the interelectrode spacing is changed. Electrode orientation (i.e., horizontal, vertical) should be maintained for repetitive stimulation at a particular site.

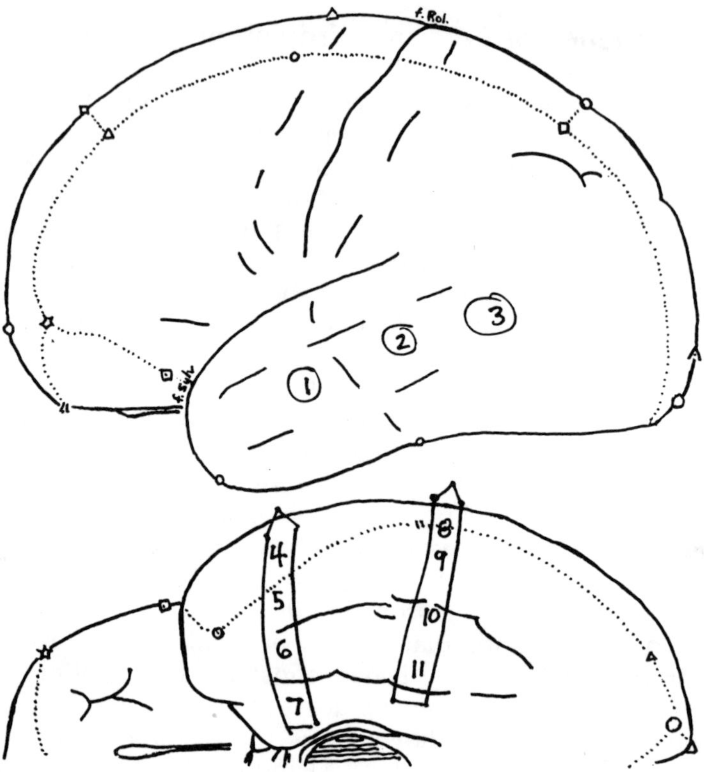

Fig. 1a. Schematic of the left cortical surface, illustrating the location of single recording electrode contacts (1, 2, and 3) and contacts along two strip electrodes slid down along the mesial and inferior surface of the left temporal lobe.

With the EEG running, an area of brain remote from the motor strip, but in the areas to be mapped for language, is stimulated for approximately 3–5 s, beginning with a current of 2 mA. The artifact of this stimulation will be readily seen on the EEG recording; if it is not, no current is reaching the cortical surface (*see below*, Complications). Afterdischarge is likely to be produced at some point following stimulation in increased steps between 2–12 mA. This afterdischarge is often at sites remote to the point of stimulation (*see* Fig. 1a,b). All stimulation studies are then conducted at a current just below the level that produces afterdischarge. It is often useful to run corticography during the period of mapping, since afterdischarge threshold may lower as stimulation proceeds. Impaired

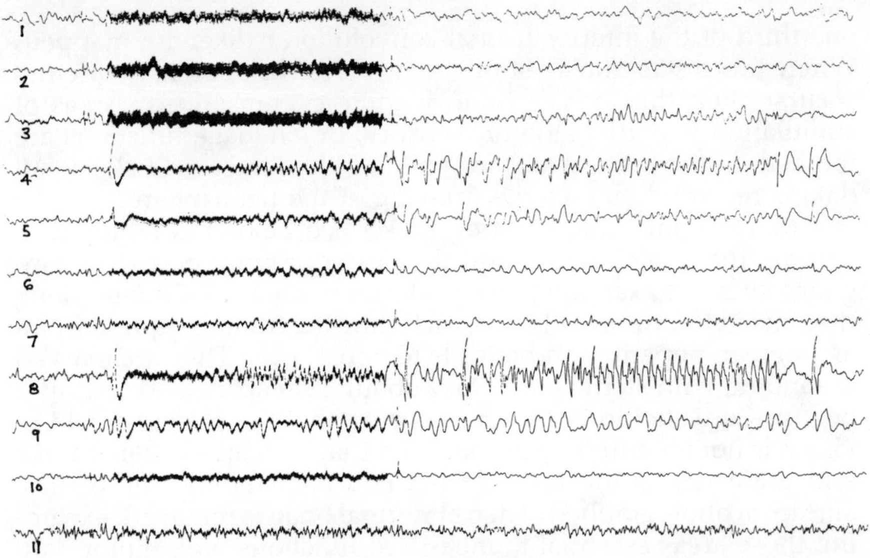

Fig. 1b. Intraoperative cortical EEG recording. Numbers relate to the electrode sites indicated in Fig. 1a. The cortex was stimulated at a level of 4 mA at site 3. Afterdischarge is seen at remote recording sites along the mesial and inferior temporal surface, most predominantly at site 8, but also at sites 4 and 5.

performance associated with such afterdischarge should be recognized and discarded from the analysis. Errors during such periods are likely to be in the form of speech arrest, and to be present across stimulation and control trials.

Motor-sensory cortex is often grossly identifiable, and is verified by stimulation in this region. Evoked movements are typically tied closely to the onset (or sometimes offset) of stimulation, and are reproducible. Most evoked movement in cases of temporal lobe stimulation will involve face, mouth, or throat followed by hand and arm. Attempts to evoke movement and/or sensory experiences should be started near the sylvian fissure in the cortical representation for the face. Both the motor and sensory homunculi may be roughly mapped in this fashion, although in truth, the sensory areas are sometimes very difficult to specifically identify, and as long as the motor strip is found, the sensory cortex may reliably be assumed to be the gyrus behind it.

When operating in the language-dominant hemisphere, the patient is then asked to begin slowly counting, and the posterior one-third of the inferior frontal convolution is likewise mapped. When Broca's region is stimulated, arrest of speech frequently occurs; often this area of brain is quite circumscribed. Areas of stimulation that affect motor, sensory, or language functions are marked on the cortical surface with sterile numbered tickets. All data is recorded on a sterile drawing of the hemisphere.

More sophisticated testing (described below) is required to identify the more anterior and the posterior temporoparietal language sites. To examine these posterior regions, 12–20 sites along the sylvian fissure are selected, including the frontal, supramarginal, angular, superior, and middle temporal gyri. They are marked at random with sterile paper tags about 2 cm apart, and testing is begun. Care should be taken to ensure that the patient's field of vision is not obscured by drapes, and that he or she is fully awake and understands the task. At the end of mapping, the cortical surface is photographed. Then abnormal brain is removed, excluding those areas essential to measured functions (i.e., motor, language, memory). Generally, the margin of the resection should not approach closer than 1 cm to identified functional cortical sites. Postresection corticography is done to confirm that no (or little) epileptiform patterns remain, and the craniotomy is closed. The entire procedure typically requires 6–8 h.

2.1. Complications

Occasionally, a seizure may be evoked in the process of the stimulation studies. In this case, appropriate intravenous drugs are immediately given, and moist abdominal sponges held firmly over the exposed cortex. This process can usually be easily controlled, but it is prudent to avoid stimulating that region again at the same current.

If stimulus artifact is absent in the EEG recording while the cortical surface is stimulated, one must troubleshoot the equipment between the electrical outlet and the stimulating electrodes. This is straightforward electronic troubleshooting, and a competent electronic engineer familiar with the equipment should be able to accomplish it.

2.2. Mapping under Special Circumstances

Mapping of motor cortex can be done in patients who are under general anesthesia. This might be the approach of choice to mapping in a child who is a candidate for epilepsy surgery, but who cannot endure the rigors of awake craniotomy. It may also be appropriate in adult tumor patients for whom tumor location does not threaten language or memory functions, but may involve motor systems. In such cases, only motor areas are mapped. The patient, although anesthetized, must not be paralyzed. Indeed, it is essential to check for reversal of anesthetically induced paralysis through peripheral-nerve stimulation. Motor stimulation mapping in this situation usually requires much higher current levels (10–20 mA) and *very* careful observation of the patient's face, hand, arm, and leg for evoked movement. This is made more difficult by the operative draping typically used with the asleep patient. The major limitation is that only motor areas can be identified, since it is impossible to get feedback regarding evoked sensation from the asleep patient, to help identify sensory areas. Also, of course, mapping of language and other cognitive functions cannot be done.

All of the principles and procedures discussed for mapping during awake craniotomy for resection of a seizure focus apply equally well to operations in tumor patients where mapping may be desired. Tumors in areas that are classically associated with a function (i.e., a tumor in Broca's area) may have caused displacement of expressive speech/language function, so that safe resection is quite possible. Much more variability in functional localization is seen in such cases than might be assumed by normal anatomy. It is impossible to know, however, unless mapping is accomplished.

Since it is important to remove as much of the tumor as possible, more detailed mapping may be necessary in tumor patients. In addition, since tumor resection often involves a deeper resection in critical areas, it is often necessary to continue behavioral mapping as the procedure moves to deep subcortical areas of the brain. If the tumor underlies a language area, the focus of mapping is often to identify areas through which a safe surgical approach might be taken, that is, through areas not indicated to be involved in language function or stimulation of which is least disruptive to language function.

3. Mapping Language Functions

Application of an alternating electrical current to cortical tissue has a variety of excitatory and inhibitory effects, both locally and at a distance from the stimulation site (Ranck, 1975). With few exceptions, the stimulation sites associated with motor and sensory responses are located in areas one might predict for them on the basis of classic neuroanatomical organization. In contrast, in the quiet patient who is not engaged in task-specific behavior, stimulation of the cortex outside these areas usually has no observable or reported effects. These areas of the cortex are said to be silent. If, however, the patient is engaged in a specific task, for example, a measure of spoken language, such as naming, application of the current to one or more sites in the silent region may disrupt performance on the ongoing task. If care is taken that the level of stimulation used is below that generating afterdischarges, recovery of normal function generally resumes the instant the current is removed. In some cases, however, the disruptive effects may persist for some seconds. If afterdischarge should be encountered, altered function is likely to persist throughout and even following the duration of afterdischarge.

This disruptive effect of stimulation on behavior has been modeled as a reversible temporary lesion, similar to the transient disruptive effect on isolated function seen in focal seizures. The exact nature and extent of functional neuronal disruption caused by the stimulating current is not well documented; empirically, the effects on behavior of stimulation at a particular site are often both repeatable and quite different from the repeated effects of stimulation at sites only a few millimeters away (Ojemann and Whitaker, 1978a). Stimulation effects thus are modeled as temporary lesions localized in both space and time.

Performance on such tasks as naming and counting is commonly disrupted in association with stimulation at discrete cortical sites on the dominant, usually left, cortex. Identification of sensorimotor cortex and of cortex important to language by the stimulation-mapping procedure allows these areas to be spared during resection, greatly increasing the margin of safety associated with cortical resection. Continued experience with stimulation-mapping of cortical function has identified minimal, if any, additional risk to surgical patients specific to cortical stimulation (Ojemann, 1983). Individual variability in the exact localization of functional sites necessitates careful mapping in each patient (Ojemann, 1979).

The strategy often adopted for intraoperative stimulation-mapping studies involves obtaining multiple samples of a number of different tasks at multiple cortical sites in an individual patient. Frequent samples of task performance on which no stimulation occurs are pseudorandomly interspersed with stimulation trials. Obtaining multiple samples of a particular task at a particular site allows for statistical evaluation of whether behaviors evoked with stimulation are significantly different from behaviors obtained in control conditions. There are usually a number of stimulation conditions for a given task at a given cortical site, commonly 3–6, and a number of nonstimulation control trials, commonly 70–80. A binomial single-sample test, for which the control performance serves as the estimate of error probability, is utilized for the statistical assessment (Siegel, 1955). A site is related to a given task only when stimulation-related errors on that task and at that site were unlikely to have occurred by chance ($p < .05$).

The larger the number of sites that can be sampled for each task, the more detailed the mapping. However, there are definite time limitations on stimulation-mapping in the operating room. Thus, there is always a trade-off among the number of samples, the number of tasks, and the number of cortical sites where stimulation effects on various tasks can be assessed. Hence, only the appropriate and relevant task should be selected for stimulation.

Stimulation studies are carried out after electrocorticographic identification of the epileptic focus and identification of sensorimotor cortex by cortical stimulation. The primary goal of these studies is to identify for the surgeon the relationship of particular tasks to the epileptic focus. Thus, the sites selected for stimulation generally encompass the posterior margins of the identified epileptic focus and sites in the nonepileptic cortex in the posterior temporal, inferior parietal, and posterior frontal cortex.

The patient must not be aware of when current is applied. Therefore, identification of stimulation sites by the surgeon should be done at the end of stimulation. Stimulation is applied at the onset of a trial or segment of a trial, and is maintained for the duration of the task, typically 4–6 s, depending on the task being tested. Patient's responses and markers indicating both trial and stimulation onset and offset are recorded on audio tape and, when appropriate, videotape for subsequent analysis.

3.1. Language and Language-Related Measures

Three of the most commonly used language tests are described below. One test measures naming, reading of simple sentences, and short-term verbal memory. This test consists of a series of consecutive trials presented visually as slides. The first segment of each trial is a slide of an object, whose name is a common word, with a carrier phrase "this is a ___" printed above it. Object names should be well known to the patient and thus of a high frequency in the language. In many of our studies, a second segment of each trial has been a slide with an 8–10-word sentence that the patient is to read aloud. This task serves two purposes—first as a distractor for the recall to follow, and second as another measure of language function. Sentence reading will elicit a longer and more linguistically complex segment of speech than naming. A wide variety of formats might be used, but results will be most interpretable if responses are not allowed to be too open-ended; target response should be well defined. We have used sentences made up of two clauses. The verb in the second clause of each sentence is left blank and is to be completed by the patient. The sentences are constructed so that they must be completed with one of a small number of inflected verb forms. This allows the patient to demonstrate not only straight reading capacity, but also the ability to generate a semantically and syntactically correct verb to complete the sentence.

The third segment of each trial has been a slide with the word "recall" printed on it. This acts as a cue for the patient to state aloud the name of the object pictured on the first slide of this trial, a name retained across the distraction produced by reading the sentences. Stimulation occurs during the naming segment on some trials, the reading segment on some trials, and the recall segment on still other trials.

Control trials on which no stimulation occurs are pseudorandomly interspersed with stimulation trials. The sequence of site and test conditions is so arranged that no site is stimulated consecutively, and stimulation at each site on each condition is distributed throughout the test period. Performance on this test is analyzed for stimulation effects on naming and reading, and for effects of stimulation at the time of input (naming), storage (reading), or retrieval (recall) on short-term verbal memory. Trials with errors in naming are excluded from analysis of memory performance to ensure that the information to be remembered has been adequately perceived.

3.2. Patterns of Language Breakdown with Cortical Stimulation

3.2.1. Arrests of Speech

The most common response to language tasks with stimulation at one or more sites in an individual patient is what is termed arrest. During an arrest, the patient appears to remain alert with eyes open and may open his/her mouth in an apparent attempt to speak, but there is no real articulatory movement and no audible vocalization. A correct response often appears immediately upon removal of the stimulating current. Arrest responses appear to be tied critically to speech motor-control systems, but cannot be further analyzed in terms of their possible linguistic role. Sites associated with arrest are typically broadly distributed in the left lateral cortex, but are always located within one gyrus of the sylvian fissure. Stimulation of a small area in the left posterior inferior frontal cortex (Broca's area) almost invariably produces speech arrest. If the arrest is associated with evoked nonverbal oral movement, it suggests stimulation of the motor strip; stimulation there is not usually applied repeatedly, since seizures can easily result.

3.2.2. Naming Errors

Naming is the language task most extensively studied with cortical stimulation-mapping. Penfield and his colleagues (Penfield and Jasper, 1954; Penfield and Roberts, 1959) were the first to report naming data from cortical stimulation. Naming errors are divided into three types:

1. Total speech arrest—during stimulation the patient is unable to produce the carrier phrase or name the object
2. Anomia—during stimulation the patient is able to produce the carrier phrase, but is unable to name the object and
3. Misnaming—during stimulation the patient is able to produce the carrier phrase, but incorrectly names the object.

Naming errors have been demonstrated with stimulation of a very broad area of the lateral dominant cortex (Ojemann and

Whitaker, 1978a; Ojemann, 1979; Ojemann and Mateer, 1979; Van Buren et al., 1978). Some of the individual sites where naming changes have been evoked extend well beyond the traditional limits of the lateral cortical language areas, but most are located in the immediate peri-sylvian cortex.

There have been a few studies involving cortical stimulation-mapping and naming in multiple languages (Ojemann and Whitaker, 1978b; Mateer and Rapport, 1982). In all cases, there have been some dissociated sites implicated in each language, i.e., cortical sites where stimulation altered naming in one language, but not in the other. This dissociation of cortical sites involving different languages is consistent with dissociated recovery of different languages seen in cases of polyglot aphasia (Paradis, 1977). One striking feature of the stimulation-mapping in two languages is that naming in the language in which the patient was least competent can be altered from a greater number of cortical sites. It has been hypothesized that larger areas of cortex must be used for object naming in the language of greater unfamiliarity and/or less automaticity.

3.2.3. Reading Errors

One of the reasons for developing the reading task was to evaluate more complex aspects of linguistic production, in order to sensitize our measure of language function. In one series of 14 patients, 26 total sites were associated with evoked naming errors (Mateer, 1982). Of these sites, 88% were also associated with significant alterations in at least one error category on the reading task. Of the 53 total sites associated with evoked changes in reading, 28 (53%) were not associated with naming errors. Thus, whereas most sites associated with naming errors were also associated with reading errors, many sites are associated only with what appears to be the more sensitive reading task. Two of the three sites involved only with naming were located in the posterior portion of the middle temporal gyrus. These findings are strikingly consistent with the lesion data. Although naming deficits are ubiquitous with almost all aphasic types and usually overlap to some extent with other kinds of linguistic disruption, anomic patients in whom the naming deficit is prominent and often isolated have been reported to have restricted lesions in this same region involving the posterior mid-temporal gyrus (Mazzochi and Vignolo, 1979).

Reading tasks provide for much more varied language performance than naming tasks, and stimulation of the dominant cortex during the reading of simple sentences has demonstrated striking patterns of linguistic alteration. Categorization of the myriad of possible changes in reading during both control and stimulation trials is a critical feature of the analysis. The major error categories associated with stimulation-related alteration include speech arrest, grammatical errors, semantic errors, and articulatory (phonetic or phonemic) errors. Errors from nonstimulation trials must be compared to errors from stimulation trials. Only errors not seen on nonstimulation trials should be considered as potentially stimulation related.

Overall, the pattern of cortical organization revealed in this analysis suggested that the motoric execution of speech as reflected in speech sound selection and production (articulatory/phonologic errors) was highly dependent on the peri-sylvian core. Both the traditional anterior "motor" area and the posterior peri-sylvian areas were critically involved. Aspects of reading relating to more linguistically based aspects of language, including grammatical and semantic selection, without any associated articulatory component, occupy more distal sites (Mateer, 1989). The concentric "ring-like" appearance of the distributions is highly reminiscent of the concentric field features associated with the primary, secondary, and tertiary association fields of other major cortical motor and sensory systems.

As seen with naming errors, there is a substantial degree of individual variability in the distribution of sites associated with stimulation-evoked alterations in reading. The areas most often involved in reading disruptions include, in order of frequency: the inferior posterior frontal zone (88%), the middle superior temporal gyrus zone (64%) and the inferior anterior frontal zone (58%), followed by the posterior mid-frontal and the posterior superior temporal gyrus zones (50% each).

The results of mapping can be plotted across groups of subjects to reveal trends in the functional distribution of language-related behavior. In Fig. 2, such a composite map is provided. Arrests of speech occur broadly over the left cortex. Phonologic errors occur only within one gyrus of the sylvian fissure in both inferior frontal, superior temporal, and inferior parietal regions. Grammatic and semantic errors occurred, in all but one case, more than one gyrus from the sylvian fissure, but in all three lobes.

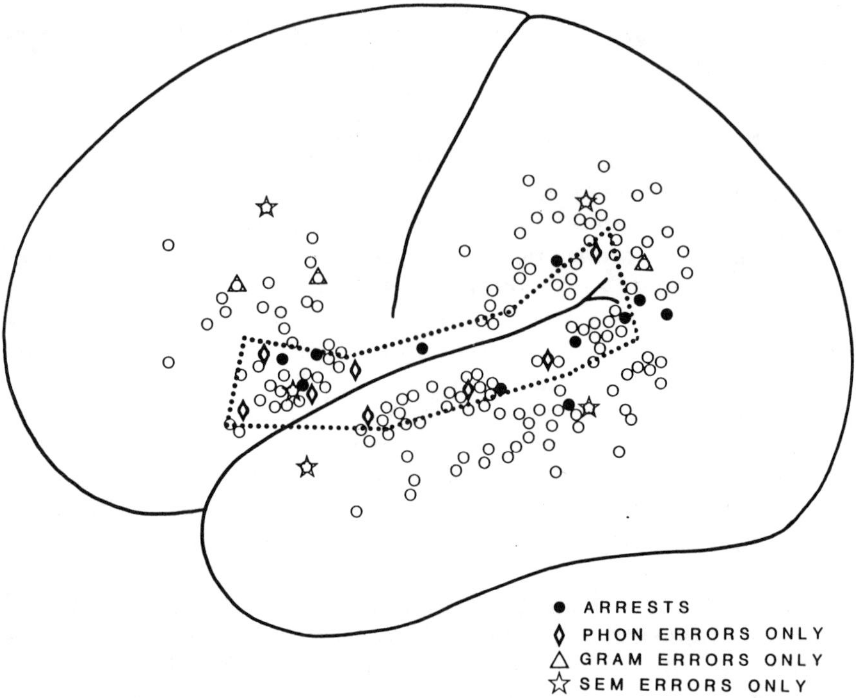

Fig. 2. Composite map of the left cortex, indicating stimulation sites and corresponding statistically significant errors on a reading task ($N = 18$ patients).

Reading stimuli have thus far been discussed in terms of providing a complex language task and a distractor for short term memory tasks. In some cases, however, single word or simple sentence reading tasks may be the stimuli of choice for evoking all language output. Some patients with restricted language skills owing to cognitive limitations may be quite unreliable on naming tasks. For these patients, very simple reading tasks are often quite helpful in providing a clear, unambiguous response which is disruptible with stimulation.

3.3. Disruption of Short-Term Verbal Memory

Short-term verbal memory (STVM) deficits are a persistent problem for patients with aphasia, suggesting that the dominant cortex plays a role in memory (Butters et al., 1970; Albert, 1976).

Milner (1967) found that resection of the dominant temporal cortex increased the verbal memory deficit almost as much as extension of the resection further into hippocampus. Selective loss of immediate and short-term verbal memory after small left-parietal lesions has been reported (Warrington and Shallice, 1969; Warrington et al., 1971; Saffran and Marin, 1975). Early observations made with the stimulation-mapping technique noted different effects of stimulation during input to or retrieval from STVM at different cortical sites. Fedio and Van Buren (1974) reported STVM changes with left, but not right cortical stimulation.

Separation of input, storage, or retrieval as parts of STVM can be obtained with a single-term memory task paradigm. Ojemann and Mateer (Ojemann, 1978a,b; 1983; Ojemann and Mateer, 1979) have used a visually presented single-term memory test during stimulation-mapping. Object naming serves as the input task. The name of the object was stored for a few seconds during a verbal distraction (reading or counting). Output of the name of the object from STVM was then cued by the word "recall." Stimulation at a given cortical site was applied during input, storage, or output on different trials of the memory task.

The locations of sites associated with STVM change in eight patients were usually at some distance from, but surrounded the peri-sylvian cortex in high- to-mid-frontal, mid-temporal, and especially parietal cortex. These memory-related sites are often adjacent to, but generally separate from, the sites where stimulation alters language, as identified by changes in naming or reading (Ojemann, 1979). Two-thirds of the sites that evoked changes in memory failed to evoke any kind of language change. Memory sites have consistently been characterized by this largely separate cortical representation across several series of patients (Ojemann, 1979, 1983; Ojemann and Mateer, 1979).

Data from a study by Ojemann (1983) suggested different roles for the frontal, temporal, and parietal cortex for STVM, based on whether memory changes were evoked by stimulation during the input, storage, or retrieval phases of the task. During the input or storage phase of the memory, stimulation of 27% of the frontal sites, 62% of the temporal sites, and 64% of the parietal sites was associated with recall errors. This represented a significantly greater role for temporal-parietal cortex relative to frontal cortex in memory input and storage. In constrast, frontal sites were significantly more often associated with errors in recall when stimulated during the recall or output phase of the task than were parietal or temporal sites.

3.4. Variability in Language Organization Relative to Gender and Verbal IQ

The degree of variability in the localization of cortical sites related to language change is great for all three linguistic behaviors: naming, reading, and memory. With the multiple linguistic function test, most patients, although not quite all, do show some kind of language change with stimulation in the traditional language zones (the inferior posterior frontal cortex and the middle to superior temporal gyrus). This suggests that the overall areas related to language functions may be relatively uniform, but with individual variability of sites related to specific language function. Such observations are consistent with the data from spontaneous lesions. Aphasias resulting from what appear to be similar cortical lesions may have quite variable linguistic characteristics (Mazzochi and Vignolo, 1979). Variability in the behavioral correlates of cortical areas is not surprising, in view of the high variability in both gross morphological structure (Rubens et al., 1976) and cytoarchitectonic patterns (Galaburda et al., 1978) in human cortex. The morphological structure of this language area is quite different from person to person. Rubens et al. (1976) noted individual variability in the gyral pattern at the end of the sylvian fissure in the dominant hemisphere. Stensass et al. (1974), after examining the total area and surface area of visual cortex in 25 normal brains, found there were variances of 300% in estimated total area and variances of 400% in surface area.

Individual variability of cortical organization for language functions found by the stimulation-mapping technique was not unexpected. We attempted to use it to further explore what may be important underlying correlates of cortical organization of language functions. Not all individuals use language with the same degree of facility, and across groups of individuals, a variety of investigative techniques yield different patterns of neurolinguistic organization. Evidence that at least some of the individual variability is not an artifact of the stimulation-mapping technique or choice of anatomical landmarks comes from the correlations that are present between the pattern of naming change in individual patients and independent measures. We correlated two independent patient characteristics, the patient's gender and preoperative Verbal IQ, with patterns of stimulation related to naming changes. We used gender as a correlate, because there are data suggesting differential patterns of aphasias in males and females following anterior vs posterior left-hemisphere spontaneous lesions (Kimura, 1980). Verbal IQ was selected as an independent measure of verbal facility.

Gender-related distribution of sites on the lateral cortex involved in naming varied significantly for a series of eight males and ten females (Mateer et al., 1982). Naming changes were evoked from more sites in men than in women. Overall, naming errors were evoked from 63% of the total sites sampled in males (32 of 51), but only from 24% of the total sites in females (16 of 68) (.025 p $<.05$). When the lateral cortex was divided into eight zones, the percentage of sites in a zone related to naming changes was significantly higher for males in two of the zones, an anterior frontal zone (80% of males vs 22% of females, $p <.05$) and a posterior parietal zone (males 57%, females 0%, $p <.05$). Proportionately, males were also at least twice as likely to demonstrate evoked naming errors with stimulation of an anterior parietal zone and two middle temporal gyrus zones, though these differences did not reach significance. Males appeared to use a broader overall area of left cortex for naming. That is, it appears that a larger area of lateral frontal and parietal cortex are involved in naming processes in males than in females. Thus, gender may to some degree determine not only interhemispheric patterns of language organization, but the extent and pattern of intrahemispheric representation of language.

Verbal IQ was also correlated with patterns of stimulation-evoked naming change (Polen, Mateer, and Ojemann, unpublished observation; Whitaker and Ojemann, 1977). A series of 21 patients, ranging in preoperative IQ from 69–115, was divided into two groups on the basis of Verbal IQ. Of ten patients with Verbal IQ at or below 96, seven demonstrated naming changes with stimulation in the posterior parietal region; only one of ten patients with Verbal IQ greater than 96 demonstrated evoked naming changes with stimulation in that area ($p <.025$). Patients with Verbal IQ above 96 were more likely to show naming changes from superior temporal gyrus stimulation than those with lower IQs (seven out of ten vs four out of eight), though this difference does not reach statistical significance. We have hypothesized that parietal representation of naming in patients with low Verbal IQs does not necessarily suggest that naming is represented more broadly in these patients. Rather, it suggests that the presence of language functions in the parietal lobe may constitute a suboptimal arrangement. Thus, the poor language function in some patients may reflect a less advantageous pattern of cortical language organization.

4. Stimulation Effects in the Nondominant Cortex

Stimulation mapping of the nondominant cortex, as established by preoperative amytal testing, does not generally alter language-related tasks, outside of motor cortex (Fried et al., 1982; Mateer, 1983b; Ojemann, 1983). Stimulation of the face motor cortex alters mimicry of single movements, with some arrests of speech. However, there is no disturbance of phoneme identification with stimulation, such as that seen with stimulation of the dominant hemisphere. There are changes in mimicry of sequential orofacial movements with stimulation of the motor cortex, but not with stimulation outside the motor cortex. Short-term verbal memory, naming, and reading changes are likewise rarely evoked from the nondominant cortex.

Stimulation of the nondominant cortex does alter various visual spatial tasks. We have previously described discrete nondominant cortical localization of evoked changes in a variety of spatial tasks: perception of and memory for faces and angles, and the identification of facial emotional expressions (Fried et al., 1982; Mateer, 1983b,c; Ojemann, 1983). In general, these studies have demonstrated a strong dissociation of cortical sites involved in separate functions across individual patients. Contrary to the notion of diffuse functional organization in the nondominant cortex, visuospatial functions in the right hemisphere appear to be as discretely localized as verbal functions in the left dominant hemisphere.

5. Conclusions

The technique of human cortical stimulation is critical to the safe conduct of a variety of neurosurgical procedures. It also provides a wealth of opportunities to investigate the detailed nature and organization of cortical information processing. Mapping should ideally answer questions of clinical utility in an individual patient. Application of principles of experimental design and repetition of procedures over time can, however, yield a unique perspective on the functional organization of cognition in the human brain.

References

Albert M. (1976) Short-term memory and aphasia. *Brain and Language,* **3,** 28–33.

Butters N., Samuel I., Goodglass H., and Brody B. (1970) Short-term visual and auditory memory disorders after parietal and frontal lobe damage. *Cortex,* **6,** 440–459.

Clarke E. and O'Malley C. (1968) *The Human Brain and Spinal Cord, Chapter IX: Brain Localization* (University of California Press, Los Angeles).

Fedio P. and Van Buren J. (1974) Memory deficits during electrical stimulation of speech cortex in conscious man. *Brain and Language,* **1,** 29–42.

Fried I., Mateer C. A., Ojemann G., Wohns R., and Fedio P. (1982). Organization of visuospatial functions in human cortex: Evidence from electrical stimulation. *Brain,* **105,** 349–371.

Galaburda A., Sanides F., and Geschwind N. (1978) Human brain: Cytoarchitectonic left-right asymmetries in the temporal speech region. *Arch. Neurol.,* **35,** 812–817.

Haymaker W. and Schiller F. (1970). *The Founders of Neurology* (Charles C. Thomas, Springfield, Illinois) pp.

Kimura D. (1980). Sex differences in intrahemispheric organization of speech. *Behav. Brain Sci.* **3,** 215–263.

Mateer C. A. (1982). Cortical organization of language: Evidence from electrical stimulation studies. *University of Washington Paper in Linguistics* **7,** 32–38.

Mateer C. A. (1983a). Motor and perceptual functions of the left hemisphere and their interaction, in *Language Functions and Brain Organization.* (Segalowitz S. J., ed.), Academic, New York.

Mateer C. A. (1983b). Functional organization of the right nondominant cortex: Evidence from electrical stimulation. *Can. J Psychol.* **37,** 36–58.

Mateer C. A. (1983c). Localization of language and visuospatial functions by electrical stimulation mapping, in *Localization in Neuropsychology* (Kertesz A., ed.), Academic, New York.

Mateer C. A. (1989) Neural bases of language, in *Neural Bases of Speech, Hearing and Language* (G. Baumgarter, ed.), Wiley, New York, pp. 259–291.

Mateer C. A. and Rapport R. (1982) Organization of language cortex in two bilinguals. Paper presented at meetings of the American Neurosurgical Society, Washington, D. C.

Mateer C. A., Polen S. B., and Ojemann G. A. (1982). Sexual variation in cortical localization of naming as determined by stimulation mapping. *Brain Behav. Sci.* **5,** 310–311.

Mazzochi R. and Vignolo L. A. (1979). Localization of lesions in aphasia: Clinical CT scan correlations in stroke patients. *Cortex,* **15,** 627–654.

Milner B. (1967). Brain mechanisms suggested by studies of temporal lobes, in *Brain Mechanisms Underlying Speech and Language* (Milliken C. and Darley F., eds.), Grune and Stratton, New York, pp. 122–145.

Ojemann G. A. (1978a) Organization of short-term verbal memory in language areas of human cortex: Evidence from electrical stimulation. *Brain and Language,* **5,** 331–348.

Ojemann G. A. (1978b). Intrahemispheric localization of language and visuospatial function: Evidence from stimulation mapping during craniotomies for epilepsy, in *Advances in Epileptoloty* (Akimoto H., Kazamatsuri H., Seino M., and Ward, A. eds.), **13,** 127–138.

Ojemann G. A. (1979) Individual variability in cortical localization of language. *J. Neurosurgery,* **50,** 164–169.

Ojemann G. A. (1983). Brain organization for language from the perspective of electrical stimulation mapping. *Behav. Brain Sci.,* **2,** 189–230.

Ojemann G. A. and Mateer C. A. (1979). Human language cortex: Localization of memory, syntax, and sequential motor-phoneme identification systems. *Science,* **250,** 1401–1403.

Ojemann G. A. and Whitaker H. A. (1978a). Language localization and variability. *Brain and Language,* **6,** 239–260.

Ojemann G. A. and Whitaker H. A. (1978b). The bilingual brain. *Arch. Neurol.* **35,** 409–412.

Paradis M. (1977) Billingualism and aphasia, in *Studies in Neurolinguistics* 3, (Whitaker H. and Whitaker H., eds.), Academic, New York, pp. 65–122.

Penfield W. and Jasper H. *Epilepsy and the Functional Anatomy of the Human Brain* (Little Brown, Boston).

Penfield W. and Perot P. (1963) The brain's record of auditory and visual experience: A final summary and discussion. *Brain* **86,** 595–696.

Penfield W. and Roberts L. (1959) *Speech and Brain Mechanisms* (Princeton University Press, Princeton, N.J.)

Ranck J., Jr. (1975) Which elements are excited in electrical stimulation of mammalian central nervous system: A review. *Brain Res.*, **98,** 417–440.

Rubens A., Mahowald M., and Hutton J. (1976). Assymmetry of the lateral (Sylvian) fissures in man. *Neurology* **26,** 620–624.

Saffran E. and Marin O. (1975). Immediate memory for word lists and sentences in a patient with deficient auditory short-term memory. *Brain and Language,* **2,** 420–433.

Siegel S. (1955). *Nonparametric Statistics for the Behavioral Sciences.* McGraw-Hill, New York.

Stensass S., Eddington D., and Dobelle W. (1974). The topography and variability of the primary visual cortex in man. *J. Neurosurgery* **40,** 747–755.

Van Buren J., Fedio P., and Frederick G. (1978) Mechanism and localization of speech in the parieto-temporal cortex. *Neurosurgery,* **2,** 233–239.

Warrington E. and Shallice T. (1969). The selective impairment of auditory verbal short-term memory. *Brain,* **29,** 855–896.

Warrington E., Logue V., and Pratt R. (1971) The anatomical localization of selective impairment of auditory verbal short-term memory. *Neuropsychologia* **9,** 377–387.

Whitaker H. and Ojemann G. A. (1977) Graded localization of naming from electrical stimulation mapping of left cerebral cortex. *Nature* **270,** 50–51.

Methods for Studying Human Laterality

John L. Bradshaw

1. Introduction

Laterality research with normal subjects is plagued by contradictory findings, stemming from the diversity of methods used. These may include reaction time (RT) or accuracy measures with an imposed speed or accuracy set, go/no-go or target/nontarget discriminations, identity, similarity or category matches, and perceptual matching of simultaneously presented stimuli vs matching a test item to a memorized target (with various retention intervals, one, many, or constantly changing targets). The stimuli may be easy or impossible to verbalize; they may be treatable as unitary wholes or as collections of discrete features or elements. The task and stimuli may be familiar or novel, easy or difficult, practiced or unpracticed, and presented in isolation or accompanied by another easy or difficult, verbal or nonverbal, concurrent task. Even superficial changes in or interpretations of instructions, or differences in methods of presentation or subject populations may alter findings. The effects of sensory modality, the nature of the task, and the problems of measurement will be considered in this chapter.

2. The Visual Modality

2.1. The Visual System

2.1.1. Exposure Durations

Stimuli flashed to one side of the fixation point, too fast for a fixational eye movement (<200 ms), project directly to the contralateral hemisphere. Thus, a stimulus in the right visual field (RVF) of either eye goes to the left hemisphere (LH), and stimuli flashed to the left visual field (LVF) go to the right hemisphere (RH). This is because optic fibers from the nasal hemiretinae cross at the optic chiasm, whereas fibers from the temporal hemiretinae project to the ipsilateral hemisphere. This need for such brief and eccentric stimulation severely limits the complexity of available stimulation. Saccadic latencies are approximately 175 ms (possibly slightly less for rightwards eye movements; Pirozzolo and Rayner, 1980). In view of individual differences around this mean, counterbalanced by a further short interval to *complete* the saccade and for perceptual sensitivity to return to normal (saccadic suppression; Volkman et al., 1968), exposure durations should not exceed approximately 150 ms.

2.1.2. Hemiretinal Division

It is possible (*see,* e.g., Gazzaniga and LeDoux, 1978) that decussation in the retinal midline may be incomplete, so that a narrow strip receives representation in both hemispheres. In monkeys, it may be 1° wide, perhaps wider in the fovea (Bunt et al., 1977; Stone et al., 1973), thus accounting for macular sparing. However, some patients exhibit perfect splitting of the fovea (Koerner and Teuber 1973), and the degree of nasotemporal hemiretinal overlap may be practically hairline (McIlwain, 1972; Westheimer and Mitchell, 1969). Thus, apart from stereopsis, it may have no functional significance, certainly in the laterality context when stimuli are usually presented at least 1.5° eccentrically. Indeed, macular sparing after occipital injury may be the result of overlapping blood supply (Harrington, 1981). In any case, Harvey (1978), Haun (1978), and Lines and Milner (1983) in a variety of tasks demonstrated visual hemifield asymmetries for foveal stimuli similar to those found for nonfoveal presentations. However, eccentricities of 1–2° do help the accuracy of fixational control, though anything beyond about 5° should be avoided because of acuity problems. If the latter effects are ignored (though *see* Sergent, 1985, and below), there is no evidence that laterality effects are much affected. Typical asymmetries (an RVF/LH advantage with verbal material, the reverse with certain types of spatial or pattern processing) occur with eccentricities between 1° and 5°.

2.1.3. Nasal/Temporal Pathway Strengths

The contralateral (nasal) pathways may possibly be more efficient than the ipsilateral (temporal) routes, particularly perhaps under normal binocular viewing or under dichoptic viewing when separate information simultaneously goes to each eye (Maddess, 1975; Osaka, 1978, though cf Harvey, 1978). Davidoff (1982) and Bradshaw and Nettleton (1983, p. 85) review the evidence; however, binocular viewing should perhaps always be used, as should subjects without marked eye dominance.

2.2. Speed and Accuracy Measures and Responding Hand

With accuracy, a vocal or manual selection response is possible, whereas with speed (RT), a manual response is more common than vocal latencies. Latency measures are typically obtained with relatively long exposures (below saccade latencies, of course) to keep errors low. This speed emphasis makes performance *resource limited* (Norman and Bobrow, 1975), whereas accuracy measures are instead typically obtained under conditions of *data limitation* (partial information, short exposures) to ensure a reasonable error rate. One drawback is the relative insensitivity of accuracy measures, floor and ceiling effects, and its susceptibility to guessing. Speed/accuracy tradeoffs can be a problem, either between subjects or, worse, when, e.g., one hemifield/hemisphere appears faster but less accurate than its fellow. Moreover is, e.g., a 10-ms superiority the same irrespective of baseline RT values? Furthermore, hand-visual field interactions and stimulus–response compatibility (Craft and Simon, 1970) can be a problem with different effects for simple and choice RTs and crossed and uncrossed arms (*see*, e.g., Berlucchi, 1978; Bradshaw and Nettleton, 1983, p. 122; Bradshaw and Umiltà, 1984). Where such effects are not of intrinsic interest, responding hand should be systematically alternated (though this can lead to asymmetric directional biases—an *entirely unimanual* response is, however, quite unsatisfactory) or a bimanual response employed where the hand responding faster stops the clock. For discriminatory responses, the two response buttons must never be *horizontally* (left/right) disposed, but, e.g., anterior/posterior out from the midline; otherwise the data can be very difficult to interpret. We recommend two laterally adjacent buttons for the two forefingers (e.g., for positive responses) and two additional buttons behind them for the middle fingers (negative responses), with this arrangement reversed for half the subjects. (Reversal *within subjects* will be confusing.) Moreover, whole-arm movements should be avoided if predominantly contralateral control is required, since fine movements of the extremities are more contralaterally organized (Brinkman and Kuypers, 1973).

To keep RT variance low, subjects must be highly practiced, and trials should be discarded or repeated where errors or aberrantly long or short RTs occur. Alternatively, such values may be replaced with an arbitrary cutoff value (e.g., two standard deviations from the subject's overall mean, or a more subjectively based estimate). Instead, medians or geometric means may be used, although assumptions of additivity may then be violated. The problem is not yet resolved, nor is the optimal length of an experimental sequence. We believe (and *see* Hamsher and Benton, 1977) that at least 32 trials should occur in each limb of the simplest possible contrast. Problems of practice, fatigue, strategy changes, and so on, prove the need to counterbalance all conditions carefully and to pseudorandomize left/right presentations and positive/negative responses, so that the subject cannot predict where the stimulus will occur or what will be the correct response. Children and clinical patients may be unable to cope with the requirements of an RT task. Other problems include how to combine positive and negative responses—a go/no-go procedure is no real solution and loses half the data. A vocal response, moreover, may be unsuitable for nonverbal tasks, since it is likely to be initiated by the LH.

2.3. Nature of Task or Decision

Sergent (1983a) distinguishes four main types of task: first, there is the detection of the presence or absence of a stimulus, either at the sensory level (e.g., gratings detection), or to infer via simple RT the duration of interhemispheric transfer. Her other three tasks all involve additional cognitive operations: relative or perceptual discrimination (simultaneous matching), recognition (delayed matching), and absolute discrimination (identification). The first two of these cognitive tasks involve similarity judgments between a target and a test stimulus presented either simultaneously or successively (i.e., thus requiring a memory component); the third requires the presentation of only a single test item for recognition as belonging to a specific and familiar category. Different cognitive processes are involved in these tasks, probably with different modes of lateralization. Moreover same/different judgments may not need consideration of all aspects of a stimulus (cf Patterson and Bradshaw, 1975), and the two judgments may involve different stringency levels with respect to criteria for correctness. Matching requires only one of two possible responses per trial, usually with a 50% probability of error; with identification, a specific response must be constructed, usually one of many possibilities. Since nonverbal stimuli are more likely to be unfamiliar than verbal, they tend not to be tested in identification tasks (though cf, e.g., familiar faces below), although this is not true of verbal stimuli. Thus, one often cannot directly compare between tasks. Target judgments tend to be faster than nontarget (Nickerson, 1978); this is perhaps because the configuration must be thoroughly checked before a nontarget response is emitted (Krueger, 1978), and perceptual noise produced by an aberrant nontarget item in a display may slow down processing (cf Sternberg's 1975 claim that target superiority occurs at the *decision* rather than the *comparison* stage). Moreover, the predicted asymmetries tend to be stronger or more consistent with target stimuli (Bradshaw and Nettleton, 1983); subjects may adopt a set to respond to targets with little awareness of what occurred with nontargets (Suberi and McKeever, 1977). Moscovitch (1972) suggests a bias to check information in both hemispheres before responding "different," but not before responding "same." Finally, Hellige et al. (1979) suggest a response bias: when in doubt, respond "different," since there are more possible ways that a stimulus can be different than it can be same. Consequently, there may be more correct guesses among correct responses on nontarget trials than on target trials. We (Bradshaw et al., 1980) found that asymmetries were stronger for nontargets than for targets when the former differed maximally from the latter and relatively little from each other.

2.4. Unilateral vs Bilateral Presentations and Fixation Controls

Few attempts have been made to simulate in vision the dichotic technique in hearing (i.e., two simultaneous though different signals, one to each ear), and for anatomical reasons there is no exact analog anyway. The most common approach, other than stimulation of one visual field at a time, randomly left or right, is to present paired stimuli, one in each visual field (bilateral presentation). Since stimuli always occur in exactly the same loci on each trial, fixation control is considerably weakened. Geffen et al. (1972)

with *unilateral* presentations compared the effects of visual field certainty and uncertainty; monitoring fixation, they found few fixation failures and no laterality differences between blocked and random presentations. With bilateral presentations, to control fixation and to reduce the probability of reporting from left to right (which would generate an LVF advantage), a digit can be placed at the fixation point, which must be reported first (McKeever and Huling, 1971). Asymmetries under these circumstances are unusually large and robust, though the procedure is unsuitable for RT measures, and may serve to continue a rightwards scan, now from the central fixation digit, and so favoring the RVF. Moreover, the verbal nature of the fixation digit may bias the processing of the lateral stimuli (*see* Beaumont, 1982; Bradshaw and Nettleton, 1983; Bryden, 1982). A solution (Piazza, 1980; Schmuller and Goodman, 1980) is to place an arrowhead at fixation to cue the item for first report or sole processing (thus permitting RT measures). Powerful asymmetries may then appear with both verbal and pictorial stimuli, although the traditional unilateral approach has proved reliable. Unilateral presentations may assess the efficiency of interhemispheric communication to permit the more specialized hemisphere to perform the task, whereas competitive bilateral presentations, by virtue of inhibiting interhemispheric communication, may provide a better means of assessing the capacity of each hemisphere individually to perform the task (Hines, 1975).

2.5. State-Limiting Variables

We may distinguish between state and process limitations (Sergent, 1983a). The former affect sensory quality, e.g., via exposure duration, size, luminance, retinal eccentricity.

2.5.1. Stimulus Duration

This may not much affect lateral asymmetries above 50 ms (*see*, e.g., Bradshaw et al., 1984). However, with shorter exposures, fine detailed information (appealing to the LH?) may disappear, thus placing greater emphasis upon global RH mechanisms. Indeed, short exposures generate LVF advantages even with verbal tasks (*see* Sergent, 1983b, for review), because of LH decrement.

2.5.2. Stimulus Eccentricity

Retinal eccentricity directly affects visual acuity. For verbal material, RVF presentations may suffer more than LVF presentations as eccentricities increase (*see* Sergent, 1983a,b, for review), although stimulus magnitude is important and exceptions have been noted. Thus, Hellige et al. (1984) found that, in a face processing task, LVF presentations were *more* detrimentally affected by increasing eccentricities. Processing requirements, stimulus duration, eccentricity, degradation, size, and luminance probably all interact, together with the relative importance of local feature detail (LH) and global holistic interrelationships (RH).

2.5.3. Stimulus Degradation

This factor may mediate all the other state-limiting variables. The RH may cope earlier or better than the LH under such circumstances (Sergent, 1983a); consequently, noise masking of verbal stimuli may cause greater performance decrement for RVF than LVF presentations (*see* Hellige et al., 1984, for review, although their own study offers an exception), leading even to a reversal of the normal RVF advantage. Thus, with stimulus degradation, the LH may be unable to analyze distinctive local features, and global or configurational information alone may be available to the holistically specialized RH (cf Bradshaw and Nettleton, 1981). Thus, clear presentations may favor the LH, and degraded the RH (Sergent, 1983a).

2.5.4. Stimulus Size

Doubling stimulus size halves its spatial frequencies (Sergent, 1983a,b) and reduces acuity requirements. Differential hemispheric sensitivities to high and low spatial frequencies (*see below*) may therefore be even more fundamental than the effects of stimulus degradation. Stimulus size must therefore be considered in conjunction with eccentricity, duration, and luminance, and the LVF advantage for the large low-frequency components present in large letters may only occur when stimuli are briefly exposed and at considerable eccentricities (Sergent, 1983b). Indeed, Pring (1981) obtained the typical RVF advantage for medium (i.e., normal) sized words, and an LVF superiority only for aberrantly sized (large *and* small) words.

2.5.5. Stimulus Luminance

This factor, along with exposure duration, determines available stimulus energy; however, it is rarely reported, and may well account for much of the conflict in the literature. Sergent (1982a) found in a face categorization task that, when all other factors were kept constant, luminance increase improved RVF performance alone, the RH thus perhaps tolerating a wider range of stimulus energies than the LH.

2.5.6. The Spatial Frequency Hypothesis and Global/Local Information

For a regular grating, its spatial frequency is the number of light and dark sinusoidal changes (bars)/degree of visual angle. The visual system may contain channels responding selectively to different spatial frequencies in a two-dimensional array, and output from these channels may compute the entire distribution of spatial frequencies as if by Fourier analysis. Harmon (1973) showed that faces, from which the relevant high-frequency spectral components had been removed by computer processing, could nevertheless be recognized from the remaining low-frequency information. A variety of methods have been used to remove or mask high or low spatial frequencies (Fiorentini et al., 1983; Inui and Miyamoto, 1984; Riley and Costall, 1980), and it appears that, although the mid-frequency range may be the most important for face recognition, this can be achieved by either high-frequency information alone (e.g., as in line drawings) or low-frequency information alone (e.g., the gross pattern of featural interrelationships only remaining).

Sergent (1983a,b) claims that the hemispheres are differentially sensitive to the outputs of spatial frequency channels, the RH to low and the LH to high. If stimulus degradation (as above) specifically impairs high-frequency information, then it should specifically decrement RVF/LH performance; conversely, since outline or schematic faces contain mostly high spatial frequencies, and as long as the cognitive discrimination is difficult, a RVF/LH superiority should—and can—be demonstrated (Fairweather et al., 1982; Patterson and Bradshaw, 1975). Moreover, long exposure durations and/or the use of similar, or familiar, faces that need processing in terms of high-frequency featural information also generate RVF/LH advantages (Jones, 1980; Marzi and Berlucchi, 1977; Sergent, 1982b). Indeed, Sergent (1985) presented familiar faces that, by digitizing, were low-pass or broad-pass filtered; the former produced LVF advantages, the latter RVF advantages for identification and categorization tasks. However, contrary results have been reported by Gazzaniga and Smylie (1983) with a commissurotomy patient, and Hellige et al. (1984) and Glass et al.

(1985) with normal subjects. A distinction should, of course, be drawn between cycles/degree of visual angle and cycles/face width. As Rhodes (1985) observes, attempts to determine hemispheric asymmetry in processing different spatial frequencies must include viewing distance and size. Moreover, the spatial-frequency approach may not explain all aspects of visual perception (Westheimer, 1984), and although it is an attractive hypothesis that the LH is better at processing local featural detail, and the RH at global holistic or integrative interrelationships, this idea need not necessarily relate at all closely to the spatial-frequency hypothesis. Indeed the global/local aspect fits very comfortably with the more cognitively based analytic/holistic dichotomy (Bradshaw and Nettleton, 1981). Thus, if a complex configuration like a face requires precise identification, then global features may be insufficient for optimal performance, and some local featural detail may be required. Any form of degradation may make it more difficult to extract this local feature information, and therefore, RH processing modes may supervene.

2.6. Process-Limiting Variables

As discussed, state-limiting variables reflect how quality of sensory information influences modes of hemispheric operation. However, processing modes imposed by task requirements, either as demanded of or perceived by the subject, are just as important. The degree of memory involvement, stimulus discriminability, and familiarity of the material all may act as process-limiting variables. Moscovitch (1979; Moscovitch et al., 1976) claims that early processes (at the sensory, iconic, or echoic level) are not lateralized, and that lateralization only occurs at later or deeper levels of processing. However, in a sense, this viewpoint merely restates the obvious (Bradshaw and Nettleton, 1983; Bryden, 1982); if language processes are lateralized, the task may have to be processed at a linguistic level for asymmetries to emerge. In any case (*see above*), laterality effects *do* occur at early sensory stages, although effects may increase if memorization is required (Oscar-Berman et al., 1978). The debate develops theoretical significance if the spatial frequency hypothesis (above), a sensory or state-limiting aspect, can be subsumed under the analytic/holistic processing dichotomy (Bradshaw and Nettleton, 1981), which has a strong cognitive flavor. The resolution between these two positions may come from recognizing that the fundamental hemispheric specialization is an LH superiority at fine, sequential feature analysis and temporal resolution—i.e., at most a dichotomy by default.

2.6.1. Difficulty

Process-limiting variables achieve significance in the contexts of discriminability (difficulty), familiarity, and memorization. Experiments with faces (the nonverbal equivalent of language tasks in complexity and meaningfulness) usually employ relatively unfamiliar (and therefore necessarily dissimilar) stimuli, whereas words are usually highly overlearned. Thus, with face discrimination tasks, there is usually an RH advantage (Bradshaw and Nettleton, 1981; Bryden, 1982, for reviews). There may be an LH superiority with difficult, very similar faces differing in only one feature (Bradshaw and Sherlock, 1982; Bradshaw et al., 1980; Fairweather et al., 1982; Patterson and Bradshaw, 1975).

2.6.2. Memory and Practice

A meaningful response inevitably involves some act of memory retrieval. Beaumont (1982) shows that the imposition of a delay between target and test stimuli in matching tasks does not necessarily increase asymmetries. Where, e.g., an LVF/RH advantage in face processing *is* so increased (Moscovitch et al., 1976), this could stem from decay in memory of inessential information and preservation of a partial representation that typically favors RH performance. With *familiar* faces, such effects may not occur, thus accounting for RVF/LH superiorities (Glass et al., 1985; Umiltà et al., 1978). In a delayed matching task with nonsense shapes, an initial LVF/RH advantage reversed, with practice, into an RVF/LH advantage (Hannay et al., 1981), possibly because of the development or increasing use of verbal codes or strategies (*see* Bradshaw and Nettleton, 1983, for review of similar findings and exceptions). Obviously, one cannot ignore contributory effects from nature of stimulus and task, task difficulty, familiarity of material, rest pauses, response compatibility, and so on. Two factors may be preeminent (Beaumont, 1982): increasing familiarity with stimuli and task, via the development of appropriate strategies, as when, e.g., children or bilinguals develop reading proficiency (Bradshaw and Nettleton, 1983), and secondly, the subject's increasing adaptation to unnatural and initially difficult lateralized presentations. Thus, task order may be important (Bradshaw and Sherlock, 1982).

2.6.3. Familiarity and Stimulus Set Size

According to Hardyck et al. (1978), reliable asymmetries in verbal tasks only occur with a limited range of amply repeated stimuli that become exhaustively familiar; there should be no asymmetries with new material on each trial. Miller and Butler (1980) confirmed this, as did Hellige (1980) with RH advantages for polygon processing, and Sullivan and McKeever (1985) in an object-naming task, although they found that repetition of stimuli led to a *reduction* in RVF/LH advantages in a *word*-naming task. However, many studies, including our own (*see* Beaumont, 1982, for review), have obtained powerful asymmetries with large ensembles of material presented once only. Sullivan and McKeever (1985) suggest that, for tasks that can be processed efficiently by *one* hemisphere, repetition *reduces* asymmetries, whereas when substantial processing is needed by both hemispheres, repetition may *increase* asymmmetries.

Thus, adopted strategies are important factors in lateral asymmetries, no less than the verbal/nonverbal nature of the actual stimuli. When members of a pair of letters are matched, an LVF/RH advantage can occur if the matching is based on physical rather than nominal qualities (e.g., AA, bb vs Aa, Bb, *see below*, and *see* also object-matching tasks). Sex differences in lateral asymmetries may even be partly because of processing strategies (Bryden, 1979, 1982), although one should resist appealing to "strategy effects" when unexpected findings eventuate. Seamon and Gazzaniga (1973) and Metzger and Antes (1976) tried to manipulate subjects' strategies directly; although there were inconsistencies, imagery instructions were associated with LVF/RH advantages, and verbal-rehearsal strategies with opposite effects in a memory task.

2.7. Problems with Tachistoscopic Presentation: Some Alternatives

The visual system did not evolve to perform shape recognition by looking at objects in the periphery for fractions of a second. As discussed above, many laterality effects may stem from such degraded and abnormal viewing. Indeed, with complex or moving patterns, with several words of text, or with slow readers or clinical patients, one often needs longer exposures.

Myers and Sperry (1982) used the limits of lateral eye turn (with head held fast) to stimulate the visual periphery. Thus, with eyes turned fully left, the temporal half of the left eye's visual field can be used for lateral input to the RH, and vice versa for input to the LH. However, the technique is uncomfortable, eye turn may interact with modes of cognitive processing (Kinsbourne, 1973), and head turn certainly affects hemispace asymmetries (*see below*). Moreover, because of the masking of one eye by the nose, the technique is essentially monocular, with all the associated drawbacks discussed above.

Dimond et al. (1975) and Dimond and Farrington (1977) used contact lenses that were opaque except for a slit, and so functioned as hemifield occluders. In an independent attempt, we found that such a technique was difficult, uncomfortable, and unreliable (because of slippage and diffraction). Recently, Sivak et al. (1985) have used a smaller hard contact lens, painted partially black to obstruct part of the visual field and inserted into a soft carrier contact lens. They claim success with it. Francks et al. (1985), instead, fitted vertical strips of opaque tape to close-fitting goggles to obscure one visual field or the other. They claim that all but 0.5° of the unwanted field was blocked out, and that appropriate asymmetries could be demonstrated.

Zaidel (1975) has developed a modified version of the stabilized image technique, which permits prolonged exposure and free scanning, since it is the image of the occluder that is stabilized and not the stimulus. However, although the system has been used successfully with commissurotomy patients (e.g., Zaidel, 1978a,b), it retains many of the disadvantages described above with hard contact lenses.

Nettleton et al. (1983) review several computer-based systems where subjects' eye movements are monitored directly as they read text on a computer display, and changes in the text are made contingent upon the eye movements. Such systems are very flexible, but require sophisticated hardware and software, and stimulus material is limited to what can be generated on the display. Crane and Kelly (1983) described a technique for accurately simulating scotomas in normal subjects, by means of a mask stabilized on a fixed retinal region by signals from an eye tracker. However, two high-speed servo-controlled mirrors are required, and as mechanical rather than electronic systems, they may possibly have been subject to inertia, response lags, or jitter.

Nettleton et al. (1983) described a system for generating a video window or mask that is yoked to a subject's eye movements.

The position and size of the window or mask is electronically derived and infinitely variable. The system uses a video monitor on which a wide range of materials can be displayed either directly via a videocamera or prerecorded on videotape, or can be computer generated. The gain and direction of the mask movement, relative to determining eye movements, can be varied in a positive or negative fashion, from an infinite range of starting positions. Moreover, the entire video screen can be blanked in response to eye movements greater than a preset limit, or in a particular direction.

2.8. Summary of Findings in the Visual Modality: (a) Verbal Processing

Single letters or numbers typically show an RVF/LH advantage (for reviews, *see* Beaumont, 1982; Bradshaw and Nettleton, 1983; Bryden, 1982), with accuracy measures, discriminatory manual RTs, or vocal naming latencies. Curiously, the asymmetries may be greater for consonants than vowels (Umiltà et al., 1972) and for stop consonants than for fricatives (Klisz, 1980), exactly as in dichotic studies (*see below*), suggesting a common LH mechanism for both sensory modes. Nonstandard typefaces may yield LVF superiorities, presumably because of visuospatial preprocessing (Bryden and Allard, 1976). When pairs of letters are simultaneously (Cohen, 1972; Davis and Schmit, 1973; Geffen et al., 1972) or successively (Wilkins and Stewart, 1974) presented for name or physical matching, RVF and LVF superiorities may respectively occur (*see also* Sergent, 1983b).

With digits, an RVF/LH advantage typically occurs for manual discrimination and vocal-naming latency (Geffen et al., 1971). Decisions of numerical magnitude are better performed to stimuli in the RVF; however, calculation and number processing may require ideographic, spatial, and semantic operations involving the RH also, especially if numbers are recoded into continuously varying analog quantities or magnitudes of a psychophysical or imaginal nature to facilitate numerical comparison (Holender and Peereman, 1987; Lamm and Gordon, 1984; Troup et al., 1983).

With words, any resultant asymmetries will be the culmination of many separate subprocesses, which may be differentially lateralized. Chiarello (1988) distinguishes three major kinds of tasks: lexical (word/nonword) decisions, semantic categorization (e.g., does the item represent an animal or plant?), and naming (where perhaps because of strongly lateralized LH mechanisms, RVF advantages may be strongest; *see*, e.g., Bradshaw and Gates, 1978; Bradshaw and Taylor, 1979). She describes three kinds of information-processing operations involved:

1. Prelexical (visual sensory), where stimulus quality is important (e.g., exposure duration, stimulus clarity and size, and word length, which are all factors that can affect LVF/RH performance). Thus, an LVF/RH superiority can occur when successively presented words are matched at a physical level (Gibson et al., 1972), though cf the RVF/LH superiority found for successively (Tomlinson-Keasey and Kelly, 1979) and simultaneously (Gross, 1972) presented words.
2. Lexical (access to the "word store": word/nonword decisions, where word frequency or class—abstract/concrete—may be important). Relevant factors (*see* Bradshaw and Nettleton, 1983, p. 86, for review) include the nature of the nonword distractor or letter string, e.g., KZFGK, GOZE, or COTE, respectively illegal, legal, and pseudohomophone (a string that *sounds* like a real word). Some studies claim an RH contribution to the processing of concrete or imageable words, though this has been disputed (Bradshaw and Nettleton, 1983, p. 152, and Lambert and Beaumont, 1983). Indeed, it is unlikely that word frequency, age of acquisition, or syntactic class greatly affect asymmetries (Beaumont, 1982). G. J. Bradshaw et al. (1979) showed that lexical decisions were possible for words exposed too briefly to be recognized, and that the resultant LVF/RH advantage shifted to a normal RVF/LH superiority with longer presentations. Chiarello (1985) employed semantic priming in a laterality context. Thus, the prime NURSE (laterally or centrally presented) is followed by the lateralized letter string or target word, which is either semantically related to it (e.g., DOCTOR) or unrelated (BREAD); the subject is timed in performing a (manual) word/nonword decision to the second item. With automatic priming (when the

prime was pattern masked and the subject was unaware of the semantic relationship between the two items), Chiarello found larger priming for *orthographically* similar stimuli (e.g., BEAK . . . BEAR) with LVF presentations, larger priming for *phonologically* similar stimuli (e.g., JUICE . . . MOOSE) with RVF presentations, and slightly larger priming for *semantically* similar stimuli (COW . . . BULL) with LVF presentations. The results varied slightly with *controlled* priming, where there was no masking of the clearly perceptible prime, and the subject knew that the considerably delayed (by 400 ms or more) target was likely to be related to the prime. Automatic priming may involve automatic hemispheric mechanisms in word processing.

3. Postlexical (postretrieval), e.g., naming or semantic category judgments. The previous level cannot, of course, be easily assessed independently of this one. Although naming shows very strong RVF/LH superiorities, semantic category judgments may generate weaker asymmetries (*see,* e.g., Bradshaw et al., 1977; Martin, 1978; Saffran, Bogyo et al., 1980; Nettleton and Bradshaw, 1983). Urcuioli et al. (1981) found an RVF/LH superiority with category *matching* (CHAIR:TABLE, both belonging to FURNITURE), but not for category membership (CHAIR belongs to FURNITURE).

To what extent may RVF advantages be an artifact of report order or a directional left-to-right scan employed in reading English? It is, of course, particularly important to control report order with bilateral presentations and accuracy measures (Beaumont, 1982). However, even with RT responses to unilaterally presented words, our Western reading habits may be a source of artifact (*see,* e.g., Bradshaw et al., 1981; Bradshaw and Nettleton, 1983, p. 87). The important first letter of a word is immediately to the right of fixation for RVF presentations. Studies before the mid-1960s comparing the effects of normally oriented and mirror-reversed English and Hebrew (which is read right-to-left) had supported the concept of a directional letter-by-letter scan in word recognition. They predicted an RVF superiority as a consequence of the scan, to and beyond the initial letter, being directionally congruent when words fell into this visual field. However, these early studies had all used long or uncontrolled exposures, no control over fixation, large and polysyllabic words, and letter-by-letter responses, all being likely to favor a sequential letter-by-letter strategy. Later studies used improved designs enabling better estimates to be made of the relative magnitudes of the RVF superiority for normal and mirror-reversed English, Hebrew, and horizontally and vertically organized English and Chinese. The absence of differential RVF superiorities among these different materials indicates that scanning is not a major artifact in RVF superiorities with horizontal, single-syllable words (*see* Bradshaw et al., 1981). The fact that asymmetries are typically weaker with sinistrals and females (Bradshaw and Nettleton, 1983; Bryden, 1982) and with certain semantic tasks (*see above*) supports the conclusion of hemispheric mediation. Indeed, the use of vertical words may well introduce further artifacts, because of their unfamiliar appearance. Although a recent study (Tramer et al., 1985) claimed to have located a scanning artifact, they used single letter report or identification. Boles (1985) manipulated orientation (horizontal/vertical), letter symmetry (e.g., A, W, X, M vs B, G, D, F, and so on), and report order, and obtained a generalized RVF superiority that did not interact with any other variable. Consequently, a horizontal format can be safely employed at least for short single-syllable words.

When color words (*red, green, blue*) are presented in color dyes, and the subject must name the dyes and ignore the words, a congruent color–word–dye relationship (e.g., *red* written in red dye) facilitates performance, and an incongruent relationship (e.g., *blue* written in a green dye) interferes with performance, relative to a baseline condition (name the dye of a meaningless letter string). Greater interference occurs when incompatible color and word information goes to the RVF/LH (Schmit and Davis, 1974; Tsao et al., 1979; Warren and Marsh, 1979). With Stroop-type letters *themselves* composed of smaller letters, Martin (1979) and Alivisatos and Wilding (1980) reported an RVF advantage for local processing of the stimulus elements, but no asymmetry for global processing of the entire letter.

Klatzky and Atkinson (1971) presented words or pictures, requiring that the initial letter of the word, or the name of the depicted object, be matched to a letter set held in memory; the picture-test stimuli produced RVF advantages and letters an LVF advantage. Thus, the former task may have involved a verbal strategy, and the latter one a physical match. With pictures, no asymmetries were reported for vocal naming latencies (Levine and Banich, 1982; McGuire et al., 1986) or in processing the "syntax" of agent-action-recipient events (Segalowitz and Hansson, 1979). For matching physically identical or merely similar animal pictures, LVF/RH advantages appeared only with reduced exposure durations and luminance (Sergent and Lorber, 1983). Hines et al. (1984) used pictures in a semantic priming task and found an RVF/LH superiority. Underwood and Whitfield (1985) found that *categorization* of pictures gave an LVF/RH advantage, whereas *naming* them gave the reverse effect (Wuillemin et al., 1982). Underwood and Whitfield also examined the effect of adding distractor words to the pictures in the picture categorization task; interference or facilitation (depending upon congruence or incongruence of the word–picture relationship) occurred only when the distractor word fell in the LVF, again indicating an RH role in picture processing. However, these effects required that a pattern mask follow the lateralized stimuli, indicating once again that stimulus degradation interferes less with RH mediation. Lupker and Sanders (1982) also employed this Stroop-like picture–word interaction; however, they found that the interactive effect required RVF presentations. Sullivan and McKeever (1985) found an RVF/LH advantage in object-picture naming latencies only when a small number of items were repeatedly presented (cf Hardyck et al., 1978, above). Seitz and McKeever (1984) used bilaterally presented pictures with a cueing arrow at fixation. For object-naming latencies, they obtained an enormous RVF/LH advantage compared to the unilateral condition; interactions with sex and familial handedness indicated a true hemispheric locus. Finally, various studies and findings (including LVF/RH superiorities) have been reported with lateralized clock faces, e.g., Berlucchi et al. (1979) and Hatta (1978).

2.9. Summary of Findings in the Visual Modality: (b) Nonverbal Processing

Verbal and nonverbal stimuli have little in common physically and may implicate different properties of the visual system (Sergent, 1983a). Verbal stimuli consist of a finite set of patterns (26 letters, 6 or 7 distinctive letter features) and are highly overlearned. The same is not generally true of faces, which are far more complex, multidimensional, and usually unfamiliar.

2.9.1. Simple Sensory Discriminations

The effects of simple sensory or low-level-perceptual manipulations on nonverbal stimuli are reviewed by Bradshaw and Nettleton (1983), Bryden (1982), Davidoff (1982), and Sergent (1983a). Better performance is reported for LVF/RH discriminations, including color (but not color naming) and brightness; LVF stimuli may even appear brighter than those on the right (Davidoff, 1982), just as lines on the left may seem longer than those on the right (Bradshaw et al., 1986). LVF superiorities include also dot detection, localization, and possibly enumeration, perception of orientation (except for verbal labelling) and curvature, stereopsis, and depth perception (though cf Julesz et al., 1976). Findings are ambivalent for motion and duration.

2.9.2. Patterns and Shapes

With more complex stimuli, e.g., polygons, effects may depend upon retention intervals, form complexity or familiarity, or nature of judgment (same or different) (*see* Bradshaw and Nettleton, 1983; Bryden, 1982; and Davidoff, 1982). Generally, an LVF/RH advantage tends to appear with intermediate levels of complexity, absence of verbal codeability, and where there is a memory interval. Opposite asymmetries may also occur if the subject is simultaneously occupied with a verbal memory load.

If the LH mediates all coded visual stimuli, including also shapes and musical notes, an RVF/LH advantage should occur for, e.g., the hand signs used by the deaf (*see* Bradshaw and Nettleton, 1983; Corballis, 1983; Davidoff, 1982). Generally, adults unable to read such signs show LVF/RH superiorities, whereas for those who *can, meaningful* signs may give an RVF/LH advantage, although there are contrary reports and even claims (Bonvillian et al., 1982; Campbell, 1986; Cranney and Ashton, 1982; Gibson and Bryden, 1984; Homan et al., 1984; Kelly and Tomlison-Keasey, 1981; Marcotte and LaBarba, 1985; Neville et al., 1982; Panou and Sewell, 1984; Suter, 1982; Weston and Weinman, 1983) that, in the absence of a normal developmental exposure to spoken language, language itself and even manual "dextrality" may be less lateralized to the LH (though cf Vargha-Khadem, 1982, and Ross, 1983, who argue that any differences may be the result of differences in strategy).

2.9.3. Faces

Faces are complex configurations *par excellence;* they also mediate person identity and transmit emotional information. They parallel language in complexity and, in so far as both mediums may require, special processors; faces may need such a special processor over and above any mechanism that may be necessary for handling other complex nonverbal configurations. We can recognize many thousands of faces during a lifetime, often after only a single brief encounter, and despite the addition of distracting paraphernalia (mustaches, beards, hats, spectacles), plastic changes with aging and differing emotions, and, frequently, the passage of many years between successive encounters (for general reviews on face recognition, *see,* e.g., Bruce, 1983; Ellis, 1981, 1983; Hay and Young, 1982).

An LVF/RH advantage (speed and/or accuracy) has been reported for processing photographed faces, cartoon drawings, Identi Kit constructions, and schematic drawings (Bradshaw and Nettleton, 1983; Bryden, 1982; Davidoff, 1982). However, task (e.g., matching or identification) is an important variable (Sergent and Bindra, 1981). In a matching task, test and target faces are compared, the latter being either memorized in long-term memory, or merely more or less immediately preceding the test face (short-term memory); memory duration may, of course, be important (Moscovitch et al., 1976). One (i.e., threshold) technique involves determining exposure durations of lateralized stimuli at which correct responses occur, e.g., when the subject subsequently chooses from an array of alternatives (Marcel and Rajan, 1975). Accuracy scores may also be obtained without threshold manipulation, with unilateral or bilateral presentations; speed measures include target/nontarget RTs (Geffen et al., 1971), go/no-go RTs (Rizzolatti et al., 1971), and same/different RTs or accuracy measures when one or both of the successive items, target and test, are lateralized, with or without various interstimulus intervals (Hilliard, 1973; Patterson and Bradshaw, 1975). Conversely, the test face may be identified as belonging to a specific (though not necessarily named) individual (Marzi and Berlucchi, 1977). Identification may promote LH advantages, either via detailed feature analysis, covert or overt naming, or both; typically, stimuli are presented singly, often from an unknown and very large set. Conversely, matching tasks may be more conducive to LVF/RH superiorities, because they promote holistic configurational processing at which the RH is superior, and the target set is known and small. Moreover, partial information may suffice for matching, unlike when full identification is required. Indeed, for "different" judgments, a decision may be reached as soon as *any* difference is detected, and even with "same" judgments, an exhaustive search is unnecessary if the subject knows how many features are relevant, and whether or not they covary (Sergent and Bindra, 1981). Some argue that the LVF/RH advantage with faces stems at least in part from an RH role in effect (Suberi and McKeever, 1977), although Ley and Bryden (1977) argue for a true independence, with the RH quite separately mediating both functions (*see also* Davidoff, 1982, and Strauss and Moscovitch, 1981). It is of course possible that face and/or emotion recognition may *themselves* be subsumable under a generalized RH mediation of complex patterns.

As discussed above, the need for memorization may increase asymmetries. Similarly, LVF/RH superiorities increased when sketched *caricatures* were matched to target photographs (Moscovitch et al., 1976). Such "deeper levels of processing" (*see also* Galper and Costa, 1980) may also explain why bigger LVF/RH superiorities occur when representations of two different *poses* of the same *person* are matched, rather than identical target *photographs* (Bertelson et al., 1979), suggesting an important RH role in the extraction of "physiognomic invariants," independent of orientation, expression, and possibly aging. However LVF/RH advantages have also been reported for processing bugs (Bradshaw and Sherlock, 1982), although not shoes (St. John, 1981), and Hay (1981) got the effect when true faces were discriminated from scrambled features.

So is there a specialist (RH?) mechanism for recognizing faces? Certainly, infants attend readily to faces (as compared to equally complex nonface stimuli), and they copy expressions, and can produce them even if blind (Young, 1986). Yin (1970) and Carey and Diamond (1977) argue for a special face processor, claiming that *inverted* faces do not generate typical asymmetries (Leehey et al., 1978; Rapaczynski and Ehrlichman, 1979), although others dispute findings, methodology, and conclusions (Bradshaw and Nettleton, 1983, p. 93; Davidoff, 1982; and Young, 1984). Inversion, of course, disturbs the perception of many classes of familiar object, and Yin (1969) compared the recognition of faces, airplanes, houses, and stick figures, upright and inverted; he found that inversion disproportionately affected faces. It may, of course, necessitate a feature-analytic strategy by the LH for processing such difficult stimuli—where difficulty here relates not to *stimulus degra-*

dation (see above), which enhances *RH* processing, but rather to *confusability*. Thus, Bradshaw and Sherlock (1982), Bradshaw et al. (1980), Fairweather et al. (1982), and Patterson and Bradshaw (1975) showed that RVF/LH advantages only occurred with difficult discriminations requiring isolation of a single feature.

Discriminating faces by sex may also induce RVF/LH advantages (Jones, 1980; Jones and Anuza, 1982), except with short exposures (Sergent, 1982b). Indeed, Glass et al. (1985) and Leehey and Cahn (1979) obtained RH mediation of *familiar* faces at short exposures.

Familiarity gained naturally over a prolonged period (e.g., from the visual media) should of course be distinguished from that acquired during an experiment; familiarity with a *person* should also be distinguished from familiarization with a *representation, photograph,* or *stimulus*. The second (experimental) form may be controlled (though it seems not to have been done) by changing materials during an experiment. *Stimulus* familiarity may be manipulated by presenting different *views* of the same (familiar or unfamiliar) person. This was done by Bertelson et al. (1979): an LVF/RH advantage occurred only for *person*-identity, not for *stimulus*-identity. Famous faces are, of course, both nameable and familiar. Umiltà et al. (1978) tried to unconfound these factors; naming speeded performance, but was not related to asymmetries and, therefore, was implicated at response selection rather than stimulus processing. Familiar faces gave an RVF/LH advantage, and unfamiliar ones the opposite. Other findings of an RVF/LH advantage with famous faces come from Marzi et al. (1974), Marzi and Berlucchi (1977), and (under certain circumstances) Glass et al. (1985). Rhodes (1985) only got the effect after long exposures and/or with verbal responses. Contrariwise, Levine and Koch-Weser (1982) got an LVF/RH advantage for famous faces; so did Leehey and Cahn (1979) with familiar faces irrespective of response (pointing or naming), and Young and Bion (1981) the same irrespective of whether the face was familiar or famous (*see also* Young et al., 1985). Whereas famous faces form a large indeterminate set, the opposite is true of familiar (colleagues') faces, thus perhaps explaining discrepancies. Indeed, Young and Bion only found an LVF superiority when subjects knew exactly whom to expect. Young (1984) with bilateral presentations also obtained an RH superiority for famous faces processed in terms of internal or external features or the whole face. How familiarity is imposed, or acquired, may therefore be important, together with task nature and difficulty, and LH involvement may demand isolation and identification of individual features.

Developmental aspects of face processing have been studied by Carey and Diamond (1977): for children aged 10 or under, inversion affected the recognition of faces and houses equally detrimentally; thereafter, faces were worse affected. Performance on upright faces improved markedly until 10, but not so upright houses and inverted faces. Carey and Diamond claim that the improvement on upright faces reflects development of a specialist RH processor for faces.

Finally, an RH involvement may occur with chimeric faces, when one investigates which side is more salient. Schwartz and Smith (1980) presented left–right composites constructed from two different original faces, with each half falling (for 36 ms exposure) in different visual fields. Subjects identified (in a subsequent whole-face matching task) more of the left halves of the chimeras, which were seen as whole faces (*see also* Finlay and French, 1978, and Milner and Dunne, 1977).

2.9.4. Right Hemisphere Preprocessing, Arousal, and Alerting

Stimulus degradation leads to or enhances RH involvement (*see above*). With unusual, florid or italic typefaces, an LVF superiority may occur for verbal material (Bryden and Allard, 1976); experience may reverse such effects (Gordon and Carmon, 1976). Cursive script may reduce RVF superiorities (Bradshaw and Mapp, 1982). With newly acquired or unfamiliar scripts or languages, RH advantages may appear, reverting with increasing familiarity to the usual RVF effects (Gordon and Carmon, 1976; Shimuzu and Endo, 1981; Silverberg et al., 1980), thus, perhaps explaining apparent bilaterality in bilinguals and polyglots (Albert and Obler, 1978; Vaid and Genesee, 1980).

Damage to the RH leads to a generalized lowering of effect, alerting, attentional capacities, and arousal, with RT decrements for either hand (Bradshaw and Nettleton, 1983; Bradshaw et al., 1986); the RH may therefore be concerned with *bilateral* manifestations of arousal. With normal subjects, LVF warning signals reduce subsequent RTs of the right hand more than RVF warnings (Van Den Abell and Heilman, 1978). Simple RTs (without warning signals) in the LVF or left ear are responded to faster than those in the RVF or right ear (for review, *see* Bradshaw and Nettleton, 1983).

Visual acuity is superior in the LVF/RH (*see* Davidoff, 1982, for review), and in vigilance tasks, the LVF/RH may be faster at maintaining sustained attention for detecting randomly and infrequently occurring signals (Dimond and Beaumont, 1973); similar effects occur in the auditory modality (Warm et al., 1980).

3. The Auditory Modality

3.1. The Auditory System

The auditory pathways are more richly interconnected than the visual ones. Several reviews discuss the relative superiority of the contralateral over the ipsilateral routes (Connolly, 1985; Geffen and Quinn, 1984; Bradshaw, 1988), particularly perhaps in the presence of competing information at the two ears, although the effects also occur physiologically and behaviorally with monaural stimulation. Indeed, the important variable may be perceptual load or difficulty, manipulable via competition *within* as well as *between* ears or hemispaces.

3.2. Behavioral Studies: Major Experimental Paradigms

Five major modes of stimulation may be distinguished:

1. Binaural: a single message to both ears (the normal experience)
2. Diotic: two different messages to the two ears
3. Dichotic: a different message to each ear
4. Monaural: one message to one ear
5. Monotic: two competitive messages both to a single ear, e.g., the full dichotic signal channeled to one ear. Sometimes called "competing monaural," it should be distinguished from pure monaural, even though it is sometimes mistakenly used for the latter procedure.

With the traditional dichotic procedure, there can be response or attentional biases. Subjects may choose to report everything from one ear first, usually the right with verbal materials. Thus, items at the second ear may be disadvantaged. When instructed to report everything, attention may be deployed in different ways, resulting in asymmetries that have little to do with cerebral lateralization. Nevertheless, the verbal LH/right ear advantage (REA) is remarkably robust, even when report is required from one ear for a block of trials, or first from one ear (pre- or postcued), or the subject must attend equally to both ears (Bryden, 1982). Bryden (1967) describes how to score only trials where all items from one ear are reported before all from the other, splitting them by which ear is reported first, and computing separate means for each ear, and each report order by ear. He can thus calculate how subjects would have performed had they always employed an ear order of report. The procedure is cumbersome, wasteful of data, and assumes (not necessarily correctly) that subjects are attending as much to the LE when they spontaneously commence with an LE item as they are to the RE when starting on that side. It does, however, provide a relatively pure measure of perceptual laterality, free of memory factors and starting or attentional biases. Freedom to deploy attention may add unwanted variability to the data, much of it attributable to subjects classifiable as biased attenders, who show asymmetry of intrusions with focused attention (Bryden et al., 1983; Geffen and Quinn, 1984); they intrude items from the RE, while attending to the LE more than vice versa. Indeed, the essence of the REA may generally be an asymmetry in distinguishing a target from background noise.

3.3. Verbal Studies (Dichotic Presentation)

Broadbent (1954) originally developed the dichotic procedure to study selective attention, an area that has developed quite independently of laterality research, although very relevant to it. Kimura (1961) adopted the technique for her classic paradigm, the presentation of three pairs of temporally aligned digits, played one digit to each ear at a rate of two pairs/s (*see* Bradshaw et al., 1986; Bryden, 1982, for an account of her physiological model to explain the resultant REA, problems associated with it, and related findings). Lists of digits or words were originally used. Backwards speech also generates REAs, though meaningless sequences may require normal syntactic ordering and intonation, especially the latter. Shankweiler and Studdert-Kennedy (1967) and Studdert-Kennedy and Shankweiler (1970) pioneered the use of *single* dichotic pairs of consonant–vowel (CV) nonsense syllables systematically differing in one or other phoneme. This procedure eliminated possible contamination from attentional and memory factors, and has largely replaced list procedures. Subjects attended to both ears and reported both items, the item they were most sure of usually first. Consonants (/b,d,g,p,t,k/) in a constant articulatory environment can be artificially synthesized, with exact onset alignment and the experimental variation of acoustic–phonetic dimensions (to which the language hemisphere presumably is most sensitive), with minimal memory and semantic involvement. Stimuli differing in terms of stop consonants generate the strongest REAs, whereas less encoded vowels that generate little or no ear differences are apparently identified by nonlateralized processors. (An alternative possibility is that the various speech sounds are differentially sensitive to degradation during callosal transfer, if hemispheric asymmetries are absolute rather than relative.) Vowels may generate REAs if embedded in noise, if shortened, or if the listener is unsure of the effective vocal tract size or voice pitch. Fricatives, liquids, and affricates produce effects intermediate between stops and vowels, though the latter may generate REAs when heard in a language context and LEAs when heard as nonspeech sounds.

The detection procedure also produces robust, reliable results; subjects report the presence of a prespecified target, and performance is measured by hit rate or d' (a bias-free measure of sensitivity). Reaction time measures are often combined with this procedure. Springer (1973) specified one CV as a target (occurring half the time, and on either ear), with five other CVs as distractors. A 14-ms REA appeared for manual RTs and for vocal-naming latencies. Bradshaw et al. (1981) and Pierson et al. (1983) have also found that vocal and manual RTs are equivalent, which suggests that the *vocal* component of verbal report in report accuracy tasks is not a major determinant of REAs. There may be less REA variability with this procedure, especially if attention is focused on one ear at a time for a block of trials, and Piazza (1980) has found the expected smaller REA for sinistrals who are known to have reduced cerebral dominance. Geffen and Gaudrey (1981) combined RTs and hit rates for each ear in a Discriminant Function Analysis procedure. They claim to have correctly classified, by language dominance, 95% of a criterion sample of 37 patients who were assessed for language lateralization by independent clinical procedures.

3.4. Other Auditory Techniques: Stroop, Delayed Auditory Feedback (DAF), and Sussman's Procedure

Cohen and Martin (1975) got their subjects to judge the pitch of pure tones (high or low) of two congruent words (the word *high* sung at a high pitch and *low* sung low) and of two incongruent words (the word *high* sung low and *low* sung high), with dichotic or monaural presentations. The Stroop effect (interference or facilitation) was greater for RE/LH presentations, showing that verbal information was automatically processed in the LH, even when irrelevant.

When Abbs and Smith (1970) directed a subject's ongoing speech, delayed, to the RE with masking noise to the LE, errors were greater than when the directions were reversed. Bradshaw et al. (1971, 1972) measured reading times with DAF to one ear and simultaneous auditory feedback, or some other irrelevant signal, to the other; times were longer with RE DAF, with opposite effects when a piano, recorder, or electronic organ was played, the last device permitting the manipulation of delayed feedback in the total absence of any simultaneous feedback. Contralateral white noise was inadequate to produce asymmetries. Irrelevant music to the LE with speech DAF to the RE *reduced* the effects of RE DAF, presumably because subjects were better able to distract themselves, by means of the music, from the disruptive verbal DAF to the RE. Roberts and Gregory (1973) got their subjects to repeat tongue twisters, and measured the point of speech breakdown as a function of DAF intensity at either ear. Intensity levels were lower in the RE for speech and in the LE for a simple tapping sequence. Both the DAF and the Stroop technique demonstrate how the LH cannot refrain from the automatic and unwanted processing of a speech signal.

A technique related to DAF examines the disruption from verbal or musical masking at one or other ear upon a separate speech or music task (Heilman et al., 1977; Manning et al., 1978). There was greater disruption in verbal processing with verbal maskers as the RE, as predicted (*see* Bradshaw, 1988, for details).

Sussman and colleagues (*see* Bradshaw and Nettleton, 1983, and Bryden, 1982, for reviews) developed a pursuit auditory tracking task: subjects heard two simultaneous tones, the target in one ear and the cursor in the other, and had to track the frequency of the randomly varying target tone with the cursor, by unidimensional movements of tongue, jaw, lips, or hand. There was better performance with target tone to LE and cursor to RE, although this asymmetry was weakest with control by respiratory or hand movements, leading to the conclusion that the speech musculature proper is more relevant to such lateralized sensorimotor integrations. However, a significant asymmetry did occur with right-hand tracking. Thus, the original effect must be due to an RE/LH superiority for cursor control, rather than an LE/RH superiority for target-tone analysis. Consequently, the LH may be specialized for the sensorimotor integration of speech-related movements and their auditory concomitants, for the complex sequential programming of changing oro–facial configurations from one target position to another, and indeed for all fine coordinated movements, including those of the limbs (Bradshaw and Nettleton, 1981; Kimura, 1977).

3.5. Nonverbal Auditory Studies (Dichotic Studies)

3.5.1. Right-Hemisphere Aspects

We have previously reviewed this area (Bradshaw, 1985; Bradshaw et al., 1986, and *see also* Bryden, 1982). Briefly, with nonmusical stimuli, LEAs appear for the dichotic report of environmental sounds, nonverbal vocalizations, for the emotional tone of speech, and for intonation patterns, although Thai speakers still show an REA when discriminating words differing only by pitch. For single notes and musical chords, LEAs are found, although the effects may be equivocal or modulated by practice. Likewise, LEAs occur for synthetic musical tones, melodies on solo orchestral instruments, or hummed melodies. Sidtis (1984) believes that the general locus of the RH's superiority in tasks of this nature lies in the processing of steady-state harmonic information and the extraction of pitch information from *complex* periodic sounds. Thus, a note produced by a musical instrument consists of a fundamental frequency and a series of partials or overtones that are integer multiples of the fundamental, constituting a harmonic series. The fundamental may be removed, leaving only the overtones; then, the subject still "hears" the (missing) fundamental. This process, the basis of timbre and pitch perception and ultimately of melody, is found by Sidtis (1984) from dichotic experiments with normal subjects, the unilaterally brain damaged, and commissurotomy patients to be mediated by the RH. Indeed, the magnitude of the

LEA was directly related to the number of overtones, and a double dissociation appeared between LH and RH damage, and verbal and complex-tone stimuli. Such an RH superiority in processing complex periodic sounds may also underly the phenomenon of LEAs for recognizing prosody, stress, pitch, and the emotional contour of speech, when such cues are paralinguistic, rather than conveying propositional information (Ross, 1983).

3.5.2. Left-Hemisphere Aspects

Leek and Brandt (1983) obtained a dichotic REA in the detection of nonverbal target stimuli that differed from nontargets in terms of the temporal order of a sequence of very brief tones. An REA has also been found in the recognition of rhythm and temporal order (for reviews, *see* Bradshaw and Nettleton, 1981, 1983; Leek and Brandt, 1983). This has led many to argue that the fundamental LH superiority underlying its mediation of speech is the processing of rapidly changing acoustic information (Bradshaw and Nettleton, 1981, 1983), although of course the opposite argument could be advanced, that the LH is better at processing the rapidly changing temporal aspects of a signal because of years of practice with speech. The sequential nature of music, therefore, poses problems for the idea of exclusive RH mediation of music. Since auditory stimuli may be integrated over time, just as visual stimuli may be integrated over space to form perceptual wholes, there have been two recent developments. One is the increasing emphasis upon the bihemispheric mediation of music, with an attempt to analyze it into its possibly discrete component elements, not all of which may be lateralized in the same way, rather than treating it as a unitary whole. In this context, we have already seen an RH mediation of complex tones, and an LH involvement in rhythm and temporal sequencing. The other consequence is the developing idea of an underlying analytic/holistic processing dichotomy (Bradshaw and Nettleton, 1981, 1983) to explain such phenomena, whether in the verbal, nonverbal, or musical domains. Thus, explanations nowadays of cerebral asymmetries are moving away from a *material*-specific to a *processing*-specific account. Instead of the blanket claim that language is the exclusive province of the LH, and nonverbal functions such as music belong entirely to the RH, some language functions are now being ascribed to the RH; moreover, songs or melodies that can be processed as nontime-dependent entities are seen as being better handled by the RH, whereas those that involve strongly temporal aspects or require step-by-step analysis are thought to demand LH mediation.

3.5.3. Tone Illusions

Deutsch (1980) used a two-tone sequence, whose members were one octave apart and were repeatedly presented in alternation. The identical sequence was presented to both ears simultaneously, except that when the RE received the high tone the LE received the low tone, and vice versa. Thus, the listener received a single, continuous two-tone chord, but the ear of input for each component switched repeatedly. However, listeners typically reported hearing a single tone switching from ear to ear, whose pitch simultaneously alternated from high to low, i.e., a single high tone in one ear alternating with a single low tone in the other. With dextrals, the high tone is usually localized in the RE and the low in the LE. Moreover, similar effects occur if loudspeakers are substituted for earphones, indicating that the illusion occurs not along pathways conveying *ear* information, but with respect to the representation of the two *sides* of auditory space (*see also* Bradshaw et al., 1986). Indeed, she holds that *what* is heard is determined partly by *where* in space the signals seem to come from.

In a somewhat related paradigm, Efron et al. (1983) note that when two pure tones differing slightly in frequency go simultaneously one to each ear via earphones, both frequencies are detectable. For most dextrals, the frequency to the LE is perceptually more salient than the RE frequency, even when the stimuli are of equal intensity or are matched for pitch.

3.6. Temporal Alignment of Dichotic Signals

A verbal signal arriving at the LE might be delayed during transcallosal transfer, with an earlier-arriving RE signal occupying the language processor. If so, making the LE lead should cancel the REA. However, this is not the case, and the REA is *increased* if the LE leads by 15–60 ms, probably because of a backward masking effect (Studdert-Kennedy et al., 1970). Indeed, onset alignment may be less important than alignment of the "psychological moments" (P-centers) of words' occurrences (Morton et al., 1976). If so, tedious tape-splicing, computer storage, or electromechanical control of multiple tape recorders for onset alignment may be unnecessary, if two practiced speakers can be trained to read materials in what they *believe* to be accurate synchrony, and the experimenter can subsequently reject perceptually unsatisfactory pairs. In any case, the same material should be replayed to the subject a second time with channel reversal between ears to cancel any unwanted asynchronies.

3.7. Monaural Asymmetries: Is Dichotic Stimulation Necessary?

According to Kimura (1967), competing temporally aligned input on the contralateral ear is necessary to suppress ipsilateral pathways and to allow auditory asymmetries to emerge. Thus, information at the "wrong" ear goes to the "wrong" hemisphere, where it is degraded, and suffers further decrement during transcallosal transmission to the "correct" hemisphere, where it then must compete with information arriving directly via the contralateral pathways. However, dichotic competition may simply serve to make the task sufficiently difficult for an ear asymmetry to emerge. Bradshaw (1988) reviews monaural asymmetries as a function of task (verbal, nonverbal, masking, DAF), measure (threshold, pure preference, report accuracy, discrimination speed, activational measures, and so on), the nature of any simultaneous input (nothing, continuous contralateral white noise or "babble," pulsed synchronous white noise, competing monaural, i.e., monotic stimulation), ear uncertainty (can the subject focus attention on a single ear, or must attention be divided), subject variables (handedness, practice, expertise), and experimental variables (between or within subject designs; use of one or both hands, simultaneously or successively, in a manual RT paradigm). With sufficient task difficulty and/or the more sensitive RT measure, monaural asymmetries can be demonstrated that are as strong and reliable as dichotic ones.

3.8. The Effects of Practice

A few studies report *decrease* in auditory asymmetries with practice (Kallman and Corballis, 1975; Murray and Richards, 1978; Spellacy, 1970). Kallman and Corballis (1975) suggest that, even with stimuli (musical sounds) normally processed by the RH, the LH may gradually increase its share of processing. Certainly, most studies indicate an increase in auditory asymmetries with practice, for verbal tasks (Fennell, et al., 1977; Lazarus-Mainka and Hörmann, 1978; Pearl and Haggard, 1975; Sidtis and Bryden, 1978; Yeni-Komshian and Gordon, 1974), and for nonverbal, tonal or musical stimuli (Sidtis and Bryden, 1978). Long-term experience, however, as with musical expertise, rather than short-term experimental practice, may bring about an RH to LH switch for processing melodic stimuli, through the switching in of analytic rather than holistic processing mechanisms (e.g., Bever and Chiarello, 1974); however, there are many contrary findings (*see* Bradshaw and Nettleton, 1981, 1983, for detailed reviews).

4. The Tactual Modality

Compared to the visual and auditory modalities, there have been far fewer tactual studies with lateralized stimuli, even though the somatosensory pathways are no less contralaterally organized (Brinkman and Kuypers, 1973; Gazzaniga and LeDoux, 1978). Stimulus construction, presentation, and control may generate special problems with touch; these are reviewed, together with the associated procedures and findings, by Bradshaw et al. (1986), and what follows is a brief summary.

4.1. General Findings

Parts of the left side of the body may have a lower pressure sensitivity threshold, though more complex tasks are normally required to demonstrate lateral asymmetries. The left hand is superior for the tactile perception of line slant, although perhaps only with pure dextrals, and the task is sensitive to right parietal lesions. Braille may be more easily read by the left hand; the simultaneous playing of music into the left ear reverses this Braille asymmetry, indicating that the differences are quantitative rather than qualitative.

Most studies, however, unlike the above, have used a tactile analog (dichhaptic stimulation) of the dichotic procedure. Witelson's (1974) subjects felt pairs of raised letters simultaneously presented one to each hand. Two pairs were presented each for 2 s, with a 1-s interval between pairs, the task being to name the four letters. Subjects (dextral boys) gave a nonsignificant right-hand superiority on this task, but a highly significant left-hand superiority with nonsense shapes. Such stimuli, presented simultaneously, in pairs one to each hand, were explored for 10 s by the index and third fingers. Subjects then pointed to the stimuli they believed had been presented, selecting them from a visual display that included four distractors. Witelson later extended these findings with nonsense shapes and found evidence of sex differences. Although her left-hand superiority for nonsense shapes has been confirmed by some other studies, sex differences have not always been found, and others have been unable to replicate her findings. Indeed, Witelson's shapes involved a much longer palpation time than 2 s, and such long exposures may have led to attentional shifts, especially since most of the studies have involved children.

In dichhaptic as in dichotic tasks, subjects may differ in the type of strategies employed, the division of attention, order of report, and how stimuli are palpated. In a study attempting to control some of these factors, subjects simultaneously palpated two unfamiliar shapes for 3.75 s (Gardner et al., 1977). Two seconds later, a light indicated the hand to be reported, and 1 s thereafter, another light indicated response mode (left or right hand pointing, or speaking). With manual responses, accuracy was greater for shapes felt by the left hand. Another study (Oscar-Berman et al., 1978) attempted to control both order of report and stimulus presentation. Two experimenters simultaneously traced pairs of letters, digits, or line orientations onto the subject's hands. Stimuli were to be reported in a particular order. There was a right-hand superiority for the letters and a left-hand superiority for the line orientation task (with no asymmetry for digits), but only for *second* hand reports. As in other modalities, *storage* processes may be more sensitive to laterality differences than measures closer in time to the actual perceptual event. Nachshon and Carmon (1975) also successfully demonstrated a double dissociation, a right-hand superiority for a sequential task, and the opposite for a simultaneous task.

Flanery and Balling (1979) improved upon the Witelson paradigm with a haptic–haptic rather than a haptic–visual matching task. Apart from demonstrating developmental effects, they concluded that dichhaptic stimulation was unnecessary, a finding supported by some but disputed by others.

Instead of active exploration, Gibson and Bryden (1983) slowly moved cutout sandpaper shapes and letters across subjects' fingertips, and demonstrated a double dissociation. Yamamoto and Hatta (1982) compared passive and active touch and a tactile thought task, and concluded that laterality differences depend upon task requirements and the neural pathways involved. Other studies have examined task difficulty, verbal association value of the stimuli, report order, different instructions, tactile–verbal paired associate learning, stimulus presentation duration, matching to sample, and field dependence. These factors are all likely to be important in determining tactile asymmetries, which have been studied in the context of maturational changes, sex differences, lateralization in the deaf, children with marked discrepancies between verbal and spatial abilities, and various clinical populations.

5. Measuring Lateralization

5.1. Measures and Indices of Lateralization

Asymmetries have traditionally been given as the difference (d) between the number of correctly reported items on the left ear, hand, or visual field (L_c) and those on the right ear, hand, or visual field (R_c), sometimes expressed as a function of total correct, i.e., $(R_c - L_c)/(R_c + L_c)$. Such an index is, of course, constrained by possible floor and ceiling effects, and can only reach a maximum when overall performance (P_o) achieves an intermediate level. Differences between subjects, tests, and experiments are likely to militate against this ideal situation. Various alternative indices have therefore been proposed and employed for both dichotic and dichhaptic situations (*see* Bradshaw et al., 1986, and Bryden, 1982, for review), with again the result that it is difficult to compare between studies; indeed, different indices can produce different interpretations of the same set of data. Two derived measures are POC, percentage of correct responses, $R_c/(R_c + L_c)$, and POE, percentage of error, $L_e/(R_e + L_e)$, where R_c and L_c here refer to the percent of correct scores at the right and left side, respectively, and R_e and L_e refer to the corresponding percent of error scores. They may be used disjunctively, depending upon P_o, in the form of the e index. However, despite claims to the contrary, measures such as d, e, and POC are all necessarily affected by P_o. Nevertheless, they may be of some limited use if very high and very low scoring subjects are first eliminated. Bryden and Sprott (1981) devised an index of lateralization based upon the log odds ratio. A log likelihood ratio is computed and an ANOVA performed upon λ values.

Data are used from trials where the subject has only a single correct response. The index has convenient statistical properties that facilitate full use of the data and permit statistical tests to be performed for individual subjects. It is claimed that λ values differ little between analyses, and that the index has no spurious correlations with accuracy. The particular method of calculating the index, and its utility, depends upon the experimental design. Jones (1983) claims that λ gives greater weight to differences at the extremes of the P_o range. Since both d and e indices correlate highly with λ when floor and ceiling effects have been avoided, this new index may be useful only if differences at the extremes of the performance range are thought to be more important than equal differences in the middle of the range. Most measures assume that laterality effects are static rather than dynamic, and attribute trial-to-trial variations in performance-to-error variance. Kuhn (1973) claims that in the traditional two-response direct-recall situation, the phi (ϕ) coefficient, where T is the number of trials, is independent of performance

$$\phi = (R_c - L_c)/[(R_c + L_c)\,(2T - R_c - L_c)^{1/2}]^{1/2}$$

level. This claim is, however, also disputed; indeed, since ϕ is the geometric mean of POC and POE, it must combine the constraints of these two indices. Levy (1983) suggests that the functional asymmetry of lateralization is specified by the correlation between performance on each trial and sensory half-field. For RT data, this yields the point biserial correlation, and for accuracy data, the ϕ coefficient.

In the absence, however, of a substantive theory of lateralization, the derivation and use of such indices may be misleading, giving a false sense of quantification. Recently, RT measures have gained popularity, since the methodology and metric are readily available. However, much information may be lost about subject distributions, and statistical analyses at the individual subject level, although in principle possible, have not proved popular. Weighted Least Squares Analyses or use of a Maximum Likelihood Estimation have been suggested, as has Discriminant Function Analysis for measures of dichotic monitoring performance. The fused-word rhyme dichotic test circumvents the need for calculating indices of lateralization. Monosyllabic CVC words begin with (a different) one of the six stops, and members of each word pair differ from each other only in terms of these six initial stops.

Subjects expect, experience, and report only one word on each trial, as the two exactly aligned rhyming words perceptually fuse. Two other innovations are recommended for general use, item and subject screening. Stimulus pairs exhibiting stimulus dominance (a tendency for one member of a pair to be consistently reported, regardless of ear of presentation) are to be dropped, as are subjects whose ear asymmetry is less than a modest criterion. With these procedures, 97% of dextrals and 30% of sinistrals gave REAs with values closely matching those predicted on the basis of other criteria for language dominance. (For full details on the above material, *see* Bradshaw et al., 1986).

5.2. Reliability and Validity of Laterality Effects

Until now, dichotic REAs have been poor predictors of language lateralization, and a doubtful noninvasive tool, grossly underestimating the 95% or so incidence of clinically determined left-hemisphere language in dextrals, with low intertest reliability. They may be a correlate of language laterality, but make poor indices, although admittedly they assess receptive aspects rather than the more strongly lateralized expressive components measured clinically. Moreover, the size and consistency of REAs often increase with much practice, not just because of reduced response variability (*see* Bradshaw et al., 1986).

Lauter's (1982) subjects identified a range (verbal to nonverbal) of complex sounds, and their ear differences were expressed along a left-to-right continuum. Verbal materials always generated ear differences to the right of nonverbal, irrespective of whether the two extremes spanned the neutral midpoint, or (in some cases) all materials generated L(or R)EAs. Subjects thus differed in terms of absolute rather than relative ear advantages. Sidtis (1982) found the same: about 75% of subjects gave verbal REAs, and the same percentage nonverbal LEAs (using his Complex Tone Test, probably the most reliable nonverbal procedure producing a lateralization equal and opposite to speech). Only 46% gave *both* the expected LEAs and REAs, but none reversed on both. A single test (i.e., verbal or nonverbal) therefore underestimates the true incidence of language lateralization, whereas two from opposite extremes may accurately index the *direction* (but not the *degree*) of language lateralization. Such lateral shifts in the lateralization continuum may be the result of factors independent of *hemispheric* asymmetry, e.g., asymmetries in the ascending auditory pathways (*see*, e.g., Majkowski et al., 1971). Only with a *symmetrical* contralateral advantage can a single (verbal or nonverbal) dichotic test accurately index hemispheric laterality, and subjects showing low test–retest reliabilities may have little contralateral advantage.

5.3. Double-Task Performance

5.3.1. Independent Processing and Capacity Limitations

The addition of secondary tasks to be performed concurrently with a task of primary interest permits "loading" of the processing system to observe activation and interference effects. Thus, Dimond and Beaumont (*see* Beaumont, 1982, for review) presented pairs of digits for recognition while subjects performed a secondary unimanual or bimanual sorting task; they observed a complex interaction between the secondary tasks and asymmetries in the primary task. The addition of a concurrent verbal task to a verbal divided-visual-field task may reduce or reverse the RVF advantage otherwise found in the primary task (Geffen et al., 1973; Hellige and Cox, 1976; Hellige et al. 1979, although *see* Beaumont, 1982; Cohen, 1979, in view of problems in controlling load, attentional division, priming, and activation). Indeed, whether primary-task asymmetries are enhanced or reversed by adding a secondary task depends upon possible priming effects, processing demands (Hellige et al., 1979; Kinsbourne and Hicks, 1978), and the available processing capacities or resources of the two hemispheres (Friedman and Polson, 1981). Generally, where two simultaneous stimuli, each requiring separate processing, have been directed to same or different hemispheres, performance has been better in the latter situation, presumably because of mutual interference and capacity reductions (*see* Bradshaw and Nettleton, 1983). Much the same occurs when two initially independent sets of information must be integrated or compared (Annett and Annett, 1979; Davis and Schmit, 1971, 1973; Dimond and Beaumont, 1972). Thus, any limitations imposed by interhemispheric intercommunication and integration must be fairly insubstantial.

5.3.2. Concurrent Task Interactions and Overflow

A particular version of the dual-task situation involves movement control (e.g., duration of dowel balancing, or speed and/or regularity of tapping, by left or right hand) during performance of a *nonlateralized* verbal task (e.g., counting backwards in threes) or spatial task (*see* Kinsbourne and Hiscock, 1983, for review). Of course, the two hands do not commence with equal proficiency, and speed and/or accuracy of *both* tasks, performed *alone* and *together*, must be obtained separately for each hand. Moreover, the *primary* task (e.g., tapping) may even interfere with the *secondary* (e.g., counting); nor do we know whether or how attention will be divided (Bryden, 1982). A commonly adopted measure for percent reduction in, e.g., tapping rate is

$$\frac{\text{performance without concurrent task–performance with it}}{\text{performance without it}} \times 100$$

In a fully parametric study with a verbal concurrent task, right-hand tapping might suffer greater decrement than left, and the verbal task itself may suffer more with right than with left-hand tapping; with a spatial concurrent task, the opposite should occur. These ideal results have rarely been fully realized.

Primary tasks include dowel balancing, and single-, dual-, or four-button tapping (in the last case using one or more fingers). Secondary verbal tasks involve backwards counting, recitation, listing, verbal or arithmetic problem solving, reading aloud, memorization, and so on. Successful nonverbal secondary tasks (i.e., resulting in a left-hand decrement) have been far fewer, e.g., finding hidden or embedded figures, performing Progressive Matrices or Wechsler Block Design tasks, or a running memory span for faces.

An orientation reflex towards the side of space contralateral to the more activated hemisphere may occur (Trevarthen, 1972). Conjugate lateral eye movements (CLEMS) are said to be a component of this contralateral orienting response or overflow of cognitive neural activity into the orientation control system within the same hemisphere (*see* Ehrlichman and Weinberger, 1978, for critical review). The direction of eye movements on interrogation is said to be fairly consistent for a given individual, and is thus believed to index his/her cognitive style (e.g. right movers are better at verbal tasks), although this is greatly disputed. (Indeed, there has been comparatively little success at correlating overall cognitive performance with any measure of lateral asymmetry). More importantly and less radically, the direction of CLEMs is also said to be a function of the *current or ongoing* verbal or nonverbal nature of the required processing. There have been many unsuccessful studies, perhaps because of using otherwise invalidated tasks, and problems in interpreting eye movements. What should count as a CLEM? *Any* movement, however small? The first movement? The longest fixation? The total number over (what?) interval? Should the eyes be "zeroed" before each task? How should corrections be made otherwise? One way of verifying a possible link between CLEMs and cognitive activation of the hemispheres would come from showing that lateral direction of the gaze (e.g., to the right) facilitates the skills of the opposite (e.g., verbal) hemisphere (*see,* e.g., Gross et al., 1978). This argument is of course quite separate from the possibility that *hemispace* rather than visual-field or ear-of-entry asymmetries determine performance (*see below*) and that head and/or eye turn may affect these asymmetries via a disturbance of spatial coordinate systems.

In the same tradition as the CLEMs studies, and indeed in the tradition of the generally well authenticated EEG and regional blood flow indices (*see* Bradshaw and Nettleton, 1983, and Bryden, 1982, for reviews) of hemispheric activation during cognitive tasks, Kimura (1976) discusses claims that, during speech, dextrals make freer hand movements with the right than the left hand, whereas self-touching movements occur equally often with either hand. These findings might again be explained in terms of activation overflow from cognitive to motor areas within a hemisphere. However, there is no good evidence of differential effects with *nonverbal* cognitive activity (but *see* Hampson and Kimura, 1984 and Moscovitch and Olds, 1982).

6. Anatomical Pathway or Hemispace Mediation of Asymmetries

The brain does not merely register events impinging upon the proximal receptor surfaces, but for good reasons, also maps events occurring out in space beyond the body, before they are directly encountered. Thus, the auditory cortex maps the contralateral sound field (Phillips and Gates, 1982), not just in species like the bat (Calford et al., 1985) where this is obviously advantageous. Polysensory areas in the tectum are involved (Harris, 1986); the auditory map in the superior colliculus is affected by gaze direction (Jay and Sparks, 1984), the superior colliculus controls eye, pinna, and head orientation (Meredith and Stein, 1985), and may be where the spatial maps from sight and hearing are combined (Meredith and Stein, 1986). In the cortex, eye and head position alter the firing of retinotopic cells (Andersen et al., 1985; Reinis et al., 1986). Lateral movements of the *hands* as well as the eyes affect the EEG (Autret et al., 1985), whereas the tactual representation of space may be affected by whether the eyes are open or closed (Blum, 1985). Bradshaw et al. (1986) demonstrated in the auditory modality, using loudspeakers and visual–auditory conflict, that it is the *perceived* position in space, rather than the ear of entry, that determines auditory asymmetries; in the visual and tactual modalities, it is where an object is located left or right of the midline that is important in judgments of length, and in the vibrotactile and kinesthetic modalities, it is not so much which *hand* receives a stimulus or performs a response, but whether it is placed to left or right of the midline. These asymmetries are disturbed by abnormal posture, e.g., lying horizontally on one or other side, or turning the head 90° to left or right. This account explains some old findings with free inspection of left–right composite faces that otherwise are difficult to explain. Thus, Gilbert and Bakan (1973) constructed photographic left-left and right-right *composites* of a person's face (i.e., they were not truly *chimeric,* since the same face was used for both halves, and in fact, was constructed from two left or two right halves appropriately reflected or reversed); the composite made from two left sides (as perceived by the viewer) seemed to resemble with free scanning or inspection an original representation more closely than one made from two right sides. (Photographic reversals showed this to be a perceptual effect, *see,* e.g., Rhodes, 1985). Similarly, Levy and Kueck (1986) found that, when subjects sought for words with a certain sound that were scattered in various orientations across the page, they tended to detect more on the right than the left side of the page. It is in this direction, free observation of spatially distributed events, that laterality research is destined to progress in the future, rather than under the artificial constraints of briefly, eccentrically, or competitively presented stimuli.

References

Abbs J. H. and Smith K. U. (1970) Laterality differences in the auditory feedback control of speech. *J. Speech Hear. Res.* **13,** 298–303.

Albert M. L. and Obler L. K. (1978) *The Bilingual Brain* (Academic Press, New York).

Alivisatos B. and Wilding J. (1980) Hemispheric differences in matching Stroop-type letter stimuli. *Bull. Br. Psychol. Soc.* **33,** 17.

Andersen R. A., Essick G. K., and Siegel P. M. (1985) Encoding of spatial location by posterior parietal neurons. *Science* **230,** 456–458.

Annett M. and Annett J. (1979) Individual differences in left and right reaction time. *Br. J. Psychol.* **70,** 393–404.

Autret A., Auvert L., Laffaut F., and Larmande P. (1985) Electroencephalographic spectral power and lateralized motor activities. *Electroencephalogr. Clin. Neurophysiol.* **60,** 228–230.

Beaumont J. G. (1982) Studies with verbal stimuli, in *Divided Visual Field Studies of Cerebral Organisation* (Beaumont J. G., ed.), Academic Press, New York. pp. 58–86.

Berlucchi G. (1978) Interhemispheric integration of simple visuomotor responses, in *Cerebral Correlates of Conscious Experience: Inserm Symposium No. 6.* (Buser P. A. and Rougeul-Buser A., eds.), Elsevier/North-Holland Biomedical Press, Amsterdam pp. 83–94.

Berlucchi G., Brizzolara D., Marzi C. A., Rizzolatti G., and Umiltà C. (1979) The role of stimulus discriminability and verbal ability in hemispheric specialization for visuospatial tasks. *Neuropsychologia,* **17,** 195–202.

Bertelson P., Vanhaelen H., and Morais J. (1979) Left hemifield superiority and the extraction of physiognomic information, in *Structure and Function of Cerebral Commissures* (Steel Russell I., Hof M. W., and Berlucchi G., eds.), MacMillan, London. pp. 400–410.

Bever T. G. and Chiarello R. J. (1974) Cerebral dominance in musicians and non musicians. *Science,* **185,** 137–139.

Blum B. (1985) Manipulation reach and visual reach neurons in the inferior parietal lobule of the rhesus monkey. *Behav. Brain Res.* **18,** 167–173.

Boles D. B. (1985) The effects of display and report order asymmetries on lateralized word recognition. *Brain Lang.,* **26,** 106–116.

Bonvillian J. D., Orlansky M. D., and Garland J. B. (1982) Handedness patterns in deaf persons. *Brain and Cognition,* **1,** 141–157.

Bradshaw G. J., Hicks R. E., and Rose B. (1979) Lexical discrimination and letter-string identification in the two visual fields. *Brain Lang.* **8,** 10–18.

Bradshaw J. L. (1985) Musik und die Funktionsteilung des Gehirns, in *Musikpsychologie: Ein Handbuch in Schlusselbegriffen* (Bruhn H., Oerter R., and Rosing H., eds.), Verlag Urban and Schwarzenberg, Munchen. pp. 70–78.

Bradshaw J. L. Monaural asymmetries (1988) in *Handbook of Dichotic Listening: Theory, Methods and Research* (Hugdahl K., ed.), Wiley, London,

Bradshaw J. L. and Gates E. A. (1978) Visual field differences in verbal tasks: Effects of task familiarity and sex of subject. *Brain Lang.* **5,** 166–187.

Bradshaw J. L. and Mapp A. (1982) Laterally presented words: Orthographic analysis and serial, parallel or holistic modes of processing. *Aust. J. Psychol.* **34,** 71–90.

Bradshaw J. L. and Nettleton N. C. (1981) The nature of hemispheric specialization in man. *Behav. Brain Sci.* **4,** 51–63.

Bradshaw J. L. and Nettleton N. C. (1983) *Human Cerebral Asymmetry.* Prentice-Hall, Englewood Cliffs, N.J.

Bradshaw J. L. and Sherlock D. (1982) Bugs and faces in the two visual fields: The analytic/holistic dichotomy and task sequencing. *Cortex,* **18,** 211–226.

Bradshaw J. L. and Taylor M. (1979) A word naming deficit in nonfamilial sinistrals? Laterality effects of vocal responses to tachistoscopically-presented letter strings. *Neuropsychologia,* **17,** 21–32.

Bradshaw J. L. and Umiltà C. (1984) A reaction time paradigm can simultaneously index spatial-compatibility and neural-pathway effects: A reply to Levy. *Neuropsychologia* **22,** 99–101.

Bradshaw J. L., Burden V., and Nettleton N. C. (1986) Dichotic and dichhaptic techniques. *Neuropsychologia,* **24,** 79–91.

Bradshaw J. L., Farrelly J., and Taylor M. J. (1981) Synonym and antonym pairs in the detection of dichotically and monaurally presented targets: Competing monaural stimulation can generate a REA. *Acta Psychol.* **47,** 189–205.

Bradshaw J. L., Gates E. A., and Nettleton N. C. (1977) Bihemispheric involvement in lexical decisions: Handedness and a possible sex difference. *Neuropsychologia* **15,** 277–286.

Bradshaw J. L., Nettleton N. C., and Geffen G. (1971) Ear differences and delayed auditory feedback: Effects on a speech and a music task. *J. Exper. Psychol.* **91,** 85–92.

Bradshaw J. L., Nettleton N. C., and Geffen G. (1972) Ear asymmetry and delayed auditory feedback: Effects of task requirements and competitive stimulation. *J. Exper. Psychol.* **94,** 269–275.

Bradshaw J. L., Nettleton N. C., and Taylor M. J. (1981) The use of laterally presented words in research into cerebral asymmetry: Is directional scanning likely to be a source of artifact? *Brain Lang.* **14,** 1–14.

Bradshaw J. L., Nettleton N. C., Wilson L., and Nathan G. (1984) A moving video window or a mask yoked to eye movement: Experiments on letters, words and biological movement with prolonged hemifield stimulation. *Int. J. Neurosci.* **25,** 81–98.

Bradshaw J. L., Taylor M. J., Patterson K., and Nettleton N. C. (1980) Upright and inverted faces, and housefronts, in the two visual fields: A right and a left hemisphere contribution. *J. Clin. Neuropsychol.* **2,** 245–257.

Bradshaw J. L., Nettleton N. C., Pierson J. M., Wilson L. E., and Nathan G. (1986) Coordinates of extracorporeal space, in *Neuropsychological and Physiological Aspects of Spatial Neglect* (Jeannerod M., ed.), North Holland, Amsterdam. pp. 41–67.

Brinkman J. and Kuypers H. G. J. M. (1973) Cerebral control of contralateral and ipsilateral arm, hand and finger movements in the split-brain Rhesus monkey. *Brain,* **96,** 653–674.

Broadbent D. (1954) The role of auditory localization in attention and memory span. *J. Exper. Psychol.* **47,** 191–196.

Bruce V. (1983) Recognizing faces. *Phil. Trans. Roy. Soc. London B,* **302,** 423–436.

Bryden M. P. (1967) An evaluation of some models of laterality effects in dichotic listening. *Acta Oto-Laryngol.* **63,** 595–604.

Bryden M. P. (1979) Evidence for sex-related differences in cerebral organization, in *Sex-Related Differences in Cognitive Functioning* (Wittig M. A. and Petersen A. C., eds.), Academic Press, New York. pp. 121–143.

Bryden M. P. (1982) *Laterality: Functional Asymmetry in the Intact Brain.* Academic Press, New York.

Bryden M. P. and Allard F. (1976) Visual hemifield differences depend upon typeface. *Brain Lang.,* **3,** 191–200.

Bryden M. P. and Sprott D. A. (1981) Statistical determination of degree of laterality. *Neuropsychologia* **19,** 571–581.

Bryden M. P., Munhall K., and Allard F. (1983) Attentional biases and the right ear effect in dichotic listening. *Brain Lang.* **18,** 236–248.

Bunt A. H., Minckler D. S., and Johanson G. W. (1977) Demonstration of bilateral projection of the central retina of the monkey with horseradish peroxidase neuronography. *J. Compar. Neurol.* **171,** 619–630.

Calford M. B., Grayden M. C., Huerta M. E., Koas J. H., and Pettigrew J. (1985) A variant on the brain somatic map in a bat. *Nature,* **313,** 477–478.

Campbell R. (1986) The lateralization of lip reading sounds: A first look. *Brain and Cognition* **5,** 1–21.

Carey S., and Diamond R. (1977) From piecemeal to configurational representation of faces. *Science,* **195,** 312–314.

Chiarello C. (1985) Hemisphere dynamics in lexical access: Automatic and controlled priming. *Brain Lang.* **26,** 146–172.

Chiarello C. (1988) Lateralization of lexical processes in the normal brain: A review of visual half field research, in *Contemporary Reviews in Neuropsychology* (Whitaker H. A. ed.), Springer-Verlag, New York, pp. 36–76.

Cohen G. (1972) Hemispheric differences in a letter classification task. *Perception and Psychophysics* **11,** 139–142.

Cohen G. (1979) Comment on "Information processing in the cerebral hemispheres: selective activation and capacity limitations" by Hellige, Cox, and Litvac. *J. Exper. Psychol.: General* **108,** 309–315.

Cohen G. and Martin M. (1975) Hemisphere differences in an auditory Stroop test. *Perception and Psychophysics* **17,** 79–83.

Connolly J. F. (1985) Stability of pathway—hemisphere differences in the auditory event related potential (ERP) for monaural stimulation. *Psychophysiology* **22,** 87–95.

Corballis M. C. (1983) *Human Laterality* (Academic Press, New York.) 1983.

Craft J. L. and Simon J. R. (1970) Processing symbolic information from a visual display: interference from an irrelevant directional cue. *J. Exper. Psychol.* **83,** 415–420.

Crane H. D. and Kelly D. H. (1983) Accurate simulation of visual scotomas in normal subjects. *Applied Optics,* **22,** 1802–1806.

Cranney J. and Ashton R. (1982) Tactile spatial ability: Lateralized performance of deaf and hearing age groups. *J. Exper. Child Psychol.* **34,** 123–134.

Davidoff J. (1982) Studies with non-verbal stimuli, in *Divided Visual Field Studies of Cerebral Organization* (Beaumont J. G., ed.), Academic Press, New York, pp. 30–55.

Davis R. and Schmit V. (1971) Timing the transfer of information between hemispheres in man. *Acta Psychol.* **35,** 335–346.

Davis R. and Schmit V. (1973) Visual and verbal coding in the interhemispheric transfer of information. *Acta Psychol.* **37,** 229–240.

Deutsch D. (1980) The octave illusion and the what-where connection, in *Attention and Performance VIII* (Nickerson R. S., ed.), Academic Press, New York, pp. 575–592.

Dimond S. J. and Beaumont J. G. (1972) Processing in perceptual integration between and within the cerebral hemispheres. *Br. J. Psychol.* **63,** 509–514.

Dimond S. J. and Beaumont J. G. (1973) Differences in the vigilance performance of the right and left hemispheres. *Cortex* **9,** 259–266.

Dimond S. J. and Farrington L. (1977) Emotional response to films shown to the right or left hemisphere of the brain measured by heart rate. *Acta Psychol.* **41,** 255–260.

Dimond S. J., Bures J., Farrington L. J., and Brouwers E. Y. M. (1975) The use of contact lenses for the lateralization of visual input in man. *Acta Psychol.* **39,** 341–349.

Efron R., Koss B., and Yund E. W. (1983) Central auditory processing IV: Ear dominance—Spatial and temporal complexity. *Brain Lang.* **19,** 264–282.

Ehrlichman H. and Weinberger A. (1978) Lateral eye movements and hemispheric asymmetry: A critical review. *Psychol. Bull.* **85,** 1080–1101.

Ellis H. D. (1981) Theoretical aspects of face recognition, in *Perceiving and Remembering Faces* (Davies G., Ellis H., and Shepherd J., eds.), Academic Press, London, pp. 171–197.

Fairweather H., Brizzolara D., Tabossi P., and Umiltà C. (1982) Functional cerebral lateralisation: Dichotomy or plurality? *Cortex* **18,** 51–66.

Fennell E. B., Bowers D., and Satz P. (1977) Within-modal and cross-modal reliabilities of two laterality tests. *Brain Lang.* **4,** 63–69.

Finlay D. C. and French J. (1978) Visual field differences in a facial recognition task using signal detection theory. *Neuropsychol.* **16,** 103–107.

Fiorentini A., Maffei L., and Sandini G. (1983) The role of high spatial frequencies in face perception. *Perception* **12,** 195–201.

Flanery R. C. and Balling J. D. (1979) Developmental changes in hemispheric specialization for tactile spatial ability. *Devel. Psychol.* **15,** 364–372.

Francks J. B., Smith S. M., and Ward T. B. (1985) The use of goggles for testing hemispheric asymmetry. *Bull. Psychonomic Soc.* **23,** 487–488.

Friedman A. and Polson M. C. (1981) Hemispheres as independent resource systems: Limited capacity processing and cerebral specialization. *J. Exper. Psychol.: Human Perception and Performance,* **7,** 1031–1058.

Galper R. E. and Costa L. (1980) Hemispheric superiority for recognizing faces depends upon how they are learned. *Cortex,* **16,** 21–28.

Gardner E. B., English A. G., Flannery B. M., Hartnett M. B., McCormick J. K., and Wilhelmy B. B. (1977) Shape-recognition accuracy and response latency in a bilateral tactile task. *Neuropsychologia* **5,** 607–616.

Gazzaniga M. S. and LeDoux J. E. (1978) *The Integrated Mind* (Plenum Press, New York).

Gazzaniga M. S. and Smylie C. S. (1983) Facial recognition and brain asymmetries: Clues to underlying mechanisms. *Annals Neurol.* **13,** 536–540.

Geffen G. and Cawdrey D. (1981) Reliability and validity of the dichotic monitoring test for language laterality. *Neuropsychologia* **19,** 413–423.

Geffen G. and Quinn K. (1984) Hemispheric specialization and ear advantages in processing speech. *Psychol. Bull.* **96,** 273–291.

Geffen G., Bradshaw J. L., and Nettleton N. C. (1972) Hemispheric asymmetry: Verbal and spatial encoding of visual stimuli. *J. Exper. Psychol.* **95,** 25–31.

Geffen G., Bradshaw J. L., and Nettleton N. C. (1973) Attention and hemispheric differences in reaction time during simultaneous audiovisual tasks. *Quart. J. Exper. Psychol.* **25,** 404–412.

Geffen G., Bradshaw J. L., and Wallace G. (1971) Interhemispheric effects on reaction time to verbal and nonverbal visual stimuli. *J. Exper. Psychol.* **87,** 415–422.

Gibson C. and Bryden M. P. (1983) Dichhaptic recognition of shapes and letters in children. *Can. J. Psychol.* **37,** 132–143.

Gibson C. J. and Bryden M. P. (1984) Cerebral lateralization in deaf and hearing children. *Brain Lang.* **23,** 1–12.

Gibson A. R., Dimond S. J., and Gazzaniga M. S. (1972) Left field superiority for word matching. *Neuropsychologia* **10,** 463–466.

Gilbert C. and Bakan P. (1973) Visual asymmetry in perception of faces. *Neuropsychologia* **11,** 355–362.

Glass C., Bradshaw J. L., Day R. H., and Umiltà C. (1985) Familiarity, spatial frequency and task determinants in processing laterally presented representations of faces. *Cortex* **21,** 513–531.

Gordon H. W. and Carmon A. (1976) Transfer of dominance in speed of verbal response to visually presented stimuli from right to left hemisphere. *Percept. Mot. Skills* **42,** 1091–1100.

Gross M. M. (1972) Hemispheric specialization for processing visually presented verbal and spatial stimuli. *Perception and Psychophysics* **12,** 357–363.

Gross Y., Franko R., and Lewin I. (1978) Effects of voluntary eye movements on hemispheric activity and choice of cognitive mode. *Neuropsychologia* **16,** 653–657.

Hampson E. and Kimura D. (1984) Hand movement asymmetries during verbal and nonverbal tasks. *Can. J. Psychol.* **38,** 102–125.

Hannay H. J., Dee H. L., Burns J. W., and Masek B. S. (1981) Experimental reversal of a left visual field superiority for forms. *Brain Lang.* **13,** 54–66.

Hardyck C., Tzeng O. J. L., and Wang W. S-Y. (1978) Cerebral lateralization of function and bilingual decision processes: Is thinking lateralized? *Brain Lang.* **5,** 56–71.

Harmon L. D. (1973) The recognition of faces. *Sci. Amer.* **229,** 71–82.

Harrington D. O. (1981) *The Visual Fields: A Textbook and Atlas of Clinical Perimetry* (Mosby, St. Louis.)

Harris W. A. (1986) Learned topography: The eye instructs the ear. *Trends Neurosci.* **9,** 97–100.

Harvey L. O. (1978) Single representation of the visual midline in humans. *Neuropsychologia* **16,** 601–610.

Hatta T. (1978) Visual field differences in a mental transformation task. *Neuropsychologia* **16,** 637–641.

Haun F. (1978) Functional dissociation of the hemispheres using foveal visual input. *Neuropsychologia* **16,** 725–733.

Hay D. C. (1981) Asymmetries in facial recognition: Evidence for a perceptual component. *Quart. J. Exper. Psychol.* **33A,** 267–274.

Hay D. C. and Young A. W. (1982) The human face, in *Normality and Pathology in Cognitive Functions* (Ellis A. W., ed.), Academic Press, New York, pp. 173–202.

Heilman K. M., Bowers D., Rasbury W. C., and Ray R. M. (1977) Ear asymmetries on a selective attentional task. *Brain Lang.* **4,** 390–395.

Hellige J. B. (1980) Visual laterality and cerebral hemisphere specialization: Methodological and theoretical considerations, in *Conditioning, Cognition, and Methodology: Contemporary Issues in Experimental Psychology* (Sidowski J. B., ed.), Erlbaum, Hillsdale, N.J.

Hellige J. B. and Cox P. J. (1976) Effects of concurrent verbal memory on recognition of stimuli from left and right visual fields. *J. Exper. Psychol.: Human Perception and Performance* **2,** 210–221.

Hellige J. B., Corwin W. H., and Jonsson J. E. (1984) Affects of perceptual quality on the processing of human faces presented to the left and right cerebral hemispheres. *J. Exper. Psychol.: Human Perception and Performance* **10,** 90–107.

Hellige J. B., Cox P. J., and Litvac L. (1979) Information processing in the cerebral hemispheres: Selective hemispheric activation and capacity limitations. *J. Exper. Psychol.: General* **108,** 251–279.

Hilliard R. D. (1973) Hemispheric laterality effects on a facial recognition task in normal subjects. *Cortex* **9,** 246–258.

Hines D. (1975) Independent functioning of the two cerebral hemispheres for recognizing bilaterally presented tachistoscopic visual half field stimuli. *Cortex,* **11,** 132–143.

Hines D., Sawyer P. K., Dura J., Gilchrist J., and Czerwinski M. (1984) Hemispheric asymmetry in use of semantic category information. *Neuropsychologia* **22,** 427–433.

Holender D. and Peereman R. (1987) Differential processing of phonographic and logographic single-digit numbers by the two hemispheres, in *Mathematical Disabilities: A Neuropsychological Perspective* (Deloche G. and Seron X., eds.), Erlbaum, Hillsdale, NJ, pp. 43–85.

Homan R. W., Criswell E., Wada J. U., and Ross E. D. (1984) Hemispheric contribution to manual communication (signing and finger spelling). *Neurology* **32,** 1020–1023.

Inui T. and Miyamoto K. (1984) The effect of changes in visible area on facial recognition. *Perception* **13,** 49–56.

Jay M. F. and Sparks D. C. (1984) Auditory receptive fields in primate superior colliculus shift with changes in eye position. *Nature* **309,** 345–347.

Jones B. (1980) Sex and handedness as factors in visual-field organization for a categorization task. *J. Exper. Psychol.: Human Perception and Performance* **6,** 494–500.

Jones B. (1983) Some problems with Bryden and Sprott's "Statistical determination of degree of lateralization." *Neuropsychologia* **21,** 295–298.

Jones B. and Anuza T. (1982) Sex differences in cerebral localization in three-and-four-year-old children. *Neuropsychologia* **20,** 347–350.

Julesz B., Breitmeyer B., and Kropfl W. (1976) Binocular disparity dependent upper lower hemifield anistropy and left right hemifield isotropy as revealed by dynamic random dot stereograms. *Perception* **5,** 129–141.

Kallman H. J. and Corballis M. C. (1975) Ear asymmetry in reaction time to musical sounds. *Perception and Psychophysics* **17,** 368–370.

Kelly R. P. and Tomlinson-Keasey C. (1981) The effect of auditory input on cerebral lateralization. *Brain and Language* **13,** 67–77.

Kimura D. (1961) Cerebral dominance and the perception of verbal stimuli. *Can. J. Psychol.* **15,** 166–171.

Kimura D. (1967) Functional asymmetry of the brain in dichotic listening. *Cortex* **3,** 163–178.

Kimura D. (1976) The neural basis of language qua gesture, in *Studies in Neurolinguistics, Vol. 1* (Whitaker H. and Whitaker H. A., eds.), Academic Press, New York, pp. 145–156.

Kimura D. (1977) Acquisition of a motor skill after left hemisphere damage. *Brain* **100,** 527–542.

Kinsbourne M. (1973) The control of attention by interaction between the cerebral hemispheres, in *Attention and Performance II* (Kornblum S., ed.), Academic Press, New York, pp. 239–257.

Kinsbourne M. and Hicks R. E. (1978) Functional cerebral space: A model for overflow transfer and interference effects in human performance: A tutorial overview, in *Attention and Performance VII* (Requin J., ed.), Lawrence Erlbaum, Hillsdale, NJ, pp. 345–362.

Kinsbourne M. and Hiscock M. (1983) Asymmetries of dual-task performance, in *Cerebral Hemisphere Asymmetry* (Hellige J. B., ed.), Praeger, New York, pp. 255–334.

Klatzky R. L. and Atkinson R. C. (1971) Specialization of the cerebral hemispheres in scanning for information in short-term memory. *Perception and Psychophysics,* **10,** 335–338.

Klisz D. K. (1980) Visual field differences between a fricative and a stop consonant. *Cortex,* **16,** 169–173.

Koerner F. and Teuber H. L. (1973) Visual field defects after missile injuries to the geniculo-striate pathway in man. *Exper. Brain Res.* **18,** 88–113.

Krueger L. E. (1978) A theory of perceptual matching. *Psychol. Rev.* **85,** 278–304.

Kuhn G. M. (1973) The Phi coefficient as an index of ear differences in dichotic listening. *Cortex* **9,** 450–457.

Lambert A. J. and Beaumont J. G (1983) Imageability does not interact with visual field in lateral word recognition with oral report. *Brain Lang.* **20,** 115–142.

Lamm O. and Gordon H. W. (1984) Right hemisphere superiority in processing new symbols for arithmetic operators. *Acta Psychol.* **57,** 29–45.

Lauter J. (1982) Dichotic identification of complex sounds: absolute and relative ear advantages. *J. Acoust. Soc. Am.* **71,** 701–707.

Lazarus-Mainka G. and Hörmann H. (1978) Strategic selection (metacontrol) of hemisphere dominance in normal human subjects. *Psychol. Res.* **40,** 15–25.

Leehey S. C. and Cahn A. (1979) Lateral asymmetries in the recognition of words, familiar faces and unfamiliar faces. *Neuropsychologia* **17,** 619–635.

Leehey S., Carey S., Diamond R., and Cahn A. (1978) Upright and inverted faces: The right hemisphere knows the difference. *Cortex,* **14,** 411–419.

Leek M. R., and Brandt J. F. (1983) Laterality of rapid auditory sequences. *Neuropsychologia,* **21,** 61–77.

Levine S. C. and Banich M. T. (1982) Lateral asymmetries in the naming of words and corresponding line drawings. *Brain Lang.* **17,** 34–45.

Levine S. C. and Koch-Weser M. P. (1982) Right hemisphere superiority in the recognition of famous faces. *Brain and Cognition* **1,** 10–22.

Levy J. (1983) Individual differences in cerebral hemisphere asymmetry: Theoretical issues and experimental considerations, in *Cerebral Hemisphere Asymmetry: Method Theory and Application* (Hellige J. B., ed.), Praeger, New York, pp. 465–497.

Levy J. and Kueck L. (1986) A right hemispatial field advantage on a verbal free-vision task. *Brain Lang.* **27,** 24–37.

Ley R. G. and Bryden M. P. (1977) Hemispheric differences in processing emotional stimuli? *Bull. Psychonomic Soc.* **10,** 261 (#239).

Lines C. R. and Milner A. D. (1983) Nasotemporal overlap in the human retina investigated by means of simple reaction time to lateralized light flash. *Exper. Brain Res.* **50,** 166–172.

Lupker S. J. and Sanders M. (1982) Visual-field differences in picture-word interference. *Brain and Cognition* **1,** 381–398.

McGuire R. A., Loren C., and Rastatter M. (1986) Naming reaction times to tachistoscopically presented pictures: Some evidence of right hemisphere encoding. *Percep. Mot. Skills* **62,** 303–306.

McIlwain J. T. (1972) Central vision: Visual cortex and superior colliculus. *Ann. Rev. Physiol.* **34,** 291–314.

McKeever W. F. and Huling M. D. (1971) Lateral dominance in tachistoscopic word recognition performances obtained with simultaneous bilateral input. *Neuropsychologia* **9,** 15–20.

Maddess R. J. (1975) Reaction time to hemiretinal stimulation. *Neuropsychologia* **13,** 213–218.

Majkowski J., Bochenek Z., Bochenek W., Knapik-Fijalkowska D., and Kopec J. (1971) Latency of averaged evoked potentials to contralateral and ipsilateral stimulation in normal subjects. *Brain Res.* **25,** 416–419.

Manning W. H., Louko L. J., and Di Salvo V. S. (1978) A right-ear effect for auditory feedback control of children's newly acquired phonemes. *J. Speech and Hearing Res.* **21,** 580–588.

Marcel T. and Rajan P. (1975) Lateral specialization for recognition of words and faces in good and poor readers. *Neuropsychologia* **13,** 489–497.

Marcotte A. J. and LaBarba R. C. (1985) Cerebral lateralization for speech in deaf and normal children. *Brain Lang.* **26,** 244–258.

Martin M. (1979) Hemispheric specialization for local and global processing. *Neuropsychologia* **17,** 33–40.

Marzi C. A. and Berlucchi G. (1977) Right visual field superiority for accuracy of recognition of famous faces in normals. *Neuropsychologia* **15,** 751–756.

Marzi C. A., Brizzolara D., Rizzolatti G., Umiltà C., and Berlucchi G. (1974) Left hemisphere superiority for the recognition of well known faces. *Brain Res.* **66,** 358.

Meredith M. A. and Stein B. E. (1985) Descending efferents from the superior colliculus relay integrated multisensory information. *Science* **227,** 657–659.

Meredith M. A. and Stein B. E. (1986) Spatial factors determine the activity of multisensory neurons in cat superior colliculus. *Brain Res.* **365,** 350–354.

Metzger R. L. and Antes J. R. (1976) Sex and coding strategy effects on reaction time to hemispheric probes. *Memory and Cognition* **4,** 167–171.

Miller L. K. and Butler D. (1980) The effect of set size on hemifield asymmetries in letter recognition. *Brain Lang.* **9,** 307–314.

Milner A. D. and Dunne J. J. (1977) Lateralized perception of bilateral chimaeric faces by normal subjects. *Nature* **268,** 175–176.

Morton J., Marcus S., and Frankish C. (1976) Perceptual centers (P-centers). *Psychol. Rev.* **83,** 405–408.

Moscovitch M. (1972) Choice reaction time study assessing the verbal behavior of the minor hemisphere in normal adult humans. *J. Compar. Physiol. Psychol.* **80,** 66–74.

Moscovitch M. (1979) Information processing and the cerebral hemispheres, in *Handbook of Behavioral Neurobiology, Vol. 2: Neuropsychology* (Gazzaniga M. S., ed.), Plenum Press, New York, pp. 379–446.

Moscovitch M. and Olds J. (1982) Asymmetries in spontaneous facial expressions and their possible relation to hemispheric specialization. *Neuropsychologia* **20,** 71–81.

Moscovitch M., Scullion D., and Christie D. (1976) Early versus late stages of processing and their relation to functional hemisphere asymmetries in face recognition. *J. Exper. Psychol.: Human Perception and Performance* **2,** 401–416.

Murray M. R. and Richards S. J. (1978) A right-ear advantage in monotic shadowing. *Acta Psychol.* **42,** 495–504.

Myers J. J. and Sperry R. W. (1982) A simple technique for lateralizing visual input that allows prolonged viewing. *Behav. Res. Methods and Instrumentation* **14,** 305–308.

Nachshon I. and Carmon A. (1975) Hand preference in sequential and spatial discrimination tasks. *Cortex* **11,** 123–131.

Nettleton N. C. and Bradshaw J. L. (1983) The effects of task, practice and sequencing upon the lateralization of semantic decisions. *Int. J. Neurosci.* **20,** 265–282.

Nettleton N. C., Wood R. G., Bradshaw J. L., Thomas C. D. L., and Donahoo, K. B. (1983) A moving video window or mask yoked to eye movements: A system to permit free ocular scanning within delimited areas of the visual field. *Behav. Res. Methods and Instrumentation* **15,** 487–496.

Neville H. J., Kutas M., and Schmidt A. (1982) Event-related potential studies of cerebral specialization during reading II: Studies of congenitally deaf adults. *Brain Lang.* **16,** 316–337.

Nickerson R. S. (1978) On the time it takes to tell things apart, in *Attention and Performance VII* (Requin J., ed.), Lawrence Erlbaum, Hillsdale, NJ, pp. 77–88.

Norman D. A. and Bobrow D. G. (1975) On data-limited and resource-limited processes. *Cognitive Psychol.* **7,** 44–64.

Osaka N. (1978) Naso-temporal differences in human reaction time in the peripheral visual field. *Neuropsychololgia* **16,** 299–303.

Oscar-Berman M., Rehbein L., Porfert A., and Goodglass H. (1978) Dichaptic hand-order effects with verbal and nonverbal tactile stimuli. *Brain Lang.* **6,** 323–333.

Panou L. and Sewell D. F. (1984) Cerebral asymmetry in congenitally deaf subjects. *Neuropsychologia* **22,** 381–383.

Patterson K. and Bradshaw J. L. (1975) Differential hemispheric mediation of nonverbal visual stimuli. *J. Exper. Psychol.: Human Perception and Performance* **1,** 246–252.

Pearl N. and Haggard M. (1975) Practice and strategy in a measure of cerebral dominance. *Neuropsychologia* **13,** 347–352.

Phillips D. P. and Gates G. R. (1982) Representation of the two ears in the auditory cortex: A re-examination. *Int. J. Neurosci.* **16,** 41–46.

Piazza D. M. (1980) The influence of sex and handedness in the hemispheric specialization of verbal and non-verbal tasks. *Neuropsychologia* **18,** 163–176.

Pierson J. M., Bradshaw J. L., and Nettleton N. C. (1983) Head and body space to left and right, front and rear 1: Unidirectional competitive auditory stimulation. *Neuropsychologia* **21,** 463–473.

Pirozzolo F. J. and Rayner K. (1980) Handedness, hemispheric specialization and saccadic eye movement latencies. *Neuropsychologia* **18,** 225–229.

Pring T. R. (1981) The effects of stimulus size and exposure duration on visual field asymmetries. *Cortex* **17,** 227–240.

Rapaczynki W. and Ehrlichman H. (1979) Opposite visual hemifield superiorities in face recognition as a function of cognitive style. *Neuropsychologia* **17,** 645–652.

Reinis S., Landolt T. P., Money K. E., Lahue R. H., and Weiss D. S. (1986) Effect of head tilt on visual cortical cell function in the intact and labyrinthectomized cat. *Exper. Neurol.* **91,** 102–106.

Rhodes G. (1985) Lateralized processes in face recognition. *Br. J. Psychol.* **76,** 249–271.

Riley D. and Costall A. (1980) Comments on "Recognition of faces in the presence of two dimensional sinusoidal masks" by Tiger and Ganz. *Perception and Psychophysics* **27,** 373–374.

Rizzolatti G., Umilta C., and Berlucchi G. (1971) Opposite superiorities of the right and left cerebral hemispheres in discriminative reaction time to physiognomical and alphabetical material. *Brain* **94,** 431–442.

Roberts L. D. and Gregory A. H. (1973) Ear differences and delayed auditory feedback: Effect on a simple verbal repetition task and a non-verbal tapping test. *J. Exper. Psychol.* **101,** 269–272.

Ross E. D. (1983) Right hemisphere lesions in disorders of affective language, in *Localization in Neuropsychology* (Kertesz A., ed.), Academic Press, New York.

Ross P. (1983) Cerebral specialization in deaf individuals, in *Language Functions and Brain Organization* (Segalowitz S. J., ed.) Academic Press, New York, pp. 287–313.

Schimuzu A. and Endo M. (1981) Tachistoscopic recognition of Kana and Hangul words: Handedness and shift of laterality differences. *Neuropsychologia* **19,** 665–677.

Schmit V. and Davis R. (1974) The role of hemispheric specialization in the analysis of Stroop stimuli. *Acta Psychol.* **38,** 149–158.

Schmuller J. and Goodman R. (1980) Bilateral tachistoscopic perception, handedness, and laterality II. Nonverbal stimuli. *Brain Lang.* **11,** 12–18.

Schwartz M. and Smith M. L. (1980) Visual asymmetries with chimeric faces. *Neuropsychologia* **18,** 103–106.

Seamon J. G. and Gazzaniga M. S. (1973) Coding strategies and cerebral laterality effects. *Cognitive Psychology* **5,** 249–256.

Segalowitz N. and Hansson P. (1979) Hemispheric functions in the processing of agent-patient information. *Brain Lang.* **8,** 51–61.

Seitz K. S. and McKeever W. F. (1984) Unilateral versus bilateral presentation methods in the reaction time paradigm. *Brain and Cognition* **3,** 413–425.

Sergent J. (1982a) Influence of luminance on hemispheric processing. *Bull. Psychonomic Soc.* **20,** 221–223.

Sergent J. (1982b) Theoretical and methodological consequences of variations in exposure duration in visual laterality studies. *Perception and Psychophysics* **31,** 451–461.

Sergent J. (1983a) Role of the input in visual hemispheric asymmetry. *Psychol. Bull.* **93,** 481–512.

Sergent J. (1983b) The effects of sensory limitations on hemispheric processing. *Can. J. Psychol.* **37,** 345–366.

Sergent J. (1985) Influence of task and input factors on hemispheric involvement in face processing. *J. Exper. Psychol.: Human Perception and Performance* **11,** 846–861.

Sergent J. and Bindra D. (1981) Differential hemispheric processing of faces: Methodological considerations and reinterpretation. *Psychol. Bull.* **89,** 541–554.

Sergent J. and Lorber E. (1983) Perceptual categorization in the cerebral hemispheres. *Brain and Cognition* **2,** 39–54.

Shankweiler D. and Studdert-Kennedy M. (1967) Identification of consonants and vowels presented to left and right ears. *Quart. J. Exper. Psychol.* **19,** 59–63.

Sidtis J. J. (1982) Predicting brain organization from dichotic listening performance: Cortical and subcortical functional asymmetries contribute to perceptual asymmetries. *Brain Lang.* **17,** 287–300.

Sidtis J. J. (1984) Music, pitch perception and the mechanisms of cortical hearing, in *Handbook of Cognitive Neurosciences* (Gazzaniga M. S., ed.), Plenum Press, New York, pp. 91–114.

Sidtis J. J. and Bryden M. P. (1978) Asymmetrical perception of language and music: Evidence for independent processing strategies. *Neuropsychologia,* **16,** 627–632.

Silverberg R., Gordon H. W., Pollack S., and Bentin S. (1980) Shift of visual field preference for Hebrew words in native speakers learning to read. *Brain Lang.* **11,** 99–105.

Sivak B., Sivak J. G., and MacKenzie C. L. (1985) Contact lens design for lateralizing visual input. *Neuropsychologia* **23,** 801–804.

Spellacy F. (1970) Lateral preferences in the identification of patterned stimuli. *J. Acoust. Soc. Am.* **47,** 574–578.

Springer S. P. (1973) Hemispheric specialization for speech opposed by contralateral noise. *Perception and Psychophysics,* **13,** 391–393.

Sternberg S. (1975) Memory scanning: New findings and current controversies, *Quart. J. Exp. Psych.* **27,** 1–32.

St. John R. C. (1981) Lateral asymmetries in face perception. *Can. J. Psychol.* **35,** 213–223.

Stone J., Leicester J., and Sherman S. M. (1973) The naso-temporal division of the monkey's retina. *J. Compar. Neurol.* **150,** 333–348.

Strauss E. and Moscovitch M. (1981) Perception of facial expressions. *Brain Lang.* **13,** 308–332.

Studdert-Kennedy M. and Shankweiler D. (1970) Hemispheric specialization for speech perception. *J. Acoust. Soc. Am.* **48,** 579–594.

Suberi M. and McKeever W. F. (1977) Differential right hemisphere memory storage of emotional and nonemotional faces. *Neuropsychologia* **15,** 757–768.

Sullivan K. F. and McKeever W. F. (1985) The roles of stimulus repetition and hemispheric activation in visual half field asymmetries. *Brain and Cognition* **4,** 413–429.

Suter S. (1982) Differences between deaf and hearing adults in task related EEG asymmetries. *Psychophysiology* **19,** 124–128.

Tomlinson-Keasey C. and Kelly R. R. (1979) A task analysis of hemispheric function. *Neuropsychologia* **17,** 345–351.

Tramer O., Butler B. E., and Mewhort D. J. K. (1985) Evidence for scanning with unilateral visual presentation of letters. *Brain Lang.* **25,** 1–18.

Trevarthen C. B. (1972) Brain bisymmetry and the role of the corpus callosum in behavior and conscious experience, in *Cerebral Interhemispheric Relations* (Cernacek J. and Podivinsky F., eds.), Publishing House of the Slovak Academy of Sciences, Bratislava, 319–333.

Troup G., Bradshaw J. L., and Nettleton N. C. (1983) The lateralization of arithmetic and number processing: A review. *Int. J. Neurosci.* **19,** 231–242.

Tsao Y. C., Feustel T., and Soseos C. (1979) Stroop interference in the left and right visual fields. *Brain Lang.* **8,** 367–371.

Umiltà C., Frost N., and Hyman R. (1972) Interhemispheric effects on choice reaction times to one-, two- and three-letter displays. *J. Exper. Psychol.* **93,** 198–204.

Umiltà C., Brizzolara D., Tabossi P., and Fairweather H. (1978) Factors affecting face recognition in the cerebral hemispheres: Familiarity and naming, in *Attention and Performance VII* (Requin J., ed.), Lawrence Erlbaum, New York, pp. 363–374.

Underwood G. and Whitfield A. (1985) Right hemisphere interactions in picture–word processing. *Brain and Cognition* **4,** 273–286.

Urcuioli P. J., Klein R. M., and Day J. (1981) Hemispheric differences in semantic processing: Category matching is not the same as category membership. *Perception and Psychophysics* **29,** 343–351.

Vaid J. and Genesee F. (1980) Neuropsychological approaches to bilingualism. *Can. J. Psychol.* **34,** 417–445.

Van Den Abell T. and Heilman K. M. (1978) Lateralized warning stimuli, phasic hemisphere arousal and reaction time. Presented before the International Neuropsychologic Association. Minneapolis.

Vargha-Khadem F. (1982) Hemispheric specialization for the processing of tactual stimuli in congenitally deaf and hearing children. *Cortex* **18,** 277–286.

Volkmann F. C., Schick A. M. L., and Riggs L. A. (1968) Time course of visual inhibition during voluntary saccades. *J. Opt. Soc. Am.* **58,** 362–369.

Warm J. S., Richter D. O., Sprague R. L., Porter P. R., and Schumsky D. A. (1980) Listening with a dual brain: Hemispheric asymmetry in sustained attention. *Bull. Psychonomic Soc.* **15,** 229–232.

Warren L. R. and Marsh G. R. (1979) Changes in event related potentials during processing of Stroop stimuli. *Int. J. Neurosci.* **9,** 217–223.

Westheimer G. (1984) Spatial vision. *Ann. Rev. Psychol.* **35,** 201–226.

Westheimer G. and Mitchell D. E. (1969) The sensory stimulus for disjunctive eye movements. *Vision Res.* **9,** 749–755.

Weston P. and Weinman J. (1983) The effects of auditory and linguistic deprivation on lateral preferences of deaf children. *Develop. Med. Child Neurol.* **25,** 207–213.

Wilkins A. and Stewart A. (1974) The time course of lateral asymmetries in visual perception of letters. *J. Exper. Psychol.* **102,** 905–908.

Witelson S. F. (1974) Hemisphere specialization for linguistic and nonlinguistic tactual perception using a dichotomous stimulation technique. *Cortex,* **10,** 3–17.

Wuillemin D. B., Krane R. V., and Richardson B. L. (1982) Hemispheric differences in picture-word interference. *Brain Lang.* **16,** 121–132.

Yamamoto M. and Hatta T. (1982) Sex, task difficulty and hemispheric differences in somatosensory function in normal children. *Psychologia,* **25,** 115–120.

Yeni-Komshian G. H. and Gordon J. (1974) The effect of memory load on the right ear advantage in dichotic listening. *Brain Lang.* **1,** 375–381.

Yin R. K. (1969) Looking at upside down faces. *J. Exper. Psychol.* **81,** 141–145.

Yin R. K. (1970) Face recognition by brain injured patients. A dissociable ability? *Neuropsychologia* **8,** 395–402.

Young A. W. (1984) Right cerebral hemisphere superiority for recognizing the internal and external features of famous faces. *Br. J. Psychology* **75,** 161–169.

Young A. W. (1986) Subject characteristics in laterality differences for face processing by normals: Age, in *The Neuropsychology of Face Perception and Facial Expression* (Bruyer R., ed.) Erlbaum, Hillsdale, NJ, pp. 167–200.

Young A. W. and Bion P. J. (1981) Accuracy of naming laterally presented known faces by children and adults. *Cortex* **17,** 97–106.

Young A. W., Hay D. C., McWeeny K. H., Ellis A. W., and Barry C. (1985) Familiarity decisions for faces presented to the left and right cerebral hemispheres. *Brain and Cognition* **4,** 439–450.

Zaidel E. (1975) A technique for presenting lateralized visual input with prolonged exposure. *Vision Res.* **15,** 283–289.

Zaidel E. (1978a) Auditory language comprehension in the right hemisphere following cerebral commissurotomy and hemispherectomy: A comparison with child language and aphasia, in *Language Acquisition and Language Breakdown: Parallels and Divergencies* (Caramazza A. and Zurif E. B., eds.), Johns Hopkins University Press, Baltimore, MD, pp. 229–275.

Zaidel E. (1978b) Concepts of cerebral dominance in the split brain, in *Cerebral Correlates of Conscious Experience* (Inserm Symposium No. 6) (Buser P. A. and Rougeul-Buser A., eds.), Elsevier, North Holland, Amsterdam, pp. 263–284.

Neuropsychological Test Batteries in Neuropsychological Assessment

R. A. Bornstein

1. Introduction

Neuropsychological assessment has become increasingly dependent upon the use of test batteries. The use of such batteries has evolved with the influence of several historical perspectives. A prominent factor has been the unequivocal elucidation in the past 30–40 yr of the complexity and diversity of brain function, and the recognition of the nature and scope of hemispheric specialization of function. Conceptualizations of brain–behavior relationships predicated on unitary models of brain dysfunction were prevalent as recently as the 1940s. For example, Kurt Goldstein's (1939) theories assumed that all brain dysfunction (regardless of location, etiology, phase of illness, or the like) resulted in a central deficit in the ability to assume the abstract attitude. Such unitary models of the effects of disordered brain function on behavior represented the theoretical basis for notions of "organicity." To some extent, these unitary models of brain dysfunction represented the historical extension of the ideas of Flourens (1824) and Lashley (1929), who argued that the effect of cerebral lesions on behavior was related to the amount of brain tissue that was damaged or destroyed.

Within the context of neuropsychological assessment, the assumption regarding a unitary model of brain dysfunction was translated into an attempt to develop and use assessment techniques that could be sensitive to this purported underlying dimension. Thus, prior to the evolution of contemporary neuropsychological models, there was considerable reliance on measures, such as the Bender Visual Motor Gestalt test or others of that ilk, that were thought to be generally sensitive to the effects of cerebral dysfunction (regardless of lateralization or localization).

Subsequent to the Second World War, there was a rapid increase in knowledge about the nature and extent of hemispheric specialization, as well as the effects of lesions in certain locations on specific aspects of behavior. Studies of the effects of partial and complete callosal section have contributed to our knowledge about the behavioral propensities of the cerebral hemispheres, and have also been instrumental in our knowledge about the integrated functioning of the hemispheres. In relation to the burgeoning data regarding hemispheric specialization, unitary models of brain dysfunction became increasingly less tenable. This disaffection for "single best tests of brain dysfunction" was an important stimulus for the development of groups of tests or test batteries that were intended to measure a broad range of brain–behavior relationships.

A second important factor in the evolution of neuropsychological tests and test batteries has been the broad professional and scientific environment in which the discipline of psychology has evolved. In North American psychology, the quantitative, empirical tradition was prevalent (Davison, 1974; Meehl, 1954). This heritage has resulted in North American neuropsychologists' emphasis on statistical measurement and comparison of groups of patients. European relationships and factors, both societal and professional, have resulted in the evolution of a different focus in neuropsychological investigation (Luria and Majovski, 1977; Goldstein, 1986). In general, European models of neuropsychological investigation tend to be more individualized, and are directed more toward detailed examination of individual patients. To some extent, and not without merit, the long history of European behavioral neurology appears to have had a much stronger influence on the direction followed by neuropsychology outside North America. Thus, the methods of neuropsychological examination are often direct parallels of the syndrome analysis, pathognomic sign approach. For example, the Luria tasks are designed as individual items in which performance is dichotomized as normal or abnormal. Similar to behavioral neurology, performance is evaluated on the constellation of tasks on which an individual's performance is deficient. This stands in distinct contrast to the psychometric tradition underlying development of tests in North American neuropsychology.

Thus, the increased knowledge of neuroscience and prevailing societal and scientific histories have been important factors in the evolution of the test batteries that are used in neuropsychological assessment. Although these test batteries have many similarities, there are also important differences. The sections that follow consider some theoretical and practical differences among the various batteries, followed by a discussion of the various test batteries. To a considerable extent, the theoretical and pragmatic issues are intertwined.

2. Fixed vs Flexible Batteries

Perhaps the most obvious difference between test batteries in contemporary use is whether they endorse "fixed" or "flexible" approaches to assessment. The Halstead Reitan Battery (HRB) represents perhaps the most widely known and used example of the former, whereas the assortment of tasks developed by Luria (1973) and further systematized by Christensen (1975) represents the most widely known example of the latter. The Luria tests should not be confused with the Luria Nebraska Battery (Golden, et al., 1980), which is in fact an example of a fixed battery. Both approaches have advantages and disadvantages that have been discussed elsewhere (Kane, 1986; Lezak, 1983; Luria and Majovski, 1977; Tarter and Edwards, 1986). The use of a fixed or flexible test battery entails theoretical as well as practical issues.

2.1. Fixed Batteries

This approach endorses the use of a predetermined set of measures that samples behavior in areas of interest. In this approach, it is intended that all patients be given the complete set of tests, regardless of the patient's presenting problems, suspected etiology, or reason for referral. Clinical reality, however, dictates that the goal of "complete" batteries is sometimes not possible. Nevertheless, where possible, the fixed battery approach strives for all tests on all patients. The fixed battery model has several distinct advantages that are pertinent to clinical and research applications. Acquisition of data on a broad range of measures in a systematic fashion permits the building of data bases and facilitates the comparison of various diagnostic groups with regard to the pattern of deficits identified. Theoretically, it would be possible for laboratories employing similar test batteries to pool data and generate large studies of specific populations (*see* the discussion below on collaborative data bases).

One important issue that has both theoretical and practical impact concerns the nature of the tests that should be included in the battery. This decision is typically based on considerations of which domains of function should be assessed and the measures available to meet those assessment requirements. For example, there are few competently trained neuropsychologists who base their examination solely on the HRB, because of the lack of adequate tests of memory and the often suggested lack of sensitivity to frontal lobe lesions. These weaknesses in the HRB have led many (if not most) investigators to supplement the battery with tests, such as the Wechsler Memory Scale and the Wisconsin Card Sorting Test, that compensate for the weaknesses of the HRB. Thus, the fixed battery is typically designed from within a theoretical framework of what the individual examiner hopes to accomplish with the battery.

2.2. Flexible Batteries

The most widely known example of the flexible battery approach is the set of tasks used by Luria (1973) and standardized by Christensen (1975). This method of neuropsychological investigation proceeds on an individualized or patient-centered model. That is, the examiner begins with no preconceptions about the specific nature of the tests to be given, but rather uses presenting complaints, reason for assessment, and initial clinical impressions as the basis for initial selection of tests. The examination proceeds with a hypothesis testing, reductionistic approach with the goal of identifying the nature of a patient's deficit(s). Satisfactory performance on initial tasks assumes satisfactory performance on all subsumed tasks. In this approach, the examination proceeds down a hierarchy of tests with the ultimate goal of arriving at a finely detailed understanding of the deficit. This type of flexible battery approach has the advantage of being tailored to the problems and symptoms presented by the individual patient. For the purposes of research, however, this individualized approach creates problems in forming groups of patients for study, because the assessment data base is composed of different tests for different patients. Use of this particular group of tasks depends, to a large extent, on endorsement of Luria's theories about brain organization (Luria, 1973).

The current realities of health-care economics have produced a different type of flexible battery approach that may represent a compromise between the methods described above. In this model, test batteries are designed to respond to specific questions that arise within particular diagnostic populations. For example, some investigators have evolved the use of batteries in evaluating patients referred for the differential diagnosis of dementia (e.g., Senile Dementia of Alzheimer's Type vs Multi–Infarct Dementia vs Pseudo Dementia of Depression). Patients referred for this particular question would be administered a more focused battery, composed of tests that have been reported to be sensitive to this discrimination. Such an approach would seem to combine some of the advantages of both fixed and flexible batteries. The examination can be tailored to a specific question(s), and simultaneously the collection of data on a consistent group of measures is conducive to research. Another compromise between fixed and flexible batteries, suggested by Tarter and Edwards (1986), involves a three-stage-decision approach to assessment. The three stages include:

1. A screening battery
2. Specialized tests or test batteries as indicated by the results of the screening battery and
3. Individualized examination of a patient's particular deficits.

The screening battery employed in this system is more comprehensive than most (Goldstein et al., 1983), but is subject to the same criticisms as other screening batteries (Spreen and Tuokko, 1982; and *see below*).

3. Theoretical, Philosophical, and Practical Issues

Apart from the rather basic issues regarding the pragmatics and mechanics of neuropsychological examination described above, fixed and flexible battery approaches can be compared with regard to the underlying theories and philosophies that, by necessity, impact on the scope and nature of the assessments that are performed. The discussion that follows will focus on the works of

Reitan and Luria, because of their prominence in the field and because the approaches represent what may be seen as opposite ends of the fixed vs flexible battery continuum.

3.1. Theoretical

The difference between these two approaches is perhaps most highlighted in consideration of the theoretical basis from which they evolved. The methods of Luria's assessment are directly linked to his theory of brain organization, whereas the Reitan method has been described as "atheoretical" (Luria and Majovski, 1977; Spreen and Tuokko, 1982; Tarter and Edwards, 1986). This description is somewhat misleading because it implies that the use of the HRB is devoid of a theoretical rationale. In the HRB model, the selection of tests is based on empirical grounds, and theoretical assumptions about brain function become operative in the interpretation of test data. The principal difference, then, is the point in the neuropsychological examination at which specific theoretical assumptions exert their influence. It would appear that flexible battery approaches (e.g., Luria) necessarily operate within the context and constructs of a particular theory. Practical decisions about which subset of tasks is to be administered are based on theoretical assumptions about brain organization, as reflected in that set of tests. On the other hand, fixed battery approaches may or may not proceed on the basis of the predications of a particular theory. Apart from accepting principles of hemispheric specialization, many fixed battery approaches select tests on pragmatic issues and make no further assumptions about organization of psychological processes (Tarter and Edwards, 1986), although there are attempts to identify the constructs that are thought to underlie the tests (Spreen and Tuokko, 1982). Since theories of brain organization are imposed (by the individual neuropsychologist) at the test interpretation (rather than test selection) phase, the same set of tests may be interpreted in relation to a variety of theories. Thus, in a certain sense, the fixed battery approach lends a greater degree of conceptual flexibility than the flexible battery approach.

The battery of tests developed by Benton and his colleagues (Benton et al., 1983) would appear to incorporate many of the strengths of both of these divergent approaches. Like theoretically based test batteries, Benton's work has been guided by his interests in classical behavioral neurology. Hence, he has developed measures that bring his careful and insightful methods to the study of such neurological symptoms as right-left disorientation, finger agnosia, constructional dyspraxia, and facial recognition. However, rather than pursuing a particular theory of cerebral organization, Benton's battery of tests is directed toward assessment of symptoms in the context of the theories and constructs of modern neurology. As a theoretically based battery, some of the very specialized tests could be administered as dictated by the presenting complaints or specific clinical issues related to an individual patient. Like broader and less theoretical batteries, the tests could be employed in a manner in which it was assumed that assessment and identification of problems in the areas addressed by these tests would be of interest in all patients.

Milner and her colleagues have developed and continue to develop a group of measures to study the behavioral effects of localized cerebral lesions. The tests have been developed individually in the context of Milner's lifelong career in the investigation of the behavioral effects of focal cortical excisions for treatment of epilepsy. Milner views her own work as experimental, and clearly has reservations about the uncritical clinical application of the measures (King, 1984). Like Benton's, the Milner battery of tests represents a composite of the numerous studies performed at the Montreal Neurological Institute over the past three decades. The development of new tests continues to evolve in the context of new discoveries in the nature of hemispheric specialization. The development of tests in this model is intimately linked to developments in experimental and comparative neuropsychology. The tests are not developed as clinical instruments, but rather as experimental techniques for studying the correlates of dysfunction in particular brain regions. To the extent that the tests developed by Milner are employed as a clinical test battery, a practice of which she might not approve (King, 1984), the basis for test development rests on sensitivity to specific brain regions.

In summary, there are several groups of tests in contemporary use, and each has been developed within a specific theoretical framework. The Halstead Reitan Battery eschews specific theoretical assumptions (other than the basic assumptions of lateralization of function) and focuses on sampling a broad range of behaviors, in order to make inferences about the integrity of cerebral function. At the opposite end of the continuum, the Luria method of assessment is intrinsically linked to his specific theories of brain function. Benton's tests have been developed to address concepts from classical neurological theory, and Milner's tests have emerged from an experimental neuropsychological approach to assessing the deficits associated with lesions in specific brain regions.

3.2. Philosophical

In addition to and related to theoretical issues, test batteries in contemporary use may be contrasted with regard to the implicit philosophy and goals of assessment. Neuropsychological examinations may be performed for a variety of reasons, including diagnosis, documentation and estimation of degree of deficit, and rehabilitation planning. The future evolution of neuropsychology in conjunction with development and availability of neuroimaging technologies will also have an impact on the questions that neuropsychologists are asked to address. The philosophy underlying the assessment necessarily influences the procedures and style of examination that are employed. To a considerable degree, these procedural differences appear to center around what is conceived to be the philosophical focus of the examination. Some styles of examination appear to emphasize an in-depth focus on deficits, whereas other approaches are more broadly focused. It is obvious, of course, that theoretical and historical influences are inextricably related to these procedural, stylistic, and philosophical questions.

Several neuropsychological batteries employ a reductionistic, hypothesis-testing approach, in which one or more general tasks form the point of departure for a detailed analysis of behavior. In general, this style follows a descending hierarchy of examination, with the ultimate goal of documenting in fine detail the inherent nature of a patient's deficit. In this model, satisfactory performance on the basic task assumes normal function on all component tasks believed to be subsumed under this more general task. This focus on the delineation of the fundamental nature of the deficit(s) reflects the lesion-finding philosophy that has its origins in neurology. That is, this style of assessment appears to emphasize a strategy that uses behavioral measures (within some theoretical structure) as a vehicle to determine what is wrong with a particular patient.

In other batteries, the implicit philosophical goal appears to be directed toward an identification of the pattern of strengths and weaknesses. Instead of a descending hierarchical approach, this style of examination typically employs a broad set of measures with the expressed intent of sampling the adequacy of a predetermined set of functions. Different tests of a particular ability may be used in particular cases, but there is a philosophical assumption that knowledge of a patient's abilities is as important as knowledge of the deficits. This style of examination emphasizes breadth (rather than depth) of assessment. The sacrifice of depth for breadth of assessment provides some potential advantage in the context of certain referral questions. For example, effective rehabilitation planning would appear to require knowledge of strengths as well as deficits, in order to maximize the potential of success. Clearly, a detailed knowledge of the deficit combined with an appreciation of a patient's residual skills will provide the optimal information, and, as above, the two philosophical approaches can (and perhaps should) be used in conjunction.

3.3. Practical

In addition to the broader theoretical and philosophical issues, there are some narrower differences in the development and utilization of neuropsychological test batteries. These differences include the manner in which test batteries are developed, the nature of the components, and the nature of the data that are evaluated. Some batteries (e.g., HRB) were developed as a group of tests, designed to be used together. Although some of Halstead's original tests had previous applications and were either adopted outright (Seashore Rhythm Test) or modified (Tactual Performance Test), the test battery as evolved through the work of Reitan and others has been developed and validated together. In contrast, the tests developed by Benton et al. (1983) and Milner (1975) have been developed individually, with validation studies documenting the utility of each clinical measure. Many of Milner's tasks have been developed in tandem within the experimen tal paradigm of double dissociation of deficit (i.e., lesion A pro duces deficit A but not B, and lesion B produces deficit B but not A).

Some batteries are comprised of individual items or tasks rather than tests (Luria, 1973; Christensen, 1975). These tasks tend to be scored in a dichotomous (pass/fail) fashion and, although conceptually related to other tasks believed to be subserved by a particular theoretical unit, the individual items are not linked together as a test of some unique function. The development of the particular tasks evolved out of Luria's experiences and his theory of brain function. Recently, there have been statistical attempts to force items together to provide some semblance of being a psychometrically based test. However, these factor analytically derived scales have little relationship to the conceptual model from which the tasks were derived.

Contemporary approaches to neuropsychological assessment also vary with regard to the unit of focus in performance evaluation. Some models focus exclusively on quantitative analysis, others include qualitative considerations, and many incorporate a combination of both. For example, the levels of test interpretation espoused by Reitan would appear to be largely quantitative, although the examination of pathognomic signs on tests, such as the Aphasia Screening Test, has some inherent qualitative interpretations and judgments. The scoring system by Russell et al. (1970) permits some standardization in the qualitative analysis of this test. Additionally, there have been some attempts to examine qualitative aspects of performance on HRB tests (Brandt and Doyle, 1983; Bornstein and Leason, 1984; Bornstein et al., 1984). Nevertheless, it is a fact that the scoring of most of the HRB tests is purely quantitative. A similar emphasis on quantitative scoring characterizes the tests developed by Milner and Benton. The tests developed by Benton, however, address qualitative aspects of performance in a systematic fashion. Much of Benton's work has demonstrated that the classical neurological symptoms are heterogeneous and may have a variety of clinical manifestations. Thus, he has developed a series of tests that explicitly examine qualitatively different aspects of performance in areas, such as finger agnosia, right-left disorientation, and facial recognition. The nature of evaluation of performance on Luria's tasks appears to be primarily qualitative. Apart from a gross characterization of pass or fail, attention to the qualitative aspects of performance is used as the basis for clinical inference.

The process approach endorsed by Kaplan and her colleagues (Milberg et al., 1986) represents a combination of both qualitative and quantitative examinations of performance (and also represents a combination of the fixed and flexible battery approaches). In this model, knowledge of how a patient performs a task is as important as the ultimate level of performance obtained. The process approach attempts to draw inferences about the cognitive substrate underlying performance by examining how a patient passes or fails on a particular task. This is accomplished by use of standardized tests that in some cases are adapted or expanded to permit a more detailed analysis of behavior. For example, the visual reproduction subtest of the Wechsler Memory Scale is expanded to include a copy trial in addition to immediate and delayed recall trials. Comparison of copy and recall trials allows for a discrimination of problems of praxis as opposed to memory. In addition, a flow chart is generated, documenting the steps used by a patient to reproduce the diagrams. Finally, the nature of errors committed or omitted can be characterized (e.g., impoverishment, simplification, distortion, disorganization, or confusion between designs). This type of careful attention to procedural, qualitative as well as quantitative aspects of performance is applied to other tests in the battery, and may lead to specific hypotheses about the integrity of cerebral function. Although there are some "standard" elaborations, the modifications of existing procedures and development of new measures are related to the specific symptoms presented and to the creativity and neuropsychological sophistication of the examiner.

The previous comments suggest that there are a broad range of theoretical and philosophical issues that influence the style and substance of neuropsychological assessment. Contemporary procedures differ considerably with regard to questions of how examinations should proceed, what should be included in the examination, and how the collected data should be analyzed and interpreted. Each of the existing models have advantages and disadvantages, and no single method or group of tests would appear to be clearly superior. The above discussion has presented some of these issues as absolute dichotomies, but the realities of neuropsychological practice indicate that various combinations of these theoretical, philosophical, and procedural methods are becoming increasingly common. The sections that follow will consider some of the empirical evidence surrounding several of the more commonly used test batteries.

4. Halstead Reitan Battery

This group of tests is probably the most widely used neuropsychological test battery in North America, and its popularity is the result of a number of factors. The HRB was probably the first group of tests developed for use as a battery of neuropsychological measures. Although the techniques of other prominent investigators (e.g., Benton and Milner) span equally long periods of time (beginning from the 1950s), their measures have changed over time. As a group of tests, the HRB therefore has the longest history of use. The longevity of the HRB is in part related to the energetic and prolific endeavors of Ralph Reitan, whose early development of Halstead's tests unequivocally established the validity of neuropsychological investigation. Reitan's numerous publications (Reitan and Wolfson, 1985) and his willingness to talk about and share his methods was (and continues to be) a major factor in the popularization of neuropsychology in general, and the use of the HRB in particular. In addition to the personal efforts of Reitan, the extensive research on the application of the HRB to a broad spectrum of disorders and referral questions has further enhanced the appeal of the battery.

Although there is little doubt of the concurrent validity of the HRB (the ability to detect brain dysfunction), other traditional psychometric aspects of reliability and validity have not been as rigorously tested. This, of course, is somewhat paradoxical in view of the fact that the HRB evolved from within the North American psychometric/empirical tradition. Nevertheless, some might argue that such mundane psychometric studies are superfluous. Indeed, it has been observed that many neuropsychological measures were developed as "standardized experiments" rather than traditional psychological tests (Davison, 1974). In that view, the amply demonstrated concurrent validity would obviate the psychometric requirements that psychologists expect of their instruments. However, this tendency to "get on with it" and find new and broader applications for the HRB may not necessarily be the best approach for further evolution of the battery. Quite the contrary, several prominent neuropsychologists have suggested that attention to this psychometric detail would be an important endeavor (Parsons and Prigatano, 1978; Satz and Fletcher, 1981).

A review of the literature regarding the sensitivity of the HRB to cerebral dysfunction would require several chapters, if not complete volumes, and thus is well beyond the scope of this chapter. The question of predictive validity is essentially untouched, but there have been a few preliminary reports of the use of neuropsychological tests in prediction of employment status (Dikmen and Morgan, 1980; Heaton et al., 1978; Heaton and Pendleton, 1981; Newnan et al., 1978).

4.1. Convergent Validity

Many studies employ independent neurodiagnostic methods, such as CAT scans, magnetic resonance imaging (MRI), or other procedures, as the basis for the formation of patient groups. In addition, there have been several studies that have directly compared the relative efficiency of neuropsychological tests and other diagnostic procedures in the identification of the presence, location, and nature of cerebral lesions (Filskov and Goldstein, 1974; Klesges et al., 1983; Snow, 1981; Swiercinsky and Leigh, 1979; Tsushima and Wedding, 1979). Most of these report good agreement between neuropsychological tests and various other neurodiagnostic methods (e.g., CAT scans, MRI, and the like). Some of these studies have suggested that the incidence of cerebral dysfunction is overestimated by neuropsychological tests. However, recent studies using MRI have shown that technique to be more sensitive than CAT scans in identification of some types of lesions (Jabbari et al., 1986; Laster et al., 1985). Thus, some of the suspected false positives associated with neuropsychological assessment in previous studies may, indeed, have been true positives. Further studies will need to be sensitive to the possible differential sensitivity of neurodiagnostic methods (e.g., CAT vs MRI) in examining the agreement between neuropsychological measures and other methodologies.

4.2. Standardization

There are potential clinical applications that can be derived from studies of the psychometric properties of the HRB. It is important to recognize, however, that psychometric data must be viewed in relation to the fact that there is more than one version of the HRB. Although the component tests may have the same name, the actual tests employed may be slightly (or not so slightly) different. For example, some versions of the Speech Perception Test and Seashore Rhythm Test have filtered out the background noise that was present on the original stimulus tapes developed by Reitan. For the purposes of standardization, Reitan has maintained the poor sound quality of the initial tapes. Other not-so-subtle changes have also emerged. One "laboratory" administers the Tactual Performance Test to adults with the board in the horizontal (as opposed to the standard vertical) position. In addition to the "innovative" approaches to the administration of the HRB, there are also a number of suppliers of test equipment, some of whom do not adhere to the original test specifications. Although there is nothing inherently wrong with change, the purveyors of such "new and improved" equipment must accept the burden of demonstrating the comparability or validity of their product. Unfortunately, this is rarely the case. Furthermore, substantial changes in equipment from the same provider may occur over time. A Finger Tapping Test recently purchased from a prominent supplier of HRB materials was noticeably different from "similar" devices purchased in the past. There was no documentation that this new device yielded comparable scores. This is not intended to suggest that one or the other version of these tests is correct, and it is unclear whether such modifications make any difference at all. Rather, this simply points out that there are differences, and the reliability and validity data, to the extent that they exist, may not be equally applicable to all versions of the tests.

4.3. Norms

Normative data for the HRB have not been established on the type of population-based representative sampling that is characteristic of some broadly used measures, such as the Wechsler Intelligence Scales. This is largely a function of the extreme costs associated with such an endeavor and the lack of a corporate sponsor to underwrite those costs. Even if such a national normative study could be undertaken, the existence of various forms of the tests would necessitate an explicit decision to use some particular form. Thus, some individuals would have to modify existing practices to conform, and thereby confront the possibility of surrendering personally desired methods of test administration. Halstead's original normal group has been criticized (Lezak, 1983), and subsequently, numerous small normative samples have been collected as control groups for specific studies. In addition, there have been a few studies performed expressly for the purpose of obtaining normative data from relatively large samples (Beardsley et al., 1978a,b; Bornstein, 1985; Fromm-Auch and Yeudall, 1983; Heaton et al., 1986; Pauker, 1980). The application of these data is constrained by the fact that these studies used limited test batteries, were based on biased samples (i.e., mean IQs of 115 or more), or were from nonrepresentative geographic areas. Three of these studies were from Canada, and there was some evidence of differences from normative scores reported from the United States. It is unclear whether this represents true geographic differences or variations in testing procedures. Nevertheless, there is as yet no well-established, demographically sophisticated, and well-stratified normative data base for the HRB. Many laboratories support local norms, but for the most part, these have not been widely disseminated.

4.4. Reliability

The reliability of various HRB tests has been examined in terms of both test–retest, as well as internal reliability. Obviously, some tests are suited for internal reliability measurement (e.g., Seashore Rhythm, Speech Perception, and Halstead Category Test), whereas tests that yield scores based on time are more suited to retest reliability studies. It is a telling commentary that the long-awaited 500-page HRB text/manual (Reitan and Wolfson, 1985) contains no information whatsoever on the psychometric properties of the tests.

The first study of internal reliability was performed by Shaw (1966) and reported a split-half correlation of .92 on the Category Test. More recently, split-half reliability and Cronbach's alpha of approximately .75 were reported for the Seashore Rhythm Test (Bornstein, 1983). Split-half correlations of .74 and .87 were reported for two samples on the Speech Sounds Perception Test (Bornstein, 1982). Several studies have examined the test–retest reliability of the HRB. Klonoff et al. (1970) reported 12-mo retest reliability in a sample of chronic schizophrenics. In that sample, the level of performance was in the range associated with brain dysfunction, and the reliability coefficients ranged from .49–.84. Matarazzo et al. (1974) examined test–retest reliability in normal young men with a mean intertest interval of 20 wk. The reliability coefficients ranged from .24 on Finger Tapping to .68 on the time score from the Tactual Performance Test (TPT). The low reliability coefficients were probably related to the general ability (mean WAIS Full Scale IQ of 118) and the associated restricted range of scores. Matarazzo et al. (1974) also reported 12 retest reliability coefficients ranging from .44 on Finger Tapping to .96 on Category Test in a sample of patients with diffuse cerebrovascular disease. In contrast to the low reliability coefficients for Finger Tapping in the samples reported by Matarazzo et al. (1974), Morrison et al. (1979) reported 1-wk retest reliability coefficients of .80 for both dominant and nondominant hand trials. Dodrill and Troupin (1975) reported reliability coefficients on four repeated assessments with 6- to 12-mo intertest intervals in a chronic seizure disorder sample. When the first assessment was used as the reference point, coefficients ranged from .30–.86 between the first two assessments, and slightly higher coefficients with the third and fourth assessments.

With the increasing use of the HRB and other neuropsychological measures to monitor short-term changes in function, studies of the retest reliability over briefer intervals are of interest. Eckardt and Matarazzo (1981) reported 3-wk reliability in detoxified alcoholics and non-alcoholic medical patients. In the nonalcoholic group, reliability coefficients ranged from .51–.90, and from .53–.74 in the alcoholic patients. In both groups, reliability coefficients for Finger Tapping were among the highest. Bornstein et al. (1987) reported 3-wk reliability in a sample of normal subjects with a mean IQ of 105. Reliability coefficients ranged from .55 on part A of the Trial Making Test to .80 for the TPT-Memory score.

The studies of internal and test–retest reliability are generally consistent, and indicate satisfactory reliability of the HRB and associated measures. However, the reliability of particular tests does appear to vary as a function of the sample studied and the intertest interval. For example, the TPT-Memory score had the highest coefficient in the normal sample reported by Bornstein et al. (1987), and the lowest coefficient reported in epileptics by Dodrill and Troupin (1975). Finger Tapping scores appear to have better reliability with shorter intertest intervals (less than 1 mo). In contrast, tests, such as TPT and Category Test, appear to have somewhat better retest reliability with longer intertest intervals. These differences in reliability with regard to duration of interval would be consistent with the suggestion that tests whose sensitivity is dependent upon novelty of the task (e.g., TPT or Category Test) are more reliable when the intertest interval is sufficiently long to overcome the effects of prior exposure. Conversely, "pure" motor tests (e.g., Finger Tapping) may be more susceptible to long-term fluctuations in performance, and therefore more reliable with shorter intervals. In summary, the available data indicate adequate reliability of the HRB in a variety of patient populations and across a range of intertest intervals.

4.5. Clinical Applications

The importance of careful psychometric examination of neuropsychological tests extends beyond simple reassurance of the reliability of our instruments. Consideration of such issues can be useful in understanding what the tests are measuring, what aspects of performance may be contributing to the sensitivity of the tests, and in further development of how the tests may be clinically employed. At present, the clinical application of the HRB measures is based on simple, rather rudimentary analyses, such as total errors or time to completion. However, there is clear potential for examining the data derived from these tests in more sophisticated ways. Analogous to use of the WAIS-R, there are few who would advocate that examination of intellectual performance should be based solely on consideration of the Full Scale IQ. Quite the contrary, interpretation typically involves a number of levels of analysis, including comparison of discrepancies between Verbal and Performance IQ, examination of intersubtest scatter, and examination of intrasubtest scatter. This type of multilevel analysis of neuropsychological tests also might well be clinically productive. Consideration of many of the HRB measures suggests numerous opportunities for analyses of patterns of performance across subtests and various other intratest patterns.

The lack of consideration of such psychometric concerns has led some investigators to look for ways to cut corners and abbreviate the HRB (Erickson et al., 1978; Golden and Anderson, 1977; Gregory et al., 1979; Kilpatrick, 1970; Ryan et al., 1982). This may be motivated by an attempt to reduce the time required, a desire to "improve" the battery, or some other motivation, but it would seem prudent to be aware of what might be lost *prior* to modification. As pointed out by Adams (1987), simple face validity of a novel idea (however divinely inspired) or innovation for its own sake is simply insufficient, and needs to be subjected to rigorous scientific tests to establish its validity. The Speech Sounds Perception Test may represent an object example of how examination of intratest patterns of performance may yield scientifically and clinically useful data.

Several studies (Bornstein, 1982; Golden and Anderson, 1977; Ryan et al., 1982) and probably countless clinicians (Adams, 1987) have recognized that the majority of errors occur on the first three subtests, and that various formulas can be derived to estimate total errors, based on these subtests. Golden and Anderson (1977) suggested that practice might account for the lower number of errors in the second half of the test, and thought that inspection of the items did not indicate obvious differences in item difficulty. Bornstein et al. (1984) altered the order of presentation of subtests and provided evidence that suggested that the items in the first two subtests were, in fact, more difficult. A subsequent study (Bornstein and Leason, 1984) reported an item analysis that confirmed the disproportionate number of more difficult items in the first two subtests. That study also reported a consideration of qualitative errors that revealed differences among groups with different types of lesions. These preliminary studies provide some basis for understanding what the test is measuring, and also provide the basis for clinical consideration of patterns of performance on the test. Similar examinations of other HRB subtests have yet to be performed, but the data suggest that abbreviation of tests without adequate scrutiny may risk the loss of potentially useful information.

In summary, the HRB is the most commonly used neuropsychological test battery that has been developed for use as a battery. Though the tests had their origin in Halstead's theories of biological intelligence (Halstead, 1947), in general, application of the battery does not occur within the constraints of that (or any other) particular theory. The HRB is generally described as a fixed battery approach, although there are probably very few trained neuropsychologists who rely solely on the HRB (Adams, 1987). Contrary to some criticisms, most of those who use the HRB have remained sensitive to changes suggested by available research. For example, several of Halstead's original measures (Time Sense Test, Critical Flicker Fusion) have been dropped from routine use because of the failure to demonstrate consistent discriminatory validity. Conversely, many HRB neuropsychologists supplement the battery with additional tests (e.g., Wisconsin Card Sorting Test) as the research supporting the utility of these tests has emerged

(Robinson et al., 1980). To the extent that they have been studied, the HRB measures appear to have adequate reliability and validity. In general, there appears to be a willingness to modify standardized test procedures or equipment without documenting the necessity for, or the independent validity of, those innovations. The HRB measures, in the current state of development, are best suited for obtaining a broad sample of an individual's abilities and deficits. Interpretations regarding the basis of particular deficits are based on examination of patterns of performance among tests, rather than a detailed analysis of performance within a test. Finally, there appears to be an increasing sensitivity to the potential contribution of careful psychometric examination of the HRB measures.

5. Benton, Milner, and Luria Batteries

Although the groups of tasks developed by these eminent neuropsychologists differ considerably in terms of the theoretical constructs in which the tests evolved, these "batteries" are much more similar with regard to their evolution and extent of psychometric study. Both the Benton and Milner tasks have been developed as individual tests. These groups of tests may be regarded as test batteries primarily to the extent that they reflect a systematic approach to brain investigation, predicated either on behavioral syndromes (Benton) or specific brain regions (Milner). As with the HRB, the focus for all of these approaches has been directed toward concurrent validity. The literature is replete with studies attesting to the ability of the Benton and Milner tasks to discriminate specific diagnostic groups. Much of the information regarding Benton's tests has been compiled into a manual of procedures (Benton et al., 1983). This brief manual provides a theoretical background for the concept being studied, a description of the procedures used, normative information, the empirical basis for deriving "cutoff" scores, and a summary of the research with various patient populations. Many of Benton's tests have relatively large normative data available. For many of the tests, there are data presented that examine the interactive effects of demographic variables (e.g., age and educational level). There are no data presented on other psychometric issues, such as reliability.

Similarly, the psychometric documentation of Milner's tests is entirely restricted to studies reporting the ability of particular tests to differentiate between groups of patients with focal lesions. There is no attempt to identify "cutoff" scores, examine reliability, or present normative data. Undoubtedly, this reflects Milner's conceptualization of herself (and, by extension, her methods) as an experimental psychologist. Much of Milner's most important work has been based on a patient undergoing focal cortical excisions for treatment of epilepsy. Although Milner explicitly recognizes the problem of generalization to other populations (King, 1984), it is unclear that others attempting to apply her techniques share that awareness. It is not uncommon to see studies employing the "Montreal Battery" in populations, such as schizophrenia (Kolb and Whishaw, 1983) or Tourette's Syndrome (Sutherland et al., 1982). It is by no means a safe assumption that the sensitivity of a test that has been validated on a specific population will necessarily generalize to all other populations. Milner's tests have been developed to a large extent on seizure disorder patients, many of whom have very longstanding cerebral dysfunction. The possibility of atypical cerebral organization exists in these patients, and Milner directly acknowledges the necessity of readapting the tests for use in other patient populations (King, 1984).

Unfortunately, there is very little in the English language literature documenting the utility of Luria's approach. However, many of his books have been published in English (Luria, 1966, 1973), and this has served to increase awareness of his theories. A discussion of the psychometric properties of the clinical application of Luria's methods is virtually a contradiction in terms. Luria did not employ statistics, and many of his prominent theoretical positions are based on single case studies. There is little doubt of the value of such studies, but many of the cases used by Luria had very large lesions. The atypical nature of these patients raises many questions about the specificity of the tasks as well as the generalizability to other patients. Of course, Luria's methods can hardly be described as a test battery, since his examination of specific patients consisted of individualized tasks based on his own clinical insight and creativity. It is clear that Luria's method is directly linked to the knowledge and clinical insight of the examiner, and the question of whether anyone other than Luria could use the approach has been raised. (Ironically, similar questions have been raised regarding Reitan's use of the HRB). In partial response to this criticism, Christensen (1975) has attempted to impose some structure and organization on the Luria method, without sacrificing the inherent value of the approach. Christensen (1975) has grouped the tasks to facilitate explanation of their relationships to each other and to Luria's theory.

The approaches of Benton and Milner differ to some extent from the HRB method, with regard to the theoretical questions underlying test development and the fact that the components of the "batteries" are developed as individual tests. Like the HRB, there is clear evidence of the concurrent validity of the tests in the vast majority of situations in which they have been applied. Luria's approach is unique and endorses a style of examination with which most North American neuropsychologists are uncomfortable. In all of these models of neuropsychological examination, there appears to be a clear rationale. The methods of assessment may be conceptually linked to brain regions (Milner), neurological syndromes (Benton), broad sampling of behavior (Reitan), or specific theories of brain organization (Luria). In addition to these approaches, there is another recently emerging approach to neuropsychological examination that attempts to combine some aspects of these established models.

6. Luria Nebraska Battery

The Luria Nebraska Battery (LNB), introduced by Golden and his associates (Golden et al., 1980), has probably been associated with the most controversy in the short history of contemporary clinical neuropsychology. Numerous studies (*see* Golden and Maruish, 1986 for a review) have attempted to document the validity and reliability of the battery and its component scales. However, the uncritical and somewhat grandiose claims of the test's proponents have been countered by powerful criticism in virtually all areas of test development and validation. Major problems with the LNB have been identified with regard to the theoretical and conceptual framework, subject selection used in validation studies, lack of construct validity of many scales, vastly overstated claims of diagnostic precision, and grossly inadequate and improper statistical analyses (Adams, 1980a, Adams, 1985; Crossen and Warren, 1982; Delis and Kaplan, 1982, 1983; Spiers, 1981, 1982,

1984; Stambrook, 1983). Many of the most important criticisms have never been satisfactorily addressed by the proponents of the LNB, and it is almost certainly the case that the vast majority of the overstated claims of the efficacy of the LNB are attributable to statistical artifact. This rancorous debate in the literature has subsided to some degree, which probably represents a crystallization of opinion (for or against). In addition, some studies have begun to appear that attempt to provide empirical comparisons of the LNB and HRB (*see* Kane, 1986 for a review).

The LNB represents an attempt to superimpose North American quantitative and statistical experimentation onto Luria's methods. The basis for this attempt was Christensen's (1975) codification of the procedures. The question of whether or not such an exercise is valid would seem to be predicated on the assumption that the endeavor is worthwhile in the first place. It is not at all clear that such is the case. The very strength of Luria's methods would appear to be the flexibility and theory-based style of examination. These are the specific aspects that are eliminated by Golden's modifications. In fact, it is in no way apparent that the battery has anything to do at all with Luria's theory. The LNB may employ tasks that Luria found useful, but the interpretation of those tasks is not necessarily the same as Luria's. Furthermore, the scales are not organized in the context of his theory and, in many cases, represent such fundamental neuropsychological knowledge that it hardly seems to need further documentation. The scales themselves are curious admixtures of items that are simply the unconstrained results of the uninspired use of multivariate statistics. That the scales have little construct validity is of no surprise. The LNB has been aggressively promoted and marketed, well in advance of the empirical base that would support such an endeavor. It is unfortunate that the apparent simplicity of administration and interpretation of the LNB (encouraged by the promoters) will likely result in sophomoric misapplications by individuals who are least qualified to critically evaluate the battery and themselves. Regardless of the flaws or potential value of the LNB, it is clear that the only relation to Luria is in the use of some of his tasks. It might be less misleading to refer to the battery as the Golden Neuropsychological Battery. Whatever the merits of this battery, it will very likely be the subject of continued empirical study and development.

7. Directions for the Future

The field of clinical neuropsychology has rapidly expanded since its contemporary emergence almost four decades ago. The initial enthusiasm for demonstrating the ability to detect behavioral changes in a variety of neurological populations has shifted to identifying neurobehavioral deficits associated with diseases in other organ systems. There also appears to be a trend toward introspection, with increasing attention being devoted to some of the basic assumptions surrounding clinical application of the tests. Some neuropsychological tests, test patterns, and "rules of thumb" are being subjected to more careful study and the empirical validity of various measures is being established.

With regard to the further development of clinical neuropsychology, it appears that there are many directions for growth and improvement. Perhaps the most urgent need relates to the development of more sophisticated and demographically sensitive normative data. The ever-increasing number of elderly subjects being referred for examination presents a clear mandate for establishing appropriate norms for these age groups. Several recent studies have reported high rates of misclassification for older subjects when conventional "cut-off" scores are employed (Bornstein, 1987; Heaton et al., 1986; Price et al., 1980). It is, therefore, imperative that proper normative studies of neuropsychological measures that are sensitive to possible subgroup differences within the elderly population be performed (e.g., consideration of educational level, sex, and age subgroups).

Another important direction for clinical neuropsychology will be in the area of predictive validity. Neuropsychologists already are asked to make judgments about the likelihood of success for an individual's return to work, ability to drive, and so forth, but there is little empirical evidence to support these recommendations. It may be necessary to pursue studies that directly (or indirectly) address these issues. For example, simple correlational studies could be performed to examine the relationship between neuropsychological measures and specific types of vocational aptitude or ability tests that have been shown to possess adequate predictive validity. Similarly, careful followup studies could examine the success (or lack thereof) of individuals who returned to previous activities, and the neuropsychological characteristics of successful individuals could be identified. In the context of the evolving role of clinical neuropsychology, these studies would be useful in demonstrating the validity of neuropsychological assessment outside of the laboratory.

It is clear that neuropsychological knowledge has not yet reached the level that indicates that there should be a moratorium on development of tests. As the neuroscience knowledge base expands, new and better tests of the important constructs of neuropsychology will be developed. The recent emergence of a number of memory tests or batteries (e.g., California Verbal Learning Test, Selective Reminding Test, Wechsler Memory Scale Revised) is obvious testimony to the vitality of the field. Simultaneously with the development of new tests, there has been a more careful examination and refinement of existing procedures. Many of the new procedures have been stimulated by current research, but in contrast to the development of many of the established test batteries, are being subjected to careful psychometric validation and study. Although the heritage of some of our measures may be as "standardized individual experiments," recent trends suggest an increasing awareness of the value of more careful studies of reliability and validity. These studies will result in a better understanding of what the tests measure and will, therefore, have direct impact on our conceptualizations regarding the fundamental nature of patients' problems. Many of the most sensitive tests are effective because of their complex nature. Further improvements in the understanding of the specifics of how (as well as how badly) patients may fail on particular tests will increase the ability to make appropriate diagnostic interpretations as well as effective remedial interventions.

References

Adams K. M. (1980a) In search of Luria's battery: a false start. *J. Cons. Clin. Psychol.* **48,** 511–516.

Adams K. M. (1980b) An end of innocence for behavioral neurology? *J. Cons. Clin. Psychol.* **48,** 522–524.

Adams K. M. (1987) Reitan's neuropsychology: take it or leave it. *J. Clin. Exp. Neuropsychol.* **9,** 235–242.

Adams R. L. (1985) Review of the Luria-Nebraska Neuropsychological Battery, in *Ninth Mental Measurements Yearbook* (Mitchell J. V., ed.), University of Nebraska Press, Lincoln, Nebraska, pp. 878–881.

Beardsley J. V., Matthews C. G., Cleeland C. S., and Harley J. P. (1978a) *Experimental t-Score Norms for C. A. 34– on the Wisconsin Neuropsychology Test Battery,* private publication.

Beardsley J. V., Matthews C. G., Cleeland C. S., and Harley J. P. (1978b) *Experimental t-Score Norms for C. A. 35+ on the Wisconsin Neuropsychology Test Battery,* private publication.

Benton A. L., Hamsher K. deS., Varney N. R., and Spreen O. (1983) *Contributions to Neuropsychological Assessment: A Clinical Manual* (Oxford, New York).

Bornstein R. A. (1982) Reliability of the Speech Sounds Perception Test. *Percept. Mot. Skill* **55,** 203–210.

Bornstein R. A. (1983) Reliability and item analysis of the Seashore Rhythm Test. *Percept. Mot. Skill* **57,** 571–574.

Bornstein R. A. (1985) Normative data on selected neuropsychological measures from a nonclinical sample. *J. Clin. Psychol.* **41,** 651–659.

Bornstein R. A. (1987) Preliminary data on classification of normal and brain-damaged elderly subjects. *The Clin. Neuropsychol.* **1,** 315–323.

Bornstein R. A. and Leason M. (1984) Item analysis of Halstead's Speech Sounds Perception Test: quantitative and qualitative analysis of errors. *J. Clin. Neuropsychol.* **6,** 205–214.

Bornstein R. A., Baker G. B., and Douglass A. B. (1987) Short term retest reliability of the Halstead-Reitan Battery in a normal sample. *J. Nerv. Ment. Dis.* **175,** 229–232.

Bornstein R. A., Weizel M., and Grant C. (1984) Error pattern and item order on Halstead's Speech Sounds Perception Test. *J. Clin. Psychol.* **40,** 266–270.

Brandt J. and Doyle L. F. (1983) Concept attainment, tracking, and shifting in adolescent polydrug abusers. *J. Nerv. Ment. Dis.* **171,** 559–563.

Christensen A. (1975) *Luria's Neuropsychological Investigation* (Spectrum, New York).

Crossen B. and Warren R. L. (1982) Use of the Luria-Nebraska Neuropsychological Battery in aphasia: a conceptual critique. *J. Cons. Clin. Psychol.* **50,** 22–31.

Davison L. A. (1974) Introduction, in *Clinical Neuropsychology: Current Status and Applications* (Reitan R. M. and Davison L. A., eds.), Winston, New York, pp. 1–18.

Delis D. and Kaplan E. (1982) The assessment of aphasia with the Luria-Nebraska Neuropsychological Battery: a case critique. *J. Cons. -lin. Psychol.* **50,** 32–39.

Delis D. C. and Kaplan E. (1983) Hazards of a standardized neuropsychological test with low content validity: comment on the Luria-Nebraska Neuropsychological Battery. *J. Cons. Clin. Psychol.* **51,** 396–398.

Dikmen S. and Morgan S. F. (1980) Neuropsychological factors related to employability and occupational status in persons with epilepsy. *J. Nerv. Ment. Dis.* **168,** 236–240.

Dodrill C. B. and Troupin A. S. (1975) Effects of repeated administrations of a comprehensive neuropsychological battery among chronic epileptics. *J. Nerv. Ment. Dis.* **161,** 185–190.

Eckardt M. J. and Matarazzo J. D. (1981) Test-retest reliability of the Halstead Impairment Index in hospitalized alcoholic and nonalcoholic males with mild to moderate neuropsychological impairment. *J. Clin. Neuropsychol.* **3,** 257–269.

Erickson R. C., Calsyn D. A., and Scheupbach C. S. (1978) Abbreviating the Halstead-Reitan Neuropsychological Test Battery. *J. Clin. Psychol.* **34,** 922–926.

Filskov S. B. and Goldstein S. G. (1974) Diagnostic validity of the Halstead-Reitan Neuropsychological battery. *J. Cons. Clin. Psychol.* **42,** 382–388.

Flourens P. (1824) *Recherches experimentales sur les Proprietes et les Fonctions du Systeme Nerveux dans les Animaux Vertebres* (Crevot, Paris).

Fromm-Auch D. and Yeudall L. T. (1983) Normative data for the Halstead-Reitan neuropsychological tests. *J. Clin. Neuropsychol.* **5,** 221–238.

Golden C. J. and Anderson S. M. (1977) Short form of the Speech Sounds Perception Test. *Percept. Mot. Skill* **45,** 485–486.

Golden C. J. and Maruish M. (1986) The Luria-Nebraska Neuropsychological Battery, in *Clinical Application of Neuropsychological Test Batteries* (Incagnoli T., Goldstein G., and Golden C. J., eds.), Plenum, New York, pp. 193–233.

Golden C. J., Hammeke T. A., and Purisch A. D. (1980) *A Manual for Administration and interpretation of the Luria-Nebraska Neuropsychological Battery* (Western Psychological Services, Los Angeles).

Golstein G. (1986) An overview of similarities and differences between the Halstead-Reitan and Luria-Nebraska Neuropsychological Batteries, in *Clinical Application of Neuropsychological Test Batteries* (Incagnoli T., Goldstein G., and Golden C. J., eds.), Plenum, New York, pp. 235–275.

Goldstein G., Tarter R., Shelly C., and Hegedus A. (1983) The Pittsburgh Initial Neuropsychological Testing System (PINTS): a neuropsychological screening battery for psychiatric patients. *J. Behav. Assess.* **5,** 227–238.

Goldstein K. (1939) *The Organism* (American Book, New York).

Gregory R. J., Paul J. J., and Morrison M. W. (1979) A short form of the Category Test for adults. *J. Clin. Psychol.* **35,** 795–798.

Halstead W. C. (1947) *Brain and Intelligence* (Univ. Chicago Press, Chicago).

Heaton R. K. and Pendleton M. G. (1981) Use of neuropsychological tests to predict adult patients' everyday functioning. *J. Cons. Clin. Psychol.* **49,** 807–821.

Heaton R. K., Chelune G. J., and Lehman R. A. W. (1978) Using neuropsychological and personality tests to assess the likelihood of patient employment. *J. Nerv. Ment. Dis.* **166,** 408–416.

Heaton R. K., Grant I., and Matthews C. G. (1986) Differences in neuropsychological test performance associated with age, education, and sex, in *Neuropsychological Assessment of Neuropsychiatric Disorders* (Grant I. and Adams K. M., eds.), Oxford, New York, pp. 100–120.

Jabbari B., Gunderson C. H., Wippold F., Citrin C., Sherman J., Bartoszek D., Daigh J. D., and Mitchell M. H. (1986) Magnetic Resonance Imaging in partial complex epilepsy. *Arch. Neurol.* **43,** 869–872.

Kane R. L. (1986) Comparison of Halstead-Reitan and Luria-Nebraska Neuropsychological Batteries: Research findings, in *Clinical Application of Neuropsychological Test Batteries* (Incagnoli T., Goldstein G., and Golden C. J., eds.), Plenum, New York. pp. 277–301.

Kilpatrick D. G. (1970) The Halstead Category test of brain dysfunction: feasibility of a short form. *Percept. Mot. Skill* **30,** 577–578.

King M. C. (1984) A conversation with Brenda Milner. *Alberta Psychology* **14,** 6–8.

Klesges R. C., Fisher L., Pheley A., Boschee P., and Vasey M. (1983) A major validational study of the Halstead-Reitan in the prediction of CAT-assessed brain damage in adults. *Int. J. Clin. Neuropsychol.* **6,** 29–34.

Klonoff H., Fibiger C. H., and Hutton G. H. (1970) Neuropsychological patterns in chronic schizophrenia. *J. Nerv. Ment. Dis.* **150,** 291–300.

Kolb B. and Whishaw I. Q. (1983) Performance of schizophrenic patients on tests sensitive to left or right frontal, temporal, or parietal function in neurological patients. *J. Nerv. Ment. Dis.* **171,** 435–443.

Lashley K. S. (1929) *Brain Mechanisms and Intelligence* (Univ. of Chicago Press, Chicago).

Laster D. W., Penry J. K., Moody D. M., Ball M. R., Witkofski R. L., and Riela A. R. (1985) Chronic seizure disorders: contribution of MR imaging when CT is normal. *Amer. J. Neuroradiol.* **6,** 177–180.

Lezak M. (1983) *Neuropsychological Assessment,* second edition (Oxford, New York).

Luria A. R. (1966) *Higher Cortical Functions in Man* (Basic Books, New York).

Luria A. R. (1973) *The Working Brain* (Basic Books, New York).

Luria A. R. and Majovski L. V. (1977) Basic approaches used in american and soviet clinical neuropsychology. *Amer. Psychol.* **32,** 959–968.

Matarazzo J. D., Wiens A. N., Matarazzo R. G., and Goldstein S. G. (1974) Psychometric and clinical test–retest reliability of the Halstead impairment index in a sample of healthy, young, normal men. *J. Nerv. Ment. Dis.* **158,** 37–49.

Meehl, P. E. (1954) *Clinical Versus Statistical Prediction* (Univ. of Minnesota Press, Minneapolis).

Milberg W. P., Hebben N., and Kaplan E. (1986) The Boston process approach to neuropsychological assessment, in *Neuropsychological Assessment of Neuropsychiatric Disorders* (Grant I. and Adams K. M., eds.) Oxford, New York, pp. 65–86.

Milner B. (1975) Psychological aspects of focal epilepsy and its neurosurgical management. *Adv. Neurol.* **8,** 299–321.

Morrison M. W., Gregory R. J., and Paul J. J. (1979) Reliability of the Finger Tapping test and a note on sex differences. *Percept. Mot. Skill* **48,** 139–142.

Newnan O. S., Heaton R. K., and Lehman R. A. W. (1978) Neuropsychological and MMPI characteristics of patients future employment characteristics. *Percept. Mot. Skill* **46,** 635–642.

Parsons O. A. and Prigatano G. P. (1978) Methodological considerations in clinical neuropsychological research. *J. Cons. Clin. Psychol.* **46,** 608–619.

Pauker J. D. (1980) Norms for the Halstead-Reitan Neuropsychological Test Battery based on a nonclinical adult sample. Paper presented at the *Canadian Psychological Association Meeting,* Calgary Alberta.

Price L. J., Fein G., and Feinberg I. (1980) Neuropsychological assessment of cognitive functioning in the elderly, in *Aging in the 1980s: Psychological Issues* (Poon L. W., ed.), Amer. Psychol. Assoc, Washington, D.C.

Reitan R. M. and Wolfson D. (1985) *The Halstead-Reitan Neuropsychological Test Battery Theory and Clinical Interpretation* (Neuropsychology Press, Tucson).

Robinson A. L., Heaton R. K., Lehman R. A. W., and Stilson D. W. (1980) The utility of the Wisconsin Card Sorting Test in detecting and localizing frontal lobe lesions. *J. Cons. Clin. Psychol.* **48,** 605–614.

Russell E. W., Neuringer C., and Goldstein G. (1970) *Assessment of Brain Damage: A Neuropsychological Key Approach* (Wiley, New York).

Ryan J. J., Larsen J., and Prifitera A. (1982) Short form of the Speech Sounds Perception Test: further considerations. *Clin. Neuropsychol.* **4,** 97–98.

Satz P. and Fletcher J. M. (1981) Emergent trends in neuropsychology: an overview. *J. Cons. Clin. Psychol.* **49,** 851–865.

Shaw D. J. (1966) The reliability and validity of the Halstead Category Test. *J. Clin. Psychol.* **22,** 176–180.

Snow W. G. (1981) A comparison of frequency of abnormal results in neuropsychological vs. neurodiagnostic procedures. *J. Clin. Psychol.* **37,** 22–28.

Spiers P. A. (1981) Have they come to praise Luria or to bury him?: the Luria-Nebraska Battery controversy. *J. Cons. Clin. Psychol.* **49,** 331–341.

Spiers P. A. (1982) The Luria-Nebraska Neuropsychological Battery revisited: a theory in practice or just practicing? *J. Cons. Clin. Psychol.* **50,** 301–306.

Spiers P. A. (1984) What more can I say? In reply to Hutchinson, one last comment from Spiers. *J. Cons. Clin. Psychol.* **52,** 546–552.

Spreen O. and Tuokko A. T. (1982) The Neuropsychological Assessment of Normal and Disordered Cognition, in *Neuropsychology and Cognition,* Vol. I, (Malatesha R. N. and Hartlage L. C., eds.), Martinus Nijhoff, The Hague, pp. 63–112.

Stambrook M. (1983) The Luria-Nebraska Neuropsychological Battery: a promise that may be partly fulfilled. *J. Clin. Neuropsychol.* **5,** 247–269.

Sutherland R. J., Kolb B., Schoel W. M., Whishaw I. Q., and Davies D. (1982) Neuropsychological assessment of children and adults with Tourette Syndrome: A comparison with learning disabilities and schizophrenia, in *Advances in Neurology,* vol. 35, *Gilles de la Tourette Syndrome* (Friedhoff A. J. and Chase T. N., eds.), Raven, New York, pp. 311–322.

Swiercinsky D. P. and Leigh G. (1979) Comparison of neuropsychological data in the diagnosis of brain impairment with computerized tomography and other neurological procedures. *J. Clin. Psychol.* **35,** 242–246.

Tarter R. E. and Edwards K. L. (1986) Neuropsychological batteries, in *Clinical Application of Neuropsychological Test Batteries* (Incagnoli T., Goldstein G., and Golden C. J., eds.), Plenum, New York, pp. 135–153.

Tsushima W. T. and Wedding D. (1979) A comparison of the Halstead-Reitan Neuropsychological Test Battery and computerized tomography in the identification of brain disorder. *J. Nerv. Ment. Dis.* **167,** 704–707.

From: *Neuromethods, Vol. 17: Neuropsychology* Edited by: A. A. Boulton, G. B. Baker, and M. Hiscock

Developmental Neuropsychological Assessment

The Systemic Approach

Jane Holmes-Bernstein and Deborah P. Waber

1. Introduction

Historically, the clinical discipline of neuropsychology evolved primarily for the purpose of evaluating the effects of brain damage in adults. Application of neuropsychological theory, techniques, and methods to the clinical evaluation and management of children, who may or may not have documentable neurological disorder, is a recent, relatively less well developed phenomenon.

As in adult neuropsychology, there exists no consensus as to how the assessment of children should be carried out, what the best measurement techniques are, or even what a particular set of findings means. The relative merit of various approaches to the neuropsychological assessment of children is and will continue to be the subject of lively debate for some time; indeed, this is a healthy and entirely normal state for a new discipline. To the clinician or student of the neurosciences attempting to gain an understanding of this field, however, the variety of opinions and approaches, how they converge, and where they diverge can be overwhelming.

Given this state of affairs, we were faced with two basic choices in formulating this chapter: we could survey the methods currently available to the practicing clinician and point out similarities and differences among them, or we could focus exclusively on our own approach and describe the assessment process as we apply it in our own practice. Simply put, we chose the latter. What follows in this chapter, therefore, is not a description of the various methods and approaches that are currently in use, or even those that are most widely accepted and prevalent. Rather, we highlight issues that we believe should be central considerations for those who perform neuropsychological assessment of children, pose questions that are most significant in this endeavor, and present the approach that we believe to be most valid.

Our presentation is therefore biased and should be viewed by the reader as such. With the test of time, our approach may be proven wrong, while others acquire increasing validation and utility. The reader is encouraged to review other approaches: the functional organization approach (Fletcher and Taylor, 1984), the battery approach (Golden, 1981; Reitan and Davison, 1974), the treatment-oriented approach (Rourke, et al., 1986), and the quantitative, hierarchical approach (Wilson and Risucci, 1986).

The approach to be presented here emphasizes the dynamic interplay of neural and behavioral systems in development. It builds on the systemic psychology of Anokhin, Vygotsky, and Luria (Luria, 1973) and extends the Wernerian process approach (Werner, 1948) elaborated by Kaplan (1983). The primary goal of the assessment process, within this framework, is not to diagnose deficits in a child, but rather to construct a *Child–World System* that characterizes the reciprocal relationship of the developing children and the world in which that child functions. Construction of the Child–World System highlights areas of "match" and "mismatch" between the child's complement of skills and the demands placed upon them. Effective management strategies address both sides of the Child–World equation, seeking to optimize the match between child and world throughout the course of development.

At the outset, we should make clear that we are presenting an *approach,* not a technique. Hence, we will not prescribe a menu of tests to be administered, nor will we provide specific diagnostic criteria or sets of recommendations. What we wish to convey is how to *think* about the problems presented in the clinical neuropsychological assessment of children and, given the instruments available for evaluation, how to go about solving these problems.

Because we do not espouse specific techniques, some readers will find our presentation frustrating, since it is not the "how to" that one might expect in the context of a volume on methods. Nevertheless, we feel strongly that *principles* endure, whereas specifics are outdated with remarkable speed. Moreover, it is the principles that we believe distinguish our approach from some others and endow it with vitality. The application and development of broad principles and theory, as well as the challenge of problem solving, can perpetuate a sense of continual discovery and learning for the practitioner. Whereas the primary goal of the assessment process is, of course, to effect change for the child, each assessment should serve as well to effect some change, be it large or small, in our understanding of neurobehavioral development.

1.1. *Assessment of Children: The Importance of Development*

Neuropsychological theory is derived primarily from the systematic observation of adults who have sustained lesions to the central nervous system. The bulk of neuropsychological theory assumes a mature nervous system, the endpoint of a complex and dynamic process. When neuropsychological theory is applied to children, although its basic structural aspects may be preserved (e.g., the affinity of language for the left hemisphere, the association of the frontal brain systems with psychological control processes), the context of a developing nervous system mandates a reformulation of the relationship between structure and function.

Given that the properties of the developing nervous system are so different from those of the adult and that, moreover, developmental considerations are so integral to the neuropsychological assessment of the child, direct application of adult methodology to the assessment of children is inappropriate. Neuropsychological assessment of children must be firmly rooted in the context of neurobehavioral development. The appropriate theory must go beyond the (relatively static) structure–function relationships that are emphasized in adult neuropsychology, and incorporate the dynamic interaction between brain and world that is essential to development.

Many issues, central in assessing children, are irrelevant to the assessment of the adult. For example, in assessing the child, the neuropsychologist is concerned not merely with the child's current status, but also how he or she has performed in the past (to the extent documentation is available) and, equally important, how he or she can be predicted to function in the future, based on our knowledge of the parameters of neural and behavioral development. A child may be performing reasonably well in the current environment, yet the assessment can reveal risk for significant difficulties as demands increase commensurate with developmental achievements typically attained later in childhood. In addition, early insults to the brain can be accompanied by dramatically different functional outcomes from those sustained later in life, or the same insult can present with diverse behavioral manifestations that are systematically related to the child's developmental status.

1.2. Models of Development

The models of development that the clinician brings to bear in approaching the assessment have direct implications for the outcome of the evaluation—the nature of the diagnosis, the organization of the report, the prescription for remediation, and, more generally, the understanding on the part of parents, teachers, and the children themselves of the academic and social difficulties that the child is experiencing. Two kinds of models are prevalent currently in the neurobehavioral assessment of children—*rate* models and *state* models.

Rate models emphasize the developmental context of the child's functioning, assuming that capacities inevitably grow and develop, and hence, change. In such a model, learning problems are indicative of developmental lags or delays that will, by implication, ultimately catch up and be righted. The goal of treatment, therefore, is to provide support for the child in identified areas of difficulty, until the normal function develops and can assume its full role.

State models, in contrast, assume an identifiable, relatively static difference that the child is unlikely to outgrow—even though the manifestations can change over time. These can include deficits in attention, in auditory or visual processing, and so forth. Whether the basis for these deficits is an early insult or some aberration in development is relatively unimportant. The aim of treatment in the context of such a model is to remediate and compensate for the identified deficit area.

Research on the long-term outcome of children with learning difficulties provides partial support for each of these models. Yet there are striking inconsistencies as well (Waber, 1989b). The bulk of the evidence indicates that catch-up is generally incomplete, and the child continues to experience problems that may range in severity from subtle to significant (Schonhaut and Satz, 1983). Closer examination of the child's overall behavioral repertoire, moreover, frequently reveals that the difficulties are not limited to the index symptom, but occur as part of broader "clusters" of behavior, as will be described below.

Contemporary research on the processes by which the nervous system develops, and on the impact of early injuries to it, is not entirely consistent with these models. In the mature adult who has sustained brain damage, recovery is essentially functional; that is, behavioral patterns reflect the function of the noncompromised areas of the nervous system (Kaplan, 1976). In the developing brain, in contrast, it is thought that damage to specific neuroanatomical structures early in life can stimulate the development of alternative neural pathways that, presumably, assume the affected function (Goldman-Rakic et al., 1985). These alternative pathways are generally less efficient than those that were damaged; nevertheless, the behavioral outcome can resemble normal function to a greater extent than that seen in adults with similar damage. Further, the nature of these alternative pathways is determined not only by the locus of the damage, but also by maturational status at the time of the insult. Thus, early damage can, in theory, stimulate the development of a variety of alternative pathways by which the same behavioral goal is accomplished. However, formation of these alternative pathways is likely to affect not only the target function, but also those for which the compensating structures were originally intended; for instance, mediation of language by right-hemisphere mechanisms after damage to the left is associated with compromise in visuospatial reasoning (Teuber and Rudel, 1962). The notion, therefore, that isolated capacities can develop or sustain damage without affecting CNS function in a more systemic way is fallacious. Nor is it reasonable to assume that children with a similar index symptom (e.g., reading difficulty) fail because of an identical physiological or cognitive mechanism.

The implications of this assertion for the clinician are by no means of purely academic interest. First, functions do not exist in isolation from one another. (Therefore, defining discrete deficits e.g., auditory discrimination or phonetic decoding, should not be the goal of evaluation.) If there has been early damage or alteration in the normal developmental course, the result is most likely the formation of alternative, less efficient pathways by which the same goal is accomplished. The existence of these alternative pathways can be expected to affect performance not only in the target domain (e.g., reading or language), but also in a variety of other behaviors that can be manifested in the learning, as well as the social, environment.

Second, in evaluating a child, it is essential to observe not only the adequacy of the product, but how he or she goes about accomplishing the task. It is only by observing the routes to success or failure on a specific task that the alternative pathways available to the child can be diagnosed and remediation determined. Remediation itself, moreover, is likely to be more effective when directed at strengthening the alternative pathways that are naturally available to the child, rather than training function according to the more usual route.

It is principles such as these, which follow directly from an appreciation of the processes by which the nervous system develops and responds to insult, that guide the systemic approach to assessment, in both its evaluative and management components. As the developmental processes that characterize the mammalian nervous system, and more specifically the human nervous system, come to be better understood, the conceptual underpinnings of this approach can be expected to change.

2. Developmental Neuropsychology: The Systemic Approach to Assessment

2.1. *The Role of Theory in Assessment*

Central to our approach is the belief that a coherent theory of cognitive and neural development is a necessary basis for assessment. Too often, in practice, the assessment procedure is not adequately guided by such a theory. Two kinds of approaches are currently dominant. One consists essentially of application to children of theory and techniques developed for the adult neurological population. Although the neuropsychological theory that is derived from adult populations is of considerable value, it does not provide an adequate basis for understanding the unique properties of the developing nervous system. The other type of approach consists of application of psychometric instruments and methods to pediatric neurological populations. The strength of this approach is that it can relate the child's behavior in an empirically valid way to that of other same-age children. Serious limitations of the psychometric approach to behavioral measurement, however, are its failure to model adequately the underlying neural substrate, to situate the behavior in the context of the specific characteristics of childhood neural pathology, and to extend our understanding of neurobehavioral development.

From our perspective, definition of a truly developmental neuropsychology entails reformulation of both of its components—"neuro" and "psychology"—for children. As discussed above, redefining the "neuro" component involves distinguishing the model of mature brain function that pertains to the adult from that of the developing system seen in childhood. There is no question that many of the characteristics of CNS function seen in the adult also appear in childhood, but the developmental considerations that must be primary in the evaluation of children are far less salient in the adult population.

On the "psychology" side, the clinical assessment of cognitive function has traditionally drawn primarily on psychometric models, comparing individual performance to normal expectation, generating profiles based on probability of occurrence, and discriminating affected from nonaffected individuals in an actuarial fashion. More congenial with the neurological basis for behavior presumed by neuropsychology, however, are structural models of cognition, drawing on European traditions as exemplified by Piaget (1985) and Merleau-Ponty (1962); these place individual domains of behavior in the context of a coherent, structural whole. This whole is also understood as highly dynamic, a perturbation in one component having extensive, far-reaching, and ultimately predictable consequences for the structure as a whole, as postulated by General Systems Theory (von Bertallanfy, 1969).

Thus, as was discussed above in relation to the CNS, alterations in specific aspects of development can be expected to affect behavior in a variety of domains, given the functional coherence of the nervous system. A structurally based theory provides a basis for characterizing development in a nonlinear fashion that better approximates phenomena seen in the developing nervous system than does an essentially linear and factorially based psychometric one. Disequilibrium induced by a single, relatively discrete change can trigger the emergence of cognitive structures that differ qualitatively from those that preceded them. Furthermore, such a model provides for a view of the child in the context of his or her entire repertoire of behaviors, not only cognitive and academic, but social and emotional as well (Vygotsky, 1978). It is more in accord with the view of the brain as a dynamic processor than is a psychometric theory. Finally, it encourages a detailed, phenomenological assessment of the processes entailed in the achievement of a final product (Werner, 1937, 1948).

2.2. Developmental Neuropsychology: The Model

In developing a theoretical formulation to guide the assessment process and, thus, the construction of the Child–World System, we have generated a model that has a set of specific components on the "neuro" and "psychology" sides. Each component on one side has its parallel or complement on the other. The reasons for our emphasis on theory will become apparent as the assessment process itself is described in greater detail, for the method stems in large part from the theoretical groundwork.

2.2.1. "Neuro"

2.2.1.1. THREE AXES. The functional parameters of the CNS are conceptualized in terms of its three primary axes, anterior–posterior, left–right, and cortical–subcortical; behavioral output is viewed as a function of the dynamic interaction among these axes. (The three main axes outlined above constitute an initial organization of the neural substrate. In theory, there is no reason why future exploration of the functional organization of the nervous system should not reveal other axes.)

The model, which is based on general characteristics of nervous system function, is essentially heuristic. Although we adhere to the three-axis basis for the model, elaboration is guided by the individual clinician's familiarity with the clinical and research literature in neuropsychology. Others may read the literature differently than we do, and arrive at a model that differs from the one presented here in terms of specific functional correlates of neuroanatomic structures. What is essential, however, is the notion of the dynamic interplay among the three axes.

The first of these is the *anterior–posterior* axis (Stuss and Benson, 1986). In simple terms, the anterior structures of the cerebral cortex (i.e., those extending from the motor strip forward to the frontal pole) are associated with executive functions. These include not only motor output, but also the "control processes" of cognitive psychology—organization, planning, and strategic behavior, as well as directed attention. Posterior structures (i.e., those extending from the sensory strip back to the occipital pole) are associated with input functions. These include not only sensory processes, but also the structure and organization of the knowledge base, semantic structure, organization of sensory input, perception of the structure of complex wholes, and so forth.

The second is the *lateral* axis. The left and right cerebral hemispheres have been variously characterized as verbal/nonverbal, sequential/parallel, and analytic/holistic, respectively. For our purposes, however, the lateral axis is viewed as functionally orthogonal to the anterior–posterior axis. Thus, distinctions between left and right concentrate not so much on process as on the specific characteristics of the stimulus input (Sergent, 1983). From a functional standpoint, the left hemisphere has an affinity for linear processing and for the fine-grained detail inherent in complex material, whereas the right has an affinity for more global and relational aspects, a functional distinction that is consistent with the patterns of neural connectivity of the respective hemispheres (Goldberg and Costa, 1981). A more detailed discussion of a hypothetical model by which the interaction between the anterior–posterior and lateral axes can be conceptualized is found in Waber (1989a).

The third is the *cortical–subcortical* axis. Cortical structures subserve higher-order behavioral functions; subcortical structures subserve vital functions, motor organization, sensory relay, attention and arousal, and the modulation of emotions and drives. From an ontogenetic standpoint, this axis parallels the neural tube, with later-developing telencephalic structures distinguished from the more primitive diencephalon and rhombencephalon.

All behavior is a function of the dynamic interaction among these three axes. For example, learning is never independent of motivation; cognitive style can be significantly modified by changes in control processes; higher-order planning skills can be compromised as a consequence of primary motor disorder.

2.2.1.2. Development Timetables. Development is not uniform across the nervous system. Cell birth, neuronal migration, synaptic generation, and "pruning" all proceed according to individual, genetically determined programs. Different regions of the brain reach maturity in a programmed sequence, changes in which can have profound implications for subsequent development (Rakic and Goldman-Rakic, 1982). There is, therefore, no basis for viewing neurological maturation as a linear, relatively undifferentiated process.

The functional consequences of these developmental timetables are largely uncharted. To the extent that knowledge is available, it pertains largely to nonhuman, especially rodent, models. What is clear from these studies, however, is that interference with these timetables has specific behavioral effects that depend upon the timing and locus of the insult. Furthermore, an apparently benign insult, having little immediate effect on behavioral function, can have a more significant impact later, as the nervous system matures and the loss of function assumes greater consequence (Goldman, 1974).

The latter point is particularly important in understanding learning difficulties. Different kinds of problems are likely to present themselves in children at different points in development, largely because it is only at those particular points that the damaged system is sufficiently challenged by environmental demands for the impact of the earlier damage or alteration in function to be of consequence (Holmes, 1986). Similarly, relatively minor characteristics of performance in a young child, which are of little consequence in terms of current functioning, can nevertheless place him or her at considerable risk, given the developmental tasks ahead.

2.2.1.3. Alternative Pathways. The developing nervous system is preeminently dynamic. As a consequence, early insults typically result in developmental alterations in the context of the overall genetic program, rather than in discrete "holes," as may be seen in the adult. Damage to a particular population of cells affects not only those specific cells, but also trophic processes that were meant to project to or from the cells. Absent their recipients, these projections will seek other, aberrant endpoints, leading to the creation of atypical circuitry. Early injuries can also result in structural abnormalities, such as malformed gyri and sulci. Finally, injuries can affect normal processes of "pruning" or refinement that are thought to render neural function more efficient.

Since these phenomena do not occur in the same way in the adult, the effects of early and late brain damage can differ significantly, despite gross similarities. For example, a cerebellar lesion that renders an adult nonambulatory can be relatively well-tolerated in a child. The child, however, may exhibit accompanying alterations, albeit mild, in multiple behavioral domains, whereas the adult's deficit is relatively restricted.

2.2.1.4. Role of Experience. Environmental stimulation is an essential requirement of the developing nervous system. Lack of appropriate experience can significantly alter the course of neural development (Hubel and Wiesel, 1962). The animal's ability to interact with the environment in the course of experience is equally critical to normal neuronal function (Held and Hein, 1963). In humans, disruption in normal experiences can have devastating functional consequences. Exposure to toxic agents; nutritional deficiency; social, emotional, and/or cognitive deprivation; and sensory or motor limitations are all potential sources of disruption of normal CNS development.

Less well understood is the impact of targeted experience on the development of the damaged nervous system. There is some indication, however, that appropriate intervention, begun early enough in life, can have a significant impact in terms of normalizing later development in a child who has sustained early brain damage (Katona, 1989). (This targeted experience should be distinguished from therapeutic attempts to recapitulate development.)

2.2.2. "Psychology"

The second part of the equation, psychology, is conceptualized so as to be compatible with the parameters of the first.

2.2.2.1. Cognitive Structures. The cognitive processes associated with the three neuroanatomical axes must be viewed in an equally dynamic fashion. Abilities or functions cannot be fully understood in isolation, but must be viewed as dynamically interactive with one another, both developmentally and at any single point. However, the preference for a structurally based theory does not, and cannot, exclude the use of traditional, psychometrically developed tests. Using these tests to compare the performance of an individual child to that of same-age peers is essential, not only for placing the child in the developmental context, but also as a means of conveying to others the child's relative functional status. The final score, however, is only one component of the total body of information yielded by the test. Careful observation of attitudes, strategies, errors, and associated behaviors also constitute essential data for analysis.

In the context of a structural theory, diagnosis proceeds by means of theory-based commonalities that transcend content areas and modalities, and provide insight into the structure of behavior as a function of the three axes. For example, a child whose left-hemisphere function is relatively inefficient compared to his or her right would be predicted to exhibit not only disordered language function, but also particular kinds of errors and strategic approaches in the visuospatial domain. These content-independent or cross-modality observations provide convergent validation for the diagnostic formulation generated within the context of the three-axis model.

2.2.2.2. Developmental Timetables. As is the case for neural development, cognitive development is not an undifferentiated maturational process, nor do individual functions develop in a linear fashion. Functions develop according to a timetable, with different functions undergoing more dramatic development at different times. Of necessity, these developmental advances affect a wide variety of other behaviors in a dynamic way, such that there are repeated structural reorganizations that build one upon another. Although there are behavioral phenomena that are tightly linked to the development of the nervous system, the more prominent changes in behavior are most likely epiphenomena of the neurobehavioral phenomena. These epiphenomena result from the child's interaction with his or her environment, given newly acquired capabilities rendered possible by neural development.

The neuropsychologist works in the context of these developmental timetables. Often, a complex skill fails to develop in the appropriate fashion because a prior function, upon which that skill builds, is not available. One task of the neuropsychologist is to define more precisely the source of the inefficiency, based on examination of not just the complex skill itself, but also other cognitive and neural correlates that could provide clues or converging evidence implicating a particular developmental event. Reading is a primary example. It is a complex skill that undoubtedly requires maturation of certain basic cognitive processes in order to progress adequately. Failure to acquire reading skills according to expectation, therefore, can stem from any one of a number of primary causes; specifying the source is prerequisite to effective intervention.

At the early grade levels, many children have difficulty with the mechanics of reading (establishing sound–symbol associations, integrating phonemic units, or segmenting). Other children, however, progress adequately in the acquisition of fundamental decoding skills, but fail to establish automaticity or generate organizational structures within which to encode meaning as the task requires, resulting in comprehension deficits (Lovett, 1987).

These developmentally based observations can be used to formulate predictions. For example, a young child who is making adequate progress in basic reading, but exhibits poor organizational skills in the context of more complex material (nonverbal as well as verbal) can be predicted to be at risk for difficulties later, as the demands for reading comprehension skills (as opposed to more basic decoding) increase.

2.2.2.3. ALTERNATIVE STRATEGIES. The qualitative nature of the behavior exhibited by children with learning problems is fundamental to the assessment process. In the context of a deficit model, the unit of analysis is the function itself, and the metric is relative to expectation. Hence, a child can be diagnosed as having a deficit in auditory processing, and the treatment plan is then focused on the deficit and on training the child to perform like his or her unaffected peers.

Although this approach may be effective in the remediation of certain specific skills, its application to more complex cognitive processes in a developmental perspective is questionable. Learning-disordered children do not exhibit the dramatic lacunae seen in brain damaged adults, but more often show a "pastel" version of the symptomology (Denckla, 1979). Rather than being incapable, they are inefficient. Close examination often reveals that the inefficiency stems from application of a less than optimal strategy that can increase processing time or lead to certain types of errors in the context of performance. The critical question then becomes whether to train the optimal stragegy (e.g., that which unaffected children typically use), or to reinforce and increase the efficiency of the alternative strategies that presumably reflect a functional or structural alteration of nervous-system development that is firmly in place.

A related question is whether attention should be directed to the inefficient system or whether efforts should be made to bypass it. Graphomotor output problems are a good example. When graphomotor skills are unusually labored (i.e., inefficient), performance of other tasks of which graphomotor skill is a component (e.g., composition) can deteriorate, since the child is forced to invest excessive effort in a skill that should be automatic, and thus, cannot focus adequately on the substance of the task. In such a case, targeted remediation to aid in the development of more fluid graphomotor output should be only one component of a treatment plan that must also include strategies for bypassing the motor system in order to develop (arguably more important) skills, such as linguistic formulation, narrative composition, or mathematics concepts.

2.2.2.4. Interaction of the Individual with the Environment. Nervous system function must be understood in the context of the environment, both the experiences that have been available to the child as well as those he or she has chosen to experience. The role of the environment must be understood at a variety of levels. First, there are the aspects of the environment that affect neural function directly, as discussed above. These can have very specific behavioral consequences, such as the effects of toxic agents on cognitive function. More complex is the nature of the stimulation to which the child is exposed. For example, children from different socioeconomic backgrounds perform differently on tasks designed to measure neuropsychological function (Waber et al., 1984). Different experiences apparently can induce children to process information differently at the most basic levels.

Educational environment is also salient. Given a particular neuropsychological profile, how well has the child been serviced in the past? A child who has yet to receive appropriate services should be regarded differently in terms of potential for achievement from one who has already received optimal teaching. Related to this is the child's self-concept in the context of his or her home and school environment, and how the child regards himself or herself as a learner. A child who experiences repeated failure will have a different motivational structure from one who has had greater opportunity for success and mastery. Home environment is of equal importance. Not only are such obvious characteristics as structure, organization, and language spoken important, but so is the family's attitude toward school and learning. The effects of these forces are by no means straightforward. A family's investment in academic learning can, on the one hand, enrich the child's exposure to relevant information, but on the other, as the primary focus of parental investment, may increase anxiety levels and have a negative impact on self-esteem.

Finally, a supportive environment can help children to develop effective compensatory strategies that may well result in appropriate levels of performance on formal tests. However, the com-

The developmental neuropsychological model

NEUROLOGY		PSYCHOLOGY
	STRUCTURE	
Three neuroanatomic axes		Cognitive structures
	DEVELOPMENT	
Developmental timetables		Developmental timetables
	PROCESSES	
Alternative pathways		Alternative strategies
	CONTEXT	
Role of experience		Environmental interaction

Fig. 1. The developmental neuropsychological model.

pensation process can itself be stressful for a child. The possibility of such stress, as manifest by fatigue, inability to complete assignments, or loss of motivation, needs to be addressed and articulated, so that more effective management and support can be offered by both family and teachers.

2.2.3. The Model

To recapitulate the above discussion, the neurodevelopmental theory within which we are working is modeled in terms of the interaction between neurological and psychological processes as the child matures and encounters his or her environment. As illustrated schematically in Fig. 1, the model posits a dynamic system incorporating four essential components: structure, development, alternative mechanisms, and context. Each component has its complement in both the neurological and psychological domains.

2.3. Implications of the Model

2.3.1. Evaluation/Diagnosis

2.3.1.1. Diagnostic Behavioral Clusters. The diversity of observations obtained in the clinical assessment is unified by the systemic neuropsychological model as it encompasses the three neuroanatomic axes and the structural approach to cognition. For example, right-hemisphere function can be linked not only to visuospatial performance, but also to social and affective cognition. Frontal brain systems are involved not only in both higher-order reasoning skills and the modulation of attention and arousal, but also in emotional and affective responses. Specific kinds of disturbances or patterns in these apparently disparate areas can be most parsimoniously understood as reflecting inefficient function in a single substrate (right cerebral hemisphere, frontal brain systems). The systemic approach permits conceptualization of these diverse areas as dynamically related to one another; it provides greater explanatory power than models based on discrete functions.

The critical unit of analysis for the diagnostic process cannot be the single behavior or function. Behaviors are never regarded in isolation, but rather, must be viewed in the context of clusters. The data for the diagnostic clusters come not only from standardized test scores, but also from a range of observations including, but not limited to, social interactional characteristics, problem-solving strategies, and historic information.

The experimental and clinical literature in neuropsychology provides the basis for cleavage of the behavioral domain into diagnostic clusters, which are increasingly referred to more formally under the rubric of "modularity" (Fodor, 1983). Modules of behavior are considered to be functionally coherent units in the context of overall neurobehavioral competence.

The relevant behaviors should be regarded as members of a "fuzzy" set. That is, the lines of demarcation as to what constitutes a cluster will not be rigid, and indeed, must vary as a function of age, experience, and/or neurological integrity. The behaviors themselves are best related to a working paradigm in the mind of the clinician, rather than to a discrete population of neurons or a strictly defined category of behavior. Because these are fuzzy sets, all the members of the set need not be present for the formulation to hold. Rather, the presence of a sufficient number of observations congruent with a specific neuropsychological construct is required to cross-validate hypotheses for interpreting single behaviors. Interpretations can shift dramatically, depending upon the context within which the behavior occurs.

2.3.1.2. DEVELOPMENTAL CONTEXT. The developmental context, including both the properties of the developing nervous system and the landmarks of cognitive development, must be evaluated critically. A particular behavioral pattern at one point in development can carry very different meaning from an apparently similar one at another. Thus, the data gathered need to be evaluated in terms of the developmental level of the nervous system, prior developmental accomplishments, and later expectations; they cannot simply be referred to age-referenced norms. The way in which a psychometrically normal score was achieved can be more important than the score itself as an indicator of future vulnerability, because of expectations for further development and contextual demands.

2.3.1.3. NATURE OF THE PROBLEM-SOLVING PROCESS. The developing nervous system, when exposed to an insult or developmental aberration, can generate alternative pathways that manifest themselves as inefficiencies or atypical strategies. Consequently, low scores on a particular test may not be indicative of a discrete deficit, but may reflect the working of an alternative mechanism that is relatively less efficient than the one that is normally programmed by the genome. The task of the neuropsychologist, therefore, is not simply to highlight areas of inefficiency, but to distinguish, where possible, the alternative mechanism used by the child and to pinpoint the source of the problem. This requires attention to strategies and component processes that can only be derived from careful observation of the child as he or she is performing the task. It is rare, moreover, that the impact of these alternative mechanisms is limited to isolated functional systems; again, converging evidence mut be elicited across domains.

2.3.1.4. CONTEXT SENSITIVITY. Given the sensitivity of the nervous system to experiential variation and the impact of environmental influences on adaptation, context variables must be examined carefully. The role of context should be considered at the level of the micro- and macroenvironment. The microenvironment is conceptualized in terms of specific task parameters and their relation to underlying skills. What kinds of presentations facilitate or interfere with performances, and how can these be specified? At the macroenvironmental level are factors like medical history, social expectations, parental attitudes and home experience, educational setting, exposure to toxic substances, and so forth.

2.3.2. Management

2.3.2.1. PREDICTIVE POWER. Where management is concerned, one of the most important characteristics of the systemic neuropsychological model proposed here lies in its predictive power, which extends far beyond that provided by psychometric approaches. In the context of a psychometric model, prediction of behavior is limited to a targeted domain. The systemic approach provides for prediction between behavioral domains as well as across developmental (in contrast to linear) time. Thus, areas of risk can be identified on the basis of a prior knowledge of future demands to which the child will be exposed, taken together with the structure of his or her competencies at the present time. For example, a child may exhibit no unusual difficulties in the area of mathematics at time 1, given the content involved, the method of teaching, the child's approach, and the compensatory strategies available. The point at which curricular demands are likely to overwhelm him or her, however, can be predicted with reasonable certainty—in response to independently predictable changes in the systemic relationship between child and environment. Preparation for stormy seas at some later date can alleviate social and emotional stresses that accompany unexplained failure, and can relieve the child of the burden to achieve at a level predicted on the basis of test scores alone.

3. The Context

Before undertaking the more detailed discussion of methodology, a brief overview of the clinical context within which the neuropsychological assessment of children occurs—the range of problems encountered in the pediatric population, and the primary questions to be addressed—is in order.

3.1. Populations

At the outset, a definition of terms is needed. Specifically, the distinction between learning disabilities and what we shall refer to as neuropsychologically based learning disorders must be highlighted. Strictly speaking, according to the programmatic definition applied by the federal government of the United States, a learning disability is defined as a selective depression of skills involving *symbolic representation* (e.g., reading, spelling, writing, or math) relative to the child's overall level of intellectual functioning. This is a strict limitation and, as one might gather from the preceding discussion, one that we feel does not do justice to the phenomenology and range of the problems encountered by children in the school setting. These kinds of problems, although often most acute in the context of written language demands, can manifest themselves in a variety of domains, for the reasons outlined above. Hence, focusing on written language or math skills alone can provide a distorted picture.

We prefer the term *neuropsychologically based learning disorder* to learning disability. A neuropsychologically based learning disorder is a failure to adapt successfully to the learning environment, and is best understood in the context of a developmental neuropsychological theory. Assuming a neurobehavioral basis, however, by no means implies brain damage. Rather, we proceed from the assumption that all behavior reflects CNS function, and thus, organize our investigations on this biological basis. From a diagnostic standpoint, learning disorders in this sense can be assigned to two main categories, those that have a documented neurological basis and those that do not.

3.1.1. Learning Disorders with Documented Neurology

In children, a finding of a static or *nonprogressive lesion* in the form of cerebral palsy or other congenital (prenatal or perinatal) insult is common. Although the lesion itself is static, it is important to bear in mind that it exists in the context of a developing organism, and the manifestations will depend not only on locus and severity, but also on the developmental context.

Progressive or *dynamic lesions* include tumors and seizure disorder. In this case, the pathology may well not be present or identifiable at birth and can, in any case, undergo change during childhood. Although tumors can be expected to affect behavior directly, it is equally important to recognize that the interventions required to treat the tumor, typically surgery and radiation, themselves cause damage to the CNS and have their own behavioral sequelae. Seizure disorders are highly dependent on the developmental state of the child: in most cases they disappear with maturation, but in some, they become progressively worse. Further, the behavioral symptoms associated with the seizure disorder can themselves change in the context of development.

The range of behavioral competence associated with *genetic disorders* is broad. Chromosomal abnormalities are frequently accompanied by mental retardation—a major exception to this observation being abnormalities of the sex chromosomes (e.g., Turner syndrome, Klinefelter syndrome) that are more often manifest as learning disorders in the context of normal IQ. There is also documentation of familial syndromes, usually involving reading and language problems, the genetics of which cannot be detected on the basis of karyotyping, which images only the gross anatomy of the chromosomes. Recombinant DNA technology, coupled with careful behavioral measurement, is, however, providing evidence of the linkage of specific learning disorders to particular gene loci (Smith et al., 1983).

Metabolic and hormonal disorders are also associated with learning problems. Phenylketonuria, or PKU, for example, is a major cause of mental retardation that is now under good control with the advent of neonatal screening and subsequent dietary therapy. Nevertheless, PKU remains a cause of learning difficulties in incompletely treated infants, as well as in the offspring of successfully treated mothers. Another example is Wilson's disease, a disorder of copper metabolism that can present with characteristic behavioral manifestations until brought under medical control. Congenital hypothyroidism and congenital adrenal hyperplasia are instances of hormonal disorders that can be accompanied by learning problems later in life. Finally, an increasingly common problem is *iatrogenic lesions*. For example, children successfully treated for leukemia with radiation and chemotherapy targeted to the nervous system frequently present with learning difficulties of varying degrees of severity later in life.

3.1.2. Learning Disorders without Documented Neurology

Perhaps the largest population in any pediatric neuropsychology clinic is children with what may be called *functional learning disorders*. These children have no documented neurological pathology (at least, prior to evaluation) and typically present with difficulties in symbol representation, attentional processes, organizational skills, social cognition, or, most typically, some combination of these.

The etiology of these disorders, if they are indeed disorders, is unclear. In some cases, they are simply representative of the spectrum of individual differences or normal variation in the population. The particular complement of skills available to these individuals is not necessarily abnormal, but rather, is not optimally adapted to the demands of the school setting. For others, there is presumably an unknown physiological cause. For instance, a significant subset of these children respond favorably to stimulant medication, suggesting that some biological condition predisposes them to the maladaptive behavior. Others may represent a failure of cell migration, an undetected lesion occurring in the pre- or perinatal period, or a familial syndrome. In some cases, the behavioral symptomatology leads to further neurological diagnostic procedures that can, in fact, establish a basis for the disorder (e.g., asymmetric motor findings, CT scan abnormalities, EEG findings). In most cases, however, the problem is best understood as a functional learning disorder.

Yet another population of children have *limited intellectual ability* without neurological findings. The etiology of the limitations is often as unclear for these children as it is for learning disordered children of higher ability levels. Even at the lower levels, however, it is generally possible to apply neuropsychological principles of behavioral clustering and delineate a pattern of strengths and weaknesses.

For children presenting with *emotional symptomatology*, a primary role of the neuropsychologist is to determine whether there is a neurobehavioral component to the disturbance. If the child is experiencing difficulties achieving in the school setting, it is important to establish whether this stems entirely from the interference of emotional issues, or whether there is an additional, constitutionally based problem. Careful delineation of the ways in which the neuropsychological components of the disorder interact with the emotional problems should be part of the description of the system.

Finally, *pervasive developmental disorder* characterizes a diverse group of children who can manifest autistic behaviors, but who do not demonstrate the full-blown syndrome of autism. They frequently exhibit unusual neuropsychological profiles, with isolated strengths embedded in a generally low level of cognitive function, pervasive language impairment, and poor social perception and interaction. One of the best studied of these disorders is hyperlexia. These children can decode printed words beyond their general level of language and cognitive functioning. Although they can decode words accurately, however, they are usually "word callers" who are unable to read for meaning.

3.2. Questions

Typically, the neuropsychologist is asked two categories of questions. The first is general and applies whether or not there is an identified source of neurological impairment. The second is more specific to children with a specific neurological or medical condition.

3.2.1. General

Subsumed in this category are questions of the following types:

Establish overall level of ability.
Define the cognitive strengths and weaknesses of the child.
Discriminate a constitutional from an emotional basis for school failure.
Formulate programmatic recommendations concerning the most appropriate educational placement.
Formulate curricular recommendations within the current or recommended placement.

3.2.2. Specific

Questions more specific to children with identified neurological or other medical conditions are as follows:

Establish a baseline level of present performance to be used for future monitoring of improvement or deterioration.
Monitor improvement or deterioration.
Estimate the prognosis for the child in terms of final level of functioning, educational potential, and so forth.
Provide localizing information that can help the neurologist in pursuing a diagnostic plan.

4. Assessment

The clinical assessment process has three components: evaluation, diagnosis, and management. The three components are unified by the developmental neuropsychological model, which provides for both construction of the Child–World System and utilization of the system as the framework for management strategies. Knowledge of the Child–World System permits a description not only of the child's characteristic style of information processing, but also of the nature and quality of his interaction with the physical and social environment. Thus, *evaluation* leads to a *diagnosis*, which, in turn, guides the formulation of the *management* plan.

The diagnostic formulation characterizes the nature of children whose failure to adapt successfully to their environments render them the subject of the clinical assessment process. This precise characterization of children's abilities, within the Child–World System, is the pivotal step that relates data gathering and interpretation to management and intervention. Without this crucial step, remedial strategies are all too likely to take on a life of their own, logically consistent in relation to individual test performance, but irrelevant to the real child in the real world (Holmes, 1988).

4.1. Evaluation

The evaluation itself has three constituents: *history, observation,* and *testing.* The history of the individual and his or her symptoms, observation of the individual (both anecdotal and direct), and the results of testing (both level and quality thereof) all provide data for analysis.

All observations and behaviors must initially be scrutinized for their fit with the diagnostic clusters predicted by the three-axis model. From the perspective of this approach, behaviors implicating motivation, concentration, persistence, and affect, which in psychometric approaches to cognitive assessment have been traditionally treated as variables that moderate cognitive performance, must themselves be considered primary indicators of neurobehavioral function—on a par with those that implicate language, reasoning, perception, memory, and sensorimotor skills. Some behaviors may eventually prove to be best understood as moderating variables, but it cannot be assumed at the outset that this will be the case.

4.1.1. History

The history is an essential prerequisite to accurate intepretation of the observation and testing data. It also constitutes, in and of itself, an important body of data for analysis in the construction of the Child–World System. Like the other components, it is organized by the developmental neuropsychological model.

Historic information is typically collected by means of interview and questionnaires that may be completed by the child, parents, teachers, or other professionals. Different features of the history will provide information relevant to either neurological or psychological components of the model, or both. Thus, cognitive and neurological *structures* must be evaluated in terms of age-appropriate exposure to relevant information as well as integrity of relevant brain systems. Delineation of *developmental timetables* presupposes a history of the child's development thus far and requires consideration of hereditary factors that may adversely influence normal maturational schedules. A history of trauma or other disruptive experiences must be evaluated for its potential impact on expected developmental *processes,* raising the possibility of alternative pathways or strategies as the basis for the presenting complaint. *Contextual* variables (social environment, cultural expectations, economic advantage, educational exposure) must be scrutinized for their role in promoting (or impeding) effective acquisition of age-appropriate skills.

In all developmental work, the first piece of information to be established is the child's *age.* Not only is this information essential in evaluating the child's knowledge base and production competencies relative to expectation, but it also permits evaluation of physical development and social interaction skills and their impact on overall functioning. *Place of residence* provides a general indication of the child's context in terms of community resources and social environment, as well as social expectations for achievement. *Parental education and occupation* may add further information in this regard.

Historic variables also speak to the child's ability to take advantage of the cultural and educational stimulation to which he or she has been exposed. Relevant information concerns family members as well as the child. A *family history* of behavioral disorder and medical or neurological conditions (e.g., neurofibromatosis, Tourette's syndrome, tuberous sclerosis, muscular dystrophies) raises the possibility that similar conditions are contributing to the child's reported difficulties. Left (or nonright) handedness, learning disorders, and a variety of autoimmune factors have been associated with an increased risk for learning difficulty in a child (Geschwind and Behan, 1982). A report of seizure disorder in a family member mandates closer examination of this condition as a possible etiology in a child referred for short attention span, daydreaming, or inconsistent learning.

Significant life events within the family can affect the child's interaction with parents or other caretakers, negatively impacting on the psychological availability of the child for learning. This is especially critical in the child who, for neurobehavioral reasons, is already vulnerable to cognitive difficulties. Loss of a significant family member; parental separation or divorce; family relocation; abuse; hospitalization of the parent or child; or substance abuse in a parent; and the social disorganization attendant on any of these will affect participation in appropriate educational experiences as well as in the normal interactions that are critical to development.

The child's own *medical history* is equally significant. *Prenatal and perinatal risk factors* must be considered initially. For example, paternal and maternal age at conception are associated with an increased risk of genetic abnormalities in the child. The mother's medical status during the pregnancy is also potentially important—quality of nutrition, hormonal condition, toxemia, and therapeutic or nontherapeutic use of drugs are all factors that can disrupt the development of the fetal brain. A history of a difficult delivery (e.g., long labor, forceps), low Apgar scores in the first several minutes, or other indications of fetal distress may signal compromise of neurobehavioral function.

Medical conditions (neurological and systemic) can have deleterious effects on neurobehavioral function at each stage throughout development. Neurological conditions include neoplasms, localized CNS trauma, seizure disorder, and other documented neurological disorders (e.g., neurofibromatosis, Tourette's syndrome); medical conditions that can be associated with compromise of CNS function include cardiac disorders, metabolic disorders, and hormonal abnormalities. The treatment for medical or neurological conditions can itself disrupt normal neurobehavioral development (e.g., radiation and chemotherapy in the young cancer patient;

neurosurgical procedures). Limitations of sensory input (e.g., repeated ear infections, strabismus), and/or sensorimotor deficits can have an adverse impact on developing functional systems, and even modify patterns of neural connectivity. Exposure to environmental toxins, of which the most common in the pediatric population is lead, can also compromise developmental progress (Needleman et al., 1980). The behavioral outcome, in terms of both the child's general adaptation to environmental demands and his or her specific responses in the context of assessment, will depend not only on the nature of the insult, but also on the age at which it was sustained (Rudel, et al., 1974; Aram and Ekelman, 1986; Shaheen, 1984).

Reported delays in *acquisition of developmental milestones* can highlight risk for later difficulties. Early sucking or feeding problems can be the first indication of oromotor apraxia in a child who later manifests articulatory difficulties and sequencing problems in both language and more general thinking and reasoning skills. Failure to acquire motor milestones within expected age limits can foreshadow later difficulties involving control of ongoing behavior, which can be attributed to frontal system function. Poorly modulated activity levels and attentional skills (relative to age expectation) also have diagnostic relevance in relation to the frontal brain systems. Delayed acquisition of language skills raises the possibility that language development is mediated by an alternate, nonoptimal pathway. This can be reflected not only in disorders of oral language, but also in difficulty in acquiring the secondary, written language skills that depend on facility in oral language. Failure to engage in age-appropriate block, puzzle, or picture play may be the first indication of fine motor or visuoconstructional deficit, potentially implicating the cortical–subcortical or lateral axes within the neuroanatomic model. Disturbed early social relations with parents, siblings, or peers, atypical separation, or eating and sleeping disturbance can each signal a potential for later emotional disturbance or for difficulties in social cognition and interpersonal skills. From the perspective of diagnosis and the construction of the Child–World System, the disturbances in development highlighted initially by delay in the acquisition of milestones do not speak simply to the cognitive skill presumed to be involved and the biological substrate that mediates it, but must also be evaluated in terms of the potential reorganization of the neural substrate and subsequent related skill development.

Educational history must be documented in detail. How much schooling has the child had, and what kind? Has he or she been maintained in the same system throughout, or have there been changes? Retention, good teachers and bad teachers, compatibility with teachers, and provision of special educational services are all critical factors that speak to disruptions in the continuity of education. They may compound the demands of critical developmental "stress points" (Holmes, 1986) and further precipitate failure. The child's academic experience and performance to date provides a context not only for interpreting observation and test data, but also for delineating the appropriate course for future instruction.

The history should also provide an indication of *attributions*, the way in which the child and his or her problems are viewed by those who live and interact most directly with the child. What do the parents, teachers, or child want and expect from the evaluation? What is their understanding of a "learning disorder/disability"? To what do the child, the parents, teachers, or a referring physician attribute the difficulty? How is the child's intellectual endowment viewed by parents and teachers; do these views accord, and are they realistic for the child? Is the child's failure to adapt primarily attributed to moral turpitude or emotional disturbance, or is the problem viewed as more constitutional in nature? Is one or the other of the parents experiencing guilt (or blaming the other) for the child's academic or social failure? How is "fault" assigned—to the child, to the family, to the school? Is there an expectation that the learning problem will be "cured" by the anticonvulsant or stimulant medication that ameliorates the presenting seizures or the hyperactivity and attentional difficulties? Answers to these questions are essential as a guide in working with the child, formulating recommendations, and, crucially, presenting the findings of the evaluation to various audiences, in both the formal written report and the interactive feedback session.

4.1.2. Observation

As in the case of the history, data derived from observation of the child must be analyzed for their congruence with the diagnostic behavioral clusters predicted by the developmental neuropsychological model. Thus, all behaviors are initially treated as dependent variables. Specific behaviors may, however, ultimately prove to be best understood as moderating variables in the construction of the Child–World System. For example, the extraneous motor activity that in one child is an indicator of poor modulation of the motor system, and thus, implicates frontal brain systems, may, in another child, reflect performance anxiety. It is the cooccurrence of the single observation with other observations that comprises a theoretically coherent diagnostic cluster that permits moderating and dependent variables to be distinguished.

Observational data can be subsumed under five categories:

1. Observations reported or elicited either in interviews with individuals familiar with the child's behavior in nonclinical (natural) contexts or from structured questionnaires
2. Direct examination of the child: size, body habitus, facial features, personal hygiene, dress
3. Direct observation of the child–parent interaction in the clinical interview
4. Analysis of the examiner–child dyad
5. Observation of the child under specific performance demands.

4.1.2.1. Nonclinical (Natural) Contexts. Specific information about the child's behavior in nonclinical, ecologically valid contexts is critical. Informants should be encouraged to provide behavioral descriptions, rather than interpretations that may be colored by *a priori* attributions. Eliciting a *typical anecdote* (a specific example of situations in which the child gets into difficulty) is important in providing data concerning the child's response given the particular demands of a situation, free (to the extent possible) of the coloration of the reporter. Information culled from these anecdotes cannot typically be obtained in the highly structured context of the test situation.

An account of *peer relationships* provides crucial information relative to the three-axis model. Children whose right hemisphere processes are inefficient often experience difficulty with interpersonal interaction, because of the social imperception that can be a component of such a profile. Language-disabled children (left hemisphere) sometimes withdraw from social groups because of their difficulty in keeping pace with fast repartee. Children with profiles consistent with frontal system inefficiency, who have difficulty organizing and modulating ongoing behavior, often get on better with younger children, whose manner and modulation are more compatible with their own skill level. Children with significant motor deficits, implicating the cortical–subcortical axis, are often unable to participate in age-appropriate physical activities and, as a result, may lack the experience needed for the development of social competence.

4.1.2.2. THE CHILD'S APPEARANCE. A child's size (height and weight) may reflect differences in developmental progress in terms of social expectations and experience, nutrition, and peer acceptance. Direct observation of facial features and body habitus is important, inasmuch as anomalies of brain development and associated neuropsychological status resulting from genetic or chromosomal disorders, systemic disease, and early brain injury or malformation can be marked by dysgenetic growth patterns. Anomalies of head shape or size (asymmetries, microcephaly, megalencephaly), facial growth planes, midline structures (cleft palate, Crouzon's syndrome), and body habitus (Turner syndrome, Klinefelter syndrome, Prader-Willi syndrome) all have diagnostic significance for overall cognitive competence.

More generally, cleanliness and tidiness signal the "good citizen." An unkempt and disheveled appearance is typical of the disorganized youngster who may (or may not) belong to a disorganized family. Style of dress also provides information about family relationships or self-concept. The child who is dressed like a younger child may be infantilized by the parents; the adolescent who dresses flamboyantly may be compensating in relatively unhealthy ways for the loss of self-esteem that so often stems from a long history of school failure.

4.1.2.3. CHILD–PARENT INTERACTION. Direct observation of the *child and parent* together, especially when they have been invited to outline the specific concerns that motivated the evaluation, can highlight significant aspects of this interaction, the role of the child within the family, the appropriacy of expectations for the child, and the mutual respect of parent and child.

4.1.2.4. EXAMINER–CHILD DYAD. Analysis of the *examiner–child* dyad in the clinical assessment is of particular importance. First, the examiner's own psychological makeup and belief system can interact both positively and negatively with that of the child. Examiner characteristics may thus lead to overcrediting the child where scoring is questionable, or to undervaluing of subtle but significant strengths.

The interaction between the child and the examiner also provides a basis for initial hypotheses about the child's abilities. The degree to which the examiner must support performance and the specific functional areas that require support highlight vulnerability in the child. Such support typically entails a change in behavior *in the examiner,* the implications of which can be quite specific: where the examiner finds it necessary to modify language or conversational style, the child is likely to have language processing problems (implicating the left hemisphere); where voice level must be raised, hearing should be examined; where exaggerated intonation contour or segmental stress is elicited from the adult, the child's attentional and interpersonal skills should be carefully considered. (*See* Holmes, 1988, pp. 154–158, for an extended discussion of the relative adult and child contributions to the "responsibility for self" that is acquired by the child with maturation.)

4.1.2.5. Specific Performance Demands. Observation of the child under specific performance demands will involve two important concepts: the problem-solving *process* and the response to *stress.* The first of these, the problem-solving process, is discussed below in section 4.1.3.

Response to stress is an important consideration in all clinical assessment of behavior, however, it is particularly valuable in the assessment of children. The precise meaning of the term "stress" in this context, however, needs definition. This can best be done by reference to the motor component of the pediatric neurological examination, in which the neurologist makes use of the "stressed gait" maneuvers. A child's ability to walk on the toes, heels, outsides, and insides of the feet is compared to age-referenced standards. With increasing age, the child is expected to perform these maneuvers, graded in difficulty, without the movement "overflowing" (that is, being mirrored by unintended posturing of the upper extremities), a phenomenon known as synkinesis (Wolff et al., 1985). Because motor system development entails increasingly precise control of movement, these synkineses disappear in a regular fashion, in order of stressfulness. In middle childhood, walking on the heels and toes should be free of overflow, whereas such movements can be expected to persist for the more stressful outside/inside maneuvers until the onset of adolescence. For children diagnosed as hyperactive, synkineses persist past the age when they have typically disappeared in normal children, reflecting the poorly developed modulation of the motor system that characterizes this population (Denckla and Rudel, 1978).

The use of stress to monitor developmental progress is equally valuable in examining performance in the context of psychological, rather than motor tasks. In the course of clinical neuropsychological evaluation, the child is stressed in a variety of ways: not only is the child expected to maintain behavioral control (reflected in motor activity level), but he or she must maintain this control in the face of a demand for performance at the frustration level, where maximum cognitive and psychological effort must be mobilized. The child is also asked questions requiring a knowledge base exceeding age expectation and is required to perform in relatively "weak" areas of function.

The child's response to such stressors contributes directly to the diagnostic process, by providing data pertinent to the identification of behavioral clusters. A child with a pervasive attentional disorder (with hyperactivity) is typically restless throughout the examination, irrespective of what task he or she is asked to perform. (Fast pacing and a highly structured test situation can, however, facilitate more effective performance and decrease the level of restlessness in such children.) Similar kinds of behavior are seen in the language-impaired child only during tasks with a high linguistic demand, and in the child with motor and/or organizational problems only when faced with drawing or constructional tasks. The differential diagnosis between these two etiologies is critical, since the latter type of child often appears hyperactive in the school setting, where he or she is maintained at a level of maximum stress. Absent an appreciation of the process, drug intervention is often inappropriately prescribed for such children.

Increased motor activity that betrays stress is not, however, restricted to restlessness and fidgeting; extraneous activity in the perioral musculature and verbal and/or vocal disinhibition (e.g., talking, humming, whistling while working) can be another manifestation. Such verbal/vocal disinhibition is often described as "verbal mediation of behavior," interpreted to indicate a strong left-hemisphere contribution to cognition. In the context of the developmental neuropsychological approach, release of the verbal/vocal system is indicative of relative *inefficiency* of left-hemisphere input to behavior; whether the child can use verbalization effectively to guide performance is a separate issue. The increasingly precise inhibitory control of motor and verbal output that is characteristic of the normally developing child is consistent with Vygotsky's (1962) formulation of the development of "inner speech."

In summary, these various sources of observational data about the child provide information necessary for the extraction of diagnostic behavioral clusters from the total corpus of material. The observed behaviors, like the formal tests of specific skills, offer information with respect to the broad range of behavioral domains that must be tapped in a comprehensive neuropsychological assessment. The relevant domains of observed behavior include:

Socioemotional: What is the child's sense of self as a learner/problem-solver? Does he or she persist effectively or give up easily? Is the child challenged by difficulty or demoralized by it? Does the child take responsibility for finding solutions or immediately seek help from the examiner? Does he or she apologize for performance or make disparaging remarks about his or her capabilities in the context of structured testing?

What is the child's affective status, and does he or she have energy to invest in learning? Is the child frightened or apprehensive, depressed, sad or unhappy, angry or oppositional, or unable to contain emotional distress or preoccupation?

How effective are his or her interpersonal skills? Is the child easy or difficult to engage, likeable, withdrawn, or distant? Is the quality of his or her interaction socially appropriate or atypical?

Cognitive: Is the child alert and attentive? Is attention appropriately focused? Does the examiner need to reengage the child constantly, or can the child regulate his or her own attentional processes adequately? Must the examiner employ encouragement and cajoling more appropriate for a younger child? Are language skills intact—in form (articulation, fluency, syntax), in content (vocabulary, semantics), and in prosody (intonation, rhythm)? Can the child make himself or herself understood in relating complex information (discourse), or does he or she require external restructuring and focusing? Is the examiner required to modify his or her own language vis-à-vis the child? Are the child's social cognitive skills intact? Does he or she "read" nonverbal cues accurately or does he or she respond inappropriately? Does the child know the boundaries of his or her own personal space and that of the examiner? Can he or she play effectively with peers and form close peer relationships (best friend)?

Motor: Is the child's motor activity level within an age-appropriate range? Is he or she too still and contained? Is the examiner required to structure the child's behavioral response or encourage and reassure him or her? Is the child clumsy and inclined to fumble with materials? Are specific motor skills (gait, reciprocal coordination, bimanual hand use, hand/finger dexterity) intact, and if not, are they asymmetrically affected (lesser use of one limb, hemiparetic posturing, asymmetric gait, asymmetries of body parts such as digits or facial features)?

Sensory: Can the child hear and listen effectively? Is voice volume well or poorly modulated? Is the child hypervigilant to nonverbal cues? Is vision adequate (with or without glasses)? Is visual scanning appropriate? Are eye movements normal? Are gaze patterns acceptable? Are there lateralizing behaviors (favoring one ear, visual field, side of space)?

4.1.3. Testing

4.1.3.1. The Role of Tests. In the context of historic and observational data, tests provide formal documentation of *performance levels* and *problem-solving processes* in specific skill areas. Test data are an important component of the diagnostic behavioral cluster, which serves to confirm or disconfirm hypotheses generated in the course of the evaluation.

4.1.3.1.1. Performance Levels. Standard test procedures are essential to neuropsychological diagnosis for a variety of reasons. They constrain interpretation, so that potential distortions resulting from bias introduced by the child–examiner interaction can be minimized. Standardized measures document the veridical nature of the knowledge base, irrespective of facile—or clumsy—production skills. For example, administration of standard tests often show that the highly verbal, articulate child from an advan taged background is not as "bright" in terms of formal reasoning skills as informally assumed. The standardized format of tests also contributes to the child-by-child comparisons that, with history and observational data, eventually form the data base of expert knowledge necessary for clinical practice. The opportunity to compare children to one another in this fashion provides a format for the neuropsychologist to form mental prototypes based on repeated instances of particular diagnostic behavioral clusters. Similarly, it provides an ongoing data base for evaluating "negative data," that is, using the absence of particular behaviors in a given protocol as evidence in support of a specific hypothesis. For example, when assessing the contribution of processes associated with the lateral axis, the absence of behaviors indicative of right-hemisphere dysfunction if further support for a hypothesized left-hemisphere deficit.

As a first step, *overall level of functioning* is ascertained from standard measures of general intelligence. Knowledge of overall ability in the context of standardized problem-solving activities is basic for both diagnostic and management decisions. The implications of overall ability level are straightforward when one is comparing the child of limited ability to the child of normal ability or better. Less obvious is its contribution to the evaluation of strength–weakness profiles and their implications for the child's adaptation. For example, a child with a left-hemisphere implicating neuropsychological profile who achieves an overall IQ score of 130 (2 standard deviations above the mean) differs in significant ways from the child with a left-hemisphere implicating profile who achieves an IQ score of 90 (on the lower end of the average range for age). These children will presumably differ markedly in their ability to compensate—an important consideration in developing management strategies. Further, the nature of the diagnostic behavioral clusters may well differ with respect to the relative contributions of quantitative vs qualitative data. For the higher-functioning child, the innate capacity to compensate will be better developed, and so diagnosis is likely to rest more heavily on qualitative observational data than on test score patterns.

Within the broader context of overall ability level, tests permit identification of *specific areas of strength and weakness*. They provide for dissection of the behavioral domain in terms of precise skills. Thus, intertest comparison is one basis for establishing the pattern of relative competencies that is critical for understanding the complementary contributions of different brain systems to specific behaviors. It is these patterns of relative strength and weakness that highlight those brain systems that are working more efficiently than others, and pinpoint the source of the child's problem vis-à-vis environmental demands.

4.1.3.1.2. Problem-Solving Processes. The level of performance with respect to overall competence and specific skills is, however, only one aspect of the information derived from testing. Rigorous documentation of the way in which individuals *proceed to the solution of a problem* is critical for accurate diagnosis (Delis and Kaplan, 1982, 1983; Goodglass and Kaplan, 1979; Milberg et al., 1986).

There are many instances in which individuals whose adaptive impairment is clearly evident perform within the average range across a variety of standard measures in the (artificial) psychological testing situation (Duncan, 1986). Without the benefit of "process" information in such cases, the nature of the underlying deficits will remain obscure. This pattern—little variation in performance level across standard tasks in the context of significant adaptive failure in real world situations—is seen in children as well as adults. Where children are concerned, it may be particularly insidious in its implications, inasmuch as the child is assumed, on the basis of the psychometric data, to have the necessary skills for academic success. Since the adaptive failure cannot be explained on the basis of scores, the child is "blamed" for social, behavioral, or moral inadequacy—an erroneous attribution that all too frequently compounds the child's problems.

Even where there is considerable inter-subtest variability, scores cannot be adequately interpreted without process analysis. For example, in the context of the Wechsler scales (Wechsler, 1974, 1981), a standard score of 10 on any subtest means that an individual performs solidly in the average range for his or her group on that particular task. The 10 provides no indication, however, of the process utilized to obtain that level of performance.

Taken at face value, a high Verbal IQ score in the context of a Block Design score of only 10 may suggest good verbal ability and less efficient nonverbal ability, thus implicating a "left-hemisphere up–right hemisphere down" neuropsychological profile. However, it could also mean just the opposite! The verbal subtests of the Wechsler scales tap verbal knowledge. Verbal knowledge, however, is as sensitive to educational exposure as it is to language competence. Individuals with a high level of educational exposure typically do well on the verbal subtests of the Wechsler scales, without this (necessarily) implying a superior level of linguistic function. Thus, a high verbal IQ may have no diagnostic significance vis-à-vis the left hemisphere, which is the biological substrate for linguistic competence. Conversely, significant left-hemisphere damage that selectively impairs the neural substrate for language, may not impair the level of performance on tasks involving overlearned verbal knowledge. Damage to the left hemisphere is, however, unlikely to spare the linear or part-oriented processing that is the left hemisphere's contribution to the appreciation of complex visuospatially represented materials—of which Block Design is a good example. Impaired linear/part-oriented processing is easily documented by careful observation of the patient's problem-solving attempts when confronted with complex visually represented materials. In such a case, the process used in tackling Block Design (and other visual materials) is a critical feature of the diagnostic behavioral cluster—and one that overrides the apparent import of the high Verbal IQ score.

More precise delineation of the process by which problems are solved can be provided by systematic *clinical limit testing*. This allows for a closer analysis of the strategies needed by the individual to solve the problem effectively. The neuropsychologist, observing a child's problem-solving attempts, generates a hypothesis about the source of the child's difficulty, provides the hypothesized "missing element," and evaluates the efficiency with which the child utilizes the new information. The aim is not only to identify the source of the child's difficulty, but also to define the potential "zone of proximal development" (Vygotsky, 1978) that is a critical consideration for management. Clinical limit testing should be undertaken with care: not only is it crucial that formal criteria for administration not be violated, but limit-testing procedures themselves should be systematically guided by the neuropsychological model.

In sum, the role of tests in the neuropsychological assessment is to provide both quantitative and qualitative data for analysis. Quantitative data are derived from age-referenced standards. Qualitative data include not only the observations detailed above in the previous section, but also the style preferred in approaching new materials and the effectiveness with which problem-solving strategies are mobilized in response to task demands. The process by which the individual arrives at a correct solution is a critical component of the diagnostic search and treatment planning.

4.1.3.2. TEST SELECTION. Tests are selected to sample a range of behaviors representative of the functions of the three neuroanatomic axes in their dynamic interaction. Given the assumption of dynamic interplay among discrete processes (rather than isolated skills as in the psychometric model), each test potentially provides data relevant to more than one axis. This has been demonstrated in the visual context, in the Block Design example described above. It is equally relevant in the verbal domain. For instance, the quality of responses given to items of the WISC-R Vocabulary subtest (e.g., precise synonyms, circumlocutory responses, empty phraseology, a need for gesture and pantomine, disorganized formulation) highlights attentional processes, word retrieval skill, syntactic competence, and knowledge of the structure of the lexicon, among other functions. "Verbal knowledge and reasoning skills," the psychometric construct commonly attached to this task, does not adequately characterize the information it can provide within a biologically based model.

Thus, individual tests, taken by themselves, do not stand in one-to-one correspondence with neuropsychological functions. Tests are selected to provide a sampling of a sufficient range of behaviors to yield data for the diagnostic formulation in terms of the interplay among the three neuroanatomic axes. The relevant behavioral constructs provide the framework for selection of tests and organization of test data. Domains of behavior that must be sampled include:

General cognitive level
Learning of new information/manipulation of old knowledge
Language and related processing
Nonverbal/visual processing
Executive control processes
Attention/vigilance
Sensorimotor capacities
Academic achievement
Emotional status/personality development.

A list (by no means exhaustive) of frequently used test instruments, organized by behavioral constructs, can be found in the Appendix of Tests.

4.2. Diagnosis

4.2.1. The Function of Diagnosis

In clinical assessment, diagnosis is the pivotal step. It serves to organize the data collected in the evaluation phase and to guide formulation of a management plan. The neuropsychological assessment is best understood as an experimental paradigm with an *N* of 1, an experiment that is carried out in the context of a specific developmental neuropsychological theory. The model derived from that theory provides the framework for generation and testing of hypotheses. In the context of clinical assessment, it is the *hypothesis testing* process that guides formulation of the diagnosis. At the heart of hypothesis testing are the *behavioral clusters* that, in a survey of the data, should stand out in relief from the less well articulated background: hypothesized diagnoses are proven or disproven by observation of, or failure to observe, specific behavioral clusters predicted by the model.

4.2.1.1. Diagnostic Behavioral Clusters. Diagnostic behavioral clusters include data from history, observation, and testing. The primary requirement for a diagnostic cluster is that there be a sufficiently large number of converging observations from diverse sources: the greater the number of observations that speak to the same brain system or systems, the more confident can one be that any single behavior has not occurred by chance (or, indeed, has been accurately observed). Aspects of the history, formal and informal observation of the child, level of performance on specific tasks, relationship between scores in different domains, quality of performance and problem-solving strategies mobilized, lateralizing signs, and so forth, must all be scrutinized for their fit with known or independently hypothesized patterns of brain–behavior relationships. These patterns, moreover, need not be limited to previous observations of childhood neuropathology. Relevant constructs come also from the study of mature adults, nonhuman species, clinical and experimental situations, and adaptive and nonfunctional behaviors. The wider context of clinical work includes the whole of neuroscience and psychology.

It is important to emphasize that the diagnostic behavioral cluster is a fuzzy set. It therefore is not, and cannot, be assumed to be fixed across individuals. What is consistent is that those behaviors that emerge as "figure" against the more generalized "ground" are internally consistent vis-à-vis the neuroanatomical axis construct. Comparable behavioral clusters from adults and children that implicate the same brain system because of their structural similarity will not include precisely the same observations. Adults and children proceed from very different competence levels and marshal very different approaches to tasks tapping the same domains. Nor will individuals of different ages necessarily mobilize the same complement of brain systems to perform the same tasks. This principle of heterology of structure for function has been extensively discussed by Goldman-Rakic (Goldman, 1974) in the context of nonhuman primates; it is equally important for human neurobehavioral development.

Behavioral clusters include observations that are subject to a variety of standards for comparison. Lezak (1983) discusses psychometrically derived norms (population average), individual baseline standards against which to assess loss of function, species-wide performance expectations (usually taken for granted, but highly diagnostic when absent), and customary standards (arbitrary ideals, such as visual acuity). Patterns of performance vis-à-vis normative standards can also be evaluated against cluster or factor analysis of such score profiles, which are themselves sensitive to age. The less clearly operationalized expectancies for performance implicit in the interactions between children and adults are discussed in Holmes (1988).

The diagnostic behavioral cluster provides the context in which a particular observation's diagnostic meaning is evaluated. For example, in a child with subtle word retrieval difficulties and discontinuities on linear-ordering tasks, a slow-for-age performance on the right side on the finger-successive condition of the *Timed Motor Examination* (Denckla, 1974) contributes to the behavioral cluster implicating inefficient left frontal system functioning. In a child with documented left hemiparesis (indicating significant damage to right-hemisphere motor systems), however, the same observation (disturbance in fine motor control on the right side) does not speak to left-hemisphere functioning: the presence of a hemiparesis indicates that the integration of the motor system *as a whole* is compromised, and thus an independent, left-hemisphere-mediated motor system dysfunction cannot be assumed. It is, of course, possible to have both left- and right-hemisphere damage (of different types) that limits motor system functioning bilaterally. Two independent diagnoses of this type would, however, require the validation of two independent diagnostic behavioral clusters. In this example, the right-side motor performance is a critical element in the behavioral cluster in the more intact child; where the damage is more extensive, its contribution is minimal.

In determining behavioral cluster membership, care must be taken to ensure that an observed behavior speaks directly to the neuroanatomic substrate and is not a product of the testing stimulus or situation. The demand characteristics (Holmes, 1988) of a given task must be assessed for their impact on an individual's performance. For example, problem-solving style as elicited in the context of visually represented materials, such as the *Rey-Osterrieth Complex Figure,* cannot be determined by means of a copy condition alone (Osterrieth, 1944; Rey, 1941). To rule out the influence of the ever-present stimulus design on the copying of the figure, the design must be produced from recall to determine what was actually encoded by the individual. Two children who accomplish an accurate copy by means of slavish, line-by-line reproduction, may differ markedly in their appreciation of the underlying organizational structure of the material. The differential ability of the two children to appreciate organizational structure (which can speak to either the anterior–posterior axis or the lateral axis) can only be determined by reference to their recall productions. In one case, the child will recall the underlying structure of the figure; in the other, the child's recall is limited to poorly organized fragments.

4.2.1.2. HYPOTHESIS TESTING. Hypothesis testing requires the identification of groups of behaviors that cluster in predictable fashion, consistent with the neuroanatomic model. Specific hypotheses are further refined by additional data or by clinical limit testing.

In the course of the assessment, the presenting complaint, historic information, and preliminary observations of the patient are the basis for initial hypotheses. Where a medical or neurological disorder is known to be present, hypothesis formulation must be guided by knowledge of the disorder and its potential neurobehavioral consequences. This is particularly important where more than one disease process is known or suspected, in which case failure to consider the potential confounding effects of multiple disorders on neurobehavioral status may render the psychometric data uninterpretable in terms of the patient's adaptive functioning.

Where there is no precise medical or neurological diagnosis, initial hypotheses formulated in light of the presenting complaint must be tested by means of data from the full complement of behavioral domains speaking to the neuroanatomic axes. This means, for example, that tests cannot simply be selected to address the primary presenting complaint: for example, a child who presents with "attentional problems" is not sufficiently evaluated by means of tests presumed to tap "attention." Phenomenological attention difficulties can be primary, in which case their constitutional basis can frequently be documented in terms of associated motor findings on the neurological examination; they can, however, be secondary—to primary problems in language processing, to preoccupation in the context of significant emotional concerns, or to cognitive overload in a disorganized or intellectually limited child. A child's performance on tests that simply tap one skill area, such as "attention," can be evaluated only in terms of psychometric standards. Failing to model the neurobehavioral substrate, they cannot adequately discriminate among the possibilities outlined above and, thus, have limited value for the diagnostic formulation that is at the core of the Child–World System.

4.2.1.3. The Diagnostic Formulation. The developmental neuropsychological theory guides all aspects of the assessment process. The diagnosis is always formulated within the framework of the three neuroanatomic axes; however, it does not have to be stated in neurological terms—and, indeed, the decision to use neurological constructs in discussing the case with parents and other nonmedical personnel must be made with care. A descriptor, such as "language and language-related deficits," may be a more effective means of highlighting a particular pattern of learning-related difficulty than "left-hemisphere dysfunction," even when the evidence for specific left-hemisphere deficit is clear. In other instances, the brain-related diagnosis may be important; emphasizing the constitutional nature of the difficulty can, in many cases, counter previous attributions of moral turpitude or emotional stress as the primary cause of the learning failure. Where a neurological (or relevant systemic) disorder is documented, a brain-based diagnostic formulation is entirely appropriate, with specific comment vis-à-vis the consistency of the findings with the known condition.

Categories of diagnostic formulation include the following:

1. A description of the child's general level of competence
2. No specific difficulty identified; skill profile consistent with individual variation within the population as a whole
3. No specific difficulty identified now, but risk predicted for the future
4. Brain damage (focal, diffuse, static, discharging)
5. Learning disability
6. Localization
7. Description of skill profile
8. Consistent with already identified disorder
9. Primary emotional disorder
10. Complex interaction of cognitive and emotional variables.

4.3. Management

The goal of the assessment process, from the systemic perspective, is not to diagnose deficits in the child at a particular point. Rather, it aims to provide a context within which to understand how a given child's total complement of talents, skills, and weaknesses interfaces with the child's world as he or she proceeds to maturity (Zigler, 1966). The Child–World System is the vehicle by means of which the understanding of the child's difficulty can be reframed—from "disorder" to "mismatch," thus shifting attribution of the difficulty from child to system. The primary focus of management is thus not to remediate deficits, but to provide a coherent description of the child, in his or her context, on which comprehensive intervention strategies are based. This description is critical to achieving the understanding of the child that can guide effective management throughout childhood and adolescence.

The function of the first two components of the assessment, evaluation and diagnosis, is to construct the Child–World System for an individual child. The nature of the *World* in this context is elaborated by weighing the contributions of expected maturational competence, medical and educational history, and family and social history against the child's performance as manifest across a variety of specific activities, some reported, some observed, and some formally elicited. The nature of the *Child* is derived from the level and quality of performance as manifest across a broad range of behavioral domains. The *Child* contribution to the system (the diagnostic formulation) is framed in terms of the three neuroanatomic axes, as specified by the developmental neuropsychological model.

The Child–World System, then, provides a comprehensive description of the interface between the individual child and his or her environment and, thus, constitutes the conceptual basis for management. It provides a systematic framework, within which the information obtained from the evaluation is used to guide, in a rigorous fashion, the development of an effective treatment plan. Recommendations developed in reference only to specific test findings or test-related deficits, as is frequently the case in neuropsychometric approaches to assessment, cannot adequately address the full scope of the child's adaptation, now and in the future. Without recognition of the Child–World System, even the diagnostic formulation itself cannot serve as the basis for treatment planning. Effective recommendations can be made only with reference to the broader context of the child's adjustment.

In evaluation and diagnosis, the focus is on delineating the nature of the child in his or her world and integrating these conceptually into a systemic whole. In addressing management, however, the focus is on the world in which the child functions. This requires analysis of the problem or problems, that is, the obstacle(s) to developmental progress, that confront the child in his or her everyday life—with a view to intervening in the system to achieve an optimal match between the child and his or her world. To accomplish this phase of the assessment requires a team approach: the neuropsychologist must work collaboratively with the child and the family to identify the goal or goals to which the findings should speak.

The concept of goal is, potentially, broad. It includes long-term educational success and socioemotional adjustment, as well as short-term mastery of particular concepts and fulfillment of specific age- or grade-referenced expectations for behavior. All recommendations must be formulated in the context of the widest ranging goal of education and child-rearing, that is, producing a competent and comfortable adult with the skills, both academic and social, to achieve economic self-sufficiency and satisfying social relationships. In this broader context, an appreciation of neurobehavioral development is as important in promoting social competence and emotional well-being as it is for facilitating cognitive performance.

4.3.1. Formulating the Management Plan

4.3.1.1. THE NATURE OF THE CHILD. *Motivation and attitude toward learning* directly influence the child's ability not only to utilize his or her skills, but also to take advantage of help offered by involved adults. Self-esteem and the sense of self as a potentially effective problem-solver are developed and maintained in the context of success and mastery. Recommendations may thus need to address motivational, emotional, or behavioral needs, in addition to specific cognitive deficits.

General competence is important, inasmuch as it is the basis for the overall level of expectations that the child will face. Treatment planning for a child with significantly limited cognitive ability will have as its primary goal the identification of an appropriate setting in which the child can function comfortably. For children of normal (or better) intelligence, the overall context may not be so critical; the management goals will emphasize better adjustment with respect to specific skill areas. Within specific skill areas, however, considerations may need to be more subtle. The same diagnostic formulation (for instance, left-hemisphere dysfunction) will have different management implications for a child with an overall IQ of 130 than for a child with an IQ in the average range. High scores on formal testing should not, however, be equated with learning competence. A high- IQ child with relatively intact reading and language skills can, nonetheless, manifest organizational difficulties that may interfere dramatically with school adjustment, especially in the junior high and high school grades, when independent, organized functioning is at a premium. In these children, the difficulty should not be attributed to "moral turpitude" simply on the basis of the IQ score.

Delineation of *specific skill profiles* is needed to achieve an optimal match between the child and the knowledge to be acquired. A child's learning style may be better matched not only to specific instructional programs, but also to particular formats for presenting information—lecture, hands-on projects, textbook, group discussion, and so forth. Certain skill profiles better facilitate progress in particular curricular subjects and, further, in specific topics within a subject. Computation and conceptual understanding are very different aspects of mathematics; a child can fail (and thus need specific help) in one or the other (as well as in both). Addition and subtraction similarly require different information-processing strategies, as do algebra and geometry. Moreover, to the extent that mathematical skills are typically taught in a linear sequence, different processing capacities will be called upon at different times. A learning style that is less than optimal at one point in a child's career may not remain so. This principle is equally applicable to the acquisition of reading skills, in which the initial task of decoding differs markedly from the later one of extracting meaning.

Consideration of optimal skill–profile matches encompasses school environments as well. Educational philosophies vary from school to school in terms of the relative importance assigned to academic skills, creativity, good citizenship, athletic requirements, individuality, and codes of behavior. The mismatch between a child's skill profile and the school environment may mandate a change of school, rather than currricular modification.

The child's ability to mobilize *problem-solving strategies* depends to a great extent on the flexibility of the educational environment. Permission to use alternate strategies, modifications of expectations (time constraints, length of assignments), and availability of "hardware" (computers, calculators) can vary enormously in different settings. Selection and use of problem-solving strategies is influenced by the nature of the material and the way in which it is taught. To the extent that teachers impose specific problem-solving strategies, a child may be "forced" to fail. Whether different problem-solving strategies can be explored may depend not only on individual teachers, however, but also on class size, availability of individual instruction, and recognition of the child's difficulty. School philosophies, administrative regulations, and financial constraints may all limit the exploration of problem-solving potential.

Medical or neurological disorders can constrain the child's ability to engage consistently in the educational process (fatigue following head injury, physical limitation attendant on trauma, hypoactivity/lethargy in endocrine disturbances). Ongoing involvement in the instructional process can also be limited by primary or secondary attentional disorders, less than optimal seizure control, or side effects of medications. Involuntary movements, obesity, or growth disturbance may cause the child to be the target of teasing or harassment.

4.3.1.2. Implications for Adaptation: Specification of Risk. An appreciation of the child's functioning in his world highlights *risk* for difficulty at those points where the match between his or her complement of skills and the demands imposed on the child is inadequate. Knowledge of the world and its demands at each stage in a child's development allows for the prediction of risk in various contexts, conceptualized in terms of micro- and macroenvironments, thus encompassing not only the acquisition of specific skills (academic and psychosocial), but also the context in which those skills must be mastered. Furthermore, risk is predicted relative to the longer-term expectations to which the child must respond, as well as the immediate demands that the child encounters. Both academic and psychosocial competencies reflect the dynamic interplay of the three neuroanatomic axes and the impact of newly available neural systems in that interplay as maturation proceeds.

Risk for the school-age child can be defined, at least initially, in terms of the (relatively uniform) expectations imposed by the school system. Assessment of risk for the specific individual, however, requires that personal goals be defined in light of that individual's particular complement of skills and preferred learning style. This issue becomes increasingly important in the older child, as the formal structure of the educational system in the grade-school years gives way to the greater freedom of choice (vocational, contextual) that typically allows for a more comfortable match of skills to specific goals.

4.3.2. Recommendations

Recommendations respond directly to the risks revealed by analysis of the Child–World System. The goal of the recommendations is to specify interventions that will reorganize the system in such a way as to achieve a new equilibrium and minimize the degree of mismatch. This may involve change in the child (that is, the child's ability to mobilize skills effectively) or modification of environmental demands; effective interventions, however, typically speak to both.

4.3.2.1. Macroenvironment. Recommendations address contextual demands in the child's social, as well as academic, world. The home environment is the foundation of the child's social world. It is here that psychosocial development, motivation, and learning are initially fostered, and where the behavioral control and responsibility for self that are crucial for social and academic adjustment are developed. Factors of the home environment that must be considered in the specification of risk, and therefore in the formulation of intervention, are varied. These include the parents' understanding of the child's difficulty; parental management skills and disciplinary styles; the family's standards for behavior and achievement; cultural expectations; emotional, financial, and medical well-being of responsible family members; and the availability of social or psychological support to the family as a group or the child in particular. As the child grows, the peer group and adults who are not family members assume an increasingly important role in his or her psychosocial development. Recommendations will need to address appropriate peer group interaction and sources of support outside the family.

Considerations vis-à-vis the child's academic world include type of school, academic and disciplinary philosophy; relevance for life goals; grade placement, need for special services, type and amount of services, adequacy of the services provided; history of education to date; opportunity for coursework selection; and match between child and teachers. All of these variables will play a role in determining for the child an optimal placement. Changes may be needed with respect to any, or all, of these.

4.3.2.2. MICROENVIRONMENT. Support for the content of a child's life must also address home environment, psychosocial development, and academic skill building. Specific recommendations for the home environment include adjustment of expectations, disciplinary guidelines, limit-setting, and rewards; more elaborate behavior management strategies with contingency planning and contracts among parents, child, and therapist; organization of homework; and explicit communication with relevant teaching personnel. Psychosocial development is promoted most effectively by appropriate educational placement and support to maximize the child's "comfort level" with respect to learning, and to foster the sense of mastery that comes with (carefully engineered) successes.

Emotional well-being and psychological development may need to be addressed directly with psychotherapeutic treatment, tailored carefully to the child's overall neuropsychological competencies. Insight-oriented psychotherapy alone is rarely helpful for the individual with a right-hemisphere-implicating profile (Weintraub and Mesulam, 1983). "Talking therapies" need to be offered carefully to adolescents with language impairment. Other psychotherapeutic decisions will involve private therapy vs school-based counseling, family therapy, child guidance, and/or group therapeutic activities. Further, extended evaluation of emotional status may be required.

Academic recommendations must address not only skill development, but also information acquisition in different contexts (listening, reading, hands-on activities) and at different ages. They include changes in specific instructional programs, combination of different types of programs, more explicit teaching of elements within programs, different instructional techniques, backup support (preteaching, postteaching, textbook review, research, and so forth) of classroom instruction, formal introduction and development of problem-solving strategies (metacognitive techniques, study and drafting skills), modification of expected assignments (amount of time, length, complexity, format), utilization of educational hardware in the guise of tape recorders, typewriters, word-processing software, and calculators (for both math and nonmath subjects).

4.3.2.3. MEDICAL MANAGEMENT. A specific set of recommendations pertains to medical management issues. The neuropsychologist is often the first nonteacher professional consulted vis-à-vis a child with learning difficulty, and thus, may be the first to query the possibility of neurological or medical conditions affecting learning potential. In the case of a child with specific learning problems, the involuntary movements, atypical sensory or motor performance, visual disturbances, headaches, interruptions in ongoing processing, loss of contact with the environment, a neuropsychological profile consistent with brain impairment, or suspected erosion of previously available skills each mandate referral to a neurologist (after contact with the child's pediatrician/family physician) for formal neurological examination and further neurodiagnostic procedures where necessary. Chromosomal analysis and genetic counseling may be needed. Pharmacologic intervention should be considered for seizure disorder, primary attentional disorders, chronic headache, movement disorders, depression, and disturbances of endocrine function. Weight control, nutritional advice, and sex education may all require medical support. The family may need encouragement to seek help or to find the appropriate physician.

4.3.3. Reporting of Findings

The goal of the assessment process is to provide a context within which the understanding of a child can be reframed and within which the *attribution* of difficulty is shifted from the child to the system. The findings of the assessment are typically communicated in two ways: a formal *written report* and a less formal *feedback session*. The report and the feedback session complement one another: they are both guided by the Child–World System, within which the child's adjustment can be understood and optimized.

4.3.3.1. ATTRIBUTION. The shift of attribution of difficulty *from child to system* is essential to the success of the assessment process. It is not only the basis for management and recommendations, but also for forming the critical alliance between clinician and family that is necessary for optimal management of children. In this context, the aim of the reporting of findings is not to label deficits, but to educate parents and teachers so that they can manage the child on an ongoing basis. The parents' understanding of the system within which their child functions gives them the tools to be active and independent problem-solvers for the child, thus allowing them to address new obstacles or developmental tasks as they arise.

4.3.3.2. THE NATURE OF THE REPORT. A primary requirement of the report is that it respond clearly and directly to the clinical questions posed or the presenting complaint. When the question is very specific, as is often the case in physician referrals, a complete description of the Child–World System may not be necessary. The Child–World System is, nonetheless, the basis for communicating findings to the family, by report or in face-to-face discussion.

Inasmuch as it documents the formal structure of the assessment process, the form of the report is more or less fixed. It is important to recognize, nonetheless, that it is likely to be read by individuals of different types of professional involvement, different levels of academic ability, and different degrees of emotional investment in the material. The words chosen and the overall tone of the language should be upbeat, the style businesslike, and "jargon" kept to a minimum (or explained). Where an (older) child is likely to read the report, it is helpful to discuss the nature of the report with the child ahead of time: adolescents are notorious for latching onto descriptions of their difficulties while completely ignoring (equally extensive) delineation of their competencies.

The written report reflects the (theoretical) structure of the assessment process. Following identification of the patient and a statement of the presenting complaint, *historic* information is outlined in terms of family variables and the child's own medical, developmental, and educational experiences. The relevant history is followed by systematic *observations* of the child's behavioral presentation, with specific categories of behavior highlighted as they respond to the neuroanatomic axes. Available skills and disrupted performance are both detailed. *Test* findings are then presented in terms of the behavioral domains listed previously (section 4.1.3.2.). Level of ability, skill profile, and problem-solving behavior—as these contribute to the diagnostic formulation—are described within each domain. The history, observation, and testing of the evaluation phase yield the *diagnostic formulation* and permit construction of the Child–World System. This may be presented in neuropsychological constructs (e.g., left-hemisphere difficulty) or in psychological constructs (e.g., language and language-related processing deficits). A representative sample of the behavioral cluster validating the neuropsychological diagnosis is listed. Available compensatory strategies and/or moderating variables are discussed as necessary.

The *nature of the risk* that this particular child faces—now and in the future—is then explored. The risks are evaluated in terms of the general *context*, both academic and social, in which this child functions and, in terms of the specific skills, the *content* (again both academic and social) that he must acquire. The Child–World System then guides the overall *management* strategy: *recommendations are formulated in direct response to the risks highlighted.* The relevance of specific recommendations is evaluated for both short-term and longer-term goals. The recommendations are organized in terms of *contextual* (macroenvironmental) demands and *content* (microenvironmental) demands. Within each of these larger categories, recommendations address psychosocial well-being, home or family expectations, and academic progress, as needed.

There are some instances in which insufficient information is obtained from formal testing: the child is unable to comply with the demands thereof. Thus, the report will not follow exactly the format outlined above, inasmuch as test data will be lacking. Rigorous use of the neuropsychological model is critical in such cases—to guide formulation of (at least) an initial diagnosis based on historic variables and close observation of relevant behaviors as the child plays or takes part in conversations. The report must still be organized in terms of the formal assessment model, with a greater emphasis on observational data in the construction of the Child–World System. Followup assessment will be needed to validate the initial diagnostic formulation as the child becomes able, with increasing maturity and/or emotional stability, to take part in the testing procedures.

4.3.3.3. THE FEEDBACK SESSION. The goal of the feedback session is to work through, with the parents, the reframing of the child's difficulty in terms of the Child–World System, and thus, to develop the alliance between clinician and family on which effective management strategies are based. The systemic approach is used in the feedback session not only to address issues pertaining to general adjustment, but also to highlight the role of specific situational demands on the child's behavior. The "typical anecdote" elicited in the course of evaluation is frequently a starting point for discussion. Helping parents to achieve an appreciation for the impact of lack of structure on a disorganized child, for example, or for the effect of high-speed conversations around the family dinner table on a child with language-processing difficulties, is a major first step in effective support of the child. Far from being deliberately difficult, most children misbehave in a predictable fashion—given both their abilities and the demands placed upon them from moment to moment. Thus, the feedback session is part of an educational process in which direct application of the Child–World System enables parents to identify potential risk factors in specific situations at school and at home.

It is in talking directly with families that the neuropsychologist can most effectively address the anxieties, fears, and misunderstandings common to both the parents of youngsters with learning disorders and the youngsters themselves. The clinician's authority can reassure a child who is—as is all too frequently the case—concerned that he or she is "stupid," "a retard," or "crazy." Even children who function, in general, at a very high intellectual level have significant difficulty maintaining self-esteem in the face of daily evidence that they are not learning to read or do math in a manner commensurate with that of their peers. Most children can appreciate the idea of "talents and untalents" and can be helped to recognize some of theirs and, perhaps, those of their siblings or parents. Many children, adolescents in particular, like the notion that they are "mismatched" with certain skills or school expectations, as posited by the systemic model.

It is equally important that parents be reassured: many parents are concerned that the need for special educational services will make the child feel different, and that this will have potentially dire consequences for his socioemotional well-being. Gently pointing out that the child is already aware of his or her difference, by virtue of failure to read and need for assessment, is usually sufficient to help the parent see the importance of addressing the "difference" as a point of departure for positive management strategies, rather than as a defect in the child.

Although feedback sessions are usually undertaken with the child and family, the same system-based explanation is useful in helping teachers gain a better appreciation of the child's difficulties in the classroom. Delineation of the Child–World System is valuable in allowing teachers to appreciate the underlying consistency of seemingly discrepant behaviors observed when the child is in different settings.

5. Finale

The systemic approach to neuropsychological assessment, outlined here, provides a powerful theoretic framework within which to understand brain–behavior relationships in the developing child. The theory offers a principled way to assemble knowledge from diverse fields—psychology (cognitive, clinical, developmental), neurobiology and neuropsychology, sociology, and education.

A systemic neuropsychology is rooted in the historic tradition of Jackson, Lashley, Vygotsky, and Luria. It is based on the principle of "extracortical organization of complex mental functions" (Vygotsky, cited in Luria, 1973), having at its core the reciprocal relationship of brain and world in human neurobehavioral development. We believe that Vygotsky's principle is not only valid, but indeed, crucial for achieving the goal of neuropsychological assessment, that is, the understanding of the individual in his or her world that is the foundation of effective intervention. Although the emphasis of this chapter has been the assessment of children, the principle is equally important as a basis for developing methodologies that will further our understanding of neurobehavioral development across the life-span.

References

Aram D. and Ekelman B. L. (1986) Cognitive profiles of children with early onset of unilateral lesions. *Developmental Neuropsychology* **2,** 147–154.

Delis D. and Kaplan E. (1982) The assessment of aphasia with the Luria-Nebraska Neuropsychological Battery: a case critique. *J. Consult. Clin. Psychol.* **50,** 32–39.

Delis D. and Kaplan E. (1983) Hazards of a standardized neuropsychological test with low content validity: Common on the Luria-Nebraska Battery. *J. Consul. Clin. Psychol.* **51,** 396–398.

Denckla M. B. (1974) Development of speed in repetitive and successive finger-movements in normal children. *Dev. Med. Child Neurol.* **15,** 635–645.

Denckla M. B. (1979) Minimal brain dysfunction, in *Education and the Brain* (Chall J. and Mirsky A., eds.) University of Chicago Press, Chicago, IL, pp. 223–268.

Denckla M. B. and Rudel R. G. (1978) Anomalies of motor development in hyperactive boys. *Ann. Neurol.* **3,** 231–233.

Duncan J. (1986) Disorganization of behavior after frontal lobe damage. *Cognitive Neuropsychology* **3,** 271–290.

Fletcher J. M. and Taylor H. G. (1984) Neuropsychological approaches to children: towards a developmental neuropsychology. *J. Clin. Neuropsychol.* **6,** 39–56.

Fodor J. (1983) *Modularity of Mind* (MIT Press, Cambridge, Massachusetts).

Geschwind N. and Behan P. (1982) Left-handedness: association with immune disease, migraine and developmental learning disorder. *Proc. Natl. Acad. Sci. USA* **79,** 5097–5100.

Goldberg E. and Costa L. D. (1981) Hemisphere differences in the acquisition and use of descriptive systems. *Brain Lang.* **14,** 144–173.

Golden C. J. (1981) A Standardized Version of Luria's Neuropsychological Tests, in *Handbook of Clinical Neuropsychology* (Filskov S. and Boll T. J., eds.) Wiley-Interscience, New York.

Goldman P. S. (1974) An alternative to developmental plasticity: Heterology of CNS structures in infants and adults, in *Plasticity and Recovery of Functions in the Central Nervous System* (Stein D. G., Rosen J., and Butters N., eds.) Academic Press, New York.

Goldman-Rakic P., Isserhoff A., Schwartz M., and Bugbee N. (1985) Neurobiology of cognitive development in non-human primates, in *Carmichael's Manual of Child Psychology* (4th Ed.) (Musser P., ed.) Wiley, New York, pp. 281–344.

Goodglass H. and Kaplan E. (1979) Assessment of cognitive deficit in the brain-injured patient, in *Handbook of Behavioral Neurobiology (Vol. 2) Neuropsychology* (Gazzaniga M. S., ed.) Plenum, New York, pp. 3–22.

Held R. and Hein A. (1963) Movement-produced stimulation in the development of visually-guided behavior. *J. Comp. Physiol. Psychol.* **56,** 607–613.

Holmes J. M. (1986) Natural histories in learning disabilities: Neuropsychological difference/environmental demand, in *Handbook of Cognitive, Social and Neuropsychological Aspects of Learning Disabilities* (Ceci S. J., ed.) Erlbaum, Hillsdale, NJ, pp. 303–319.

Holmes J. M. (1988) Neuropsychological assessment of children, in *Assessment of Developmental Learning Disorders: A Neuropsychological Approach* (Rudel R. G., Holmes J. M., and Pardes J., eds.) Basic Books, New York.

Hubel D. H. and Wiesel T. N. (1962) Receptive fields, binocular interaction and functional architecture in the cat's visual cortex. *J. Physiol. (Lond)* **160,** 106–154.

Kaplan E. (1976) The role of the non-compromised hemisphere in patients with local brain disease, in (Teuber, H-L. chair) *Alterations in Brain Functioning and Changes in Cognition.* Symposium presented at the meeting of the American Psychological Association, Washington, D.C., August, 1976.

Kaplan E. (1983) Process and achievement revisited, in *Toward a Holistic Developmental Psychology* (Wapner S. and Kaplan B., eds.) Erlbaum, Hillsdale, NJ, pp. 143–156.

Katona F. (1989) Early identification and neurohabilitation of brain-injured infants, in *Challenges to Paradigms in Development* (Barr R. and Zelazo P., eds.), Erlbaum, Hillsdale, New Jersey.

Lezak M. D. (1983) *Neuropsychological Assessment* (2nd Ed.) (Oxford University Press, New York).

Lovett M. (1987) A developmental approach to reading disability: accuracy and speed criteria of normal and deficient reading skills. *Child Dev.* **58,** 234–260.

Luria A. R. (1973) *The Working Brain* (The Penguin Press, London).

Merleau-Ponty M. (1962) *The Phenomenology of Perception* (Humanities Press, New York.)

Milberg W. P., Hebben N., and Kaplan E. (1986) The Boston process approach to neuropsychological assessment, in *Neuropsychological*

Assessment of Neuropsychiatric Disorders (Grant I. and Adams K. M., eds.) Oxford University Press, New York, pp. 65–86.

Needleman H. (ed.) (1980) *Low Level Lead Exposure: The Clinical Implications of Current Research* (Raven Press, New York.)

Osterrieth P. A. (1944) Le test de copie d'une figure complexe. *Archives de Psychologie* **30,** 206–356.

Piaget J. (1985) *The Equilibration of Cognitive Structures* (University of Chicago Press, Chicago.)

Rakic P. and Goldman-Rakic P. S. (1982) Development and modifiability of the cerebral cortex. *Neuroscience Research Program Bulletin* **20,** 429–611.

Reitan R. M. and Davison L. A. (1974) *Clinical Neuropsychology: Current Status and Applications* (Hemisphere, New York.)

Rey A. (1941) L'examen psychologique dans le cas d'encephalopathie traumatique. *Archives de Psychologie* **28,** 286–340.

Rourke B. P., Fisk J. L., and Strang J. D. (1986) *Neuropsychological Assessment of Children—A Treatment-Oriented Approach* (Guilford, New York.)

Rudel R., Teuber H-L., and Twitchell T. (1974) Levels of impairment of sensorimotor function in children with early brain damage. *Neuropsychologia* **12,** 95–108.

Schonhaut S. and Satz P. (1983) Prognosis for children with learning disabilities: a review of follow-up studies, in *Developmental Neuropsychiatry* (Rutter M., ed.) Guilford Press, New York, pp. 542–563.

Sergent J. (1983) Role of the input in visual hemispheric asymmetries. *Psychol. Bull.* **93,** 481–512.

Shaheen S. J. (1984) Neuromaturation and behavioral development: The case of childhood lead poisoning. *Dev. Psychol.* **20,** 542–550.

Smith S. D., Kimberling W. J., Pennington B. F., and Lubs M. A. (1983) Specific reading disability: Identification of an inherited form through linkage analysis. *Science* **219,** 1345–1347.

Stuss D. T. and Benson D. F. (1986) *The Frontal Lobes* (Raven Press, New York.)

Teuber H-L, and Rudel R. G. (1962) Behavior after cerebral lesions in children and adults. *Dev. Med. Child Neurol.* **4,** 3–20.

von Bertallanfy L. (1969) *General Systems Theory* (G. Braziller, New York).

Vygotsky L. S. (1962) *Thought and Language* (MIT Press, Cambridge, MA.)

Vygotsky L. S. (1978) *Mind in Society* (Harvard University Press, Cambridge, Massachusetts).

Waber D. P. (1989a) The biological boundaries of cognitive style: a neuropsychological analysis, in *Cognitive Style and Cognitive Development* (Globerson T. and Zelniker T., eds.) Ablex, New York, in press.

Waber D. P. (1989b) Rate and state: A critique of models underlying the assessment of learning disabled children, in *Challenges to Paradigms in Development* (Barr R. and Zelazo P., eds.) Erlbaum, Hillsdale, New Jersey, in press.

Waber D. P., Carlson D., Mann M., Merola J., and Moylan P. (1984) SES-related aspects of neuropsychological performance. *Child Dev.* **55,** 1878–1886.

Wechsler D. (1974) *Wechsler Intelligence Scale for Children—Revised* (The Psychological Corporation, New York).

Wechsler D. (1981) *Wechsler Adult Intelligence Scale—Revised* (The Psychological Corporation, New York).

Weintraub S. and Mesulam M.-M. (1983) Developmental learning disabilities of the right hemisphere: emotional, interpersonal and cognitive components. *Arch. Neurol.* **40,** 463–468.

Werner H. (1937) Process and achievement: a basic problem of education and developmental psychology. *Harvard Educational Review* **1,** 353–368.

Werner H. (1948) *Comparative Psychology of Mental Development* (International Universities Press, New York).

Wilson B. C. and Risucci D. A. (1986) A model for clinical-quantitative classification. Application to language-disordered preschool children. *Brain Lang.* **27,** 282–309.

Wolff P. H., Gunnoe C., and Cohen C. (1985) Neuromotor maturation and psychological performance: a developmental study. *Dev. Med. Child Neurol.* **27,** 344–354.

Zigler E. (1966) Mental retardation: current issues and approaches. *Review of Child Development Research Volume 2.* (Hoffman L. W. and Hoffman M. L., eds.) Russell Sage, New York, pp. 107–168.

Appendix of Tests

General Cognitive

- Kaufman Assessment Battery for Children
- Raven's Progressive Matrices
- Stanford-Binet Intelligence Scale (4th Edition)
- Wechsler Primary and Preschool Scale of Intelligence
- Wechsler Intelligence Scale for Children—Revised
- Wechsler Adult Intelligence Scale—Revised

"Memory"

- Benton Visual Retention/Visual Recognition
- Logical Memory for Children

Rey Auditory Verbal Learning/California Verbal Learning
Rey-Osterrieth Complex Figure
Sentence Repetition
Wechsler Memory Scale

Language and Related Processing
Auditory Discrimination
Automatized Series
Boston Naming
Expressive One-Word Picture Vocabulary Test
Logical Memory for Children/Logical Memory (WMS)
Peabody Picture Vocabulary Test—Revised
Token Test
Word Fluency (F-A-S)

Visual/Nonverbal Processing
Embedded Figures Test
Face Recognition
Judgment of Line Orientation
Locomotor Mazes
Mooney Closure Test
Porteus Mazes
Rey-Osterrieth Complex Figure
Developmental Test of Visual Motor Integration

Executive Control Processes
Booklet Category Test
Rapid Automatized Naming/Series
Stroop Color-Word Interference
Trail Making
Wisconsin Card Sort

Sensory/Motor
Parietal Lobe Battery (BDAE)
Pegboard
Tactual Performance Test
Timed Motor Examination

Achievement
Analytic Reading Inventory
Gilmore Oral Reading Test
Gray Oral Reading Test—Revised
KeyMath

Mathematics Diagnostic/Prescriptive Inventory
Test of Written Language
Wide Range Achievement Test—Revised
Woodcock Reading Mastery Tests

Projectives
Projective Drawings
Rorschach
Sentence Completion
Tasks of Emotional Development
Thematic Apperception Test/Children's Apperception Test

Questionnaire/Rating Scale
Child Behavior Checklist
Rating Scale for Hyperactivity

Appendix

Achenbach T. M. and Edelbrock C. S. (1982) *Manual for the Child Behavior Checklist and Child Behavior Profile* Child Psychiatry (University of Vermont, Burlington, Vermont).

Beery K. E. (1982) *Revised Manual for the Development Test of Visual Motor Integration* (Modern Curriculum Press, Cleveland).

Benton A. L. (1974) *The Revised Visual Retention Test* (4th Ed.) (Psychological Corporation, New York).

Benton A. L., Hamsher K. deS., Varney N. R., and Spreen O. (1983) *Facial recognition* (Oxford University Press, New York.)

Chelune G. J. and Baer R. Z. (1986) Developmental norms for the Wisconsin card sorting test. *J. Clin. Exper. Neuropsychol.* **8,** 219–228.

Cohen H. and Weil G. R. (1975) *Tasks of Emotional Development* (TED Associates, Brookline, Massachusetts).

Conners C. K. (1970) Symptom patterns in hyperkinetic, neurotic and normal children. *Child Dev.* **41,** 667–682.

Conners C. K. (1979) A teacher rating scale for use in drug studies with children. *Am. J. Psychiatry* **126,** 884–888.

Connoly A. J., Nachtman W., and Pritchett E. M. (1976) *Keymath Diagnostic Arithmetic Test* (American Guidance Service, Circle Pines, Michigan).

Davidson P. S. and Marolda M. R. (1978) *Mathematics Diagnostic/Prescriptive Inventory* Cusenaire Company of America, New Rochelle (in preparation).

DeFilippis N. A. and McCampbell E. (1979) *The Booklet Category Test*, Vol II. (Psychological Assessment Resources, Inc., Odessa, Florida).

Delis D. C., Kramer J. H., Kaplan E., and Ober B. A. (1986) *The California Verbal Learning Test-Children's Version* (The Psychological Corporation, San Antonio, Texas).

Denckla M. B. (1974) Development of speed in repetitive and successive finger movements in normal children. *Dev. Med. Child Neurol.* **15,** 635–645.

Denckla M. B., and Rudel R. G. (1976) Rapid Automatized Naming (R.A.N.): Dyslexia differentiated from other learning disabilities. *Neuropsychologia,* **14,** 471–479.

Denckla M. B., Rudel R. G., and Broman M. (1980) The development of a spatial orientation skill in normal, learning-disabled and neurologically-impaired children, in *Biological Studies of Mental Processes,* (Caplan D., ed.), M.I.T. Press, Cambridge, Massachusetts, pp. 44–59.

DeRenzi E. and Vignolo L. A. (1962) The Token Test: a sensitive test to detect disturbances in aphasics. *Brain* **85,** 665–678.

Dunn L. M. (1981) *Peabody Picture Vocabulary Test–revised* American Guidance Service, Circle Pines, Michigan).

Exner J. E. (1974) *The Rorschach* (Wiley-Interscience, New York).

Gardner M. F. (1979) *Expressive One-Word Picture Vocabulary Test* (Academic Therapy Publications, Novato, California).

Gilmore J. V. and Gilmore, E. C. (1968) *Gilmore Oral Reading Test* (Harcourt Brace Jovanovich, New York).

Golden C. J. (1978) *Manual for the Stroop Color and Word Test* (Stoelting, Chicago.)

Goodglass H. and Kaplan E. (1972) *Assessment of Aphasia and Related Disorders* (Lea and Febiger, Philadelphia.)

Hammill D. D. and Larsen S. C. (1983) *The Test of Written Language* (Pro-Ed, Austin, Texas).

Harris A. J. (1958) Harris tests of lateral dominance, in *Manual of Directions for Administration and Interpretation* (Psychological Corporation, New York.)

Heaton R. K. (1981) *Manual for the Wisconsin Card Sorting Test* (Psychological Assessment Resources, Odessa, Florida).

Ivinskis A., Allen S., and Shaw E. (1971) An extension of Weschler Memory Scale norms to lower age groups. *J. Clin. Psychol.* **27,** 354–357.

Jastak J. F. and Jastak S. R. (1984) *The Wide Range Achievement Test Manual—Revised* (Jastak Associates, Inc, Wilmington).

Kaplan E. F., Goodglass H., and Weintraub S. (1983) *The Boston Naming Test* (Lea and Febiger, Philadelphia).

Kaufman A. S. and Kaufman N. L. (1983) *Kaufman Assessment Battery for Children* (American Guidance Service, Circle Pines, Michigan).

Knights R. M. and Moule A. D. (1968) Normative data on the Motor Steadiness Battery for children. *Percept Mot. Skills* **26,** 643–650.

McArthur, D. S. and Roberts, G. E. (1982) *Roberts Apperception Test for Children* (Western Psychological Services, Los Angeles).

Mooney C. M. (1957) Age in the development of closure ability in children. *Can. J. Psychol.* **2,** 219–226.

Mooney C. M. and Furguson G. A. (1951) A new closure test. *Can. J. Psychol.* **5,** 129–133.

Oldfield R. C. (1971) The assessment and analysis of handedness: the Edinburgh inventory. *Neuropsychologia* **9,** 19–113.

Osterrieth P. A. (1944) Le test de copie d'une figure complexe. *Archives de Psychologie* **30,** 206–356.

Porteus S. D. (1965) *Porteus Maze Test: 50 Years' Application* (Pacific Books, Palo Alto, California).

Raven J. C. (1965) *Guide to Using the Coloured Matrices* (Lewis H. K., ed., (Psychological Corporation, London).

Raven J. C. (1960) *Guide to the Standard Progressive Matrices* (Lewis H. K., ed.) (Psychological Corporation, London).

Reitan R. M. and Davison L. A. (1974) *Clinical Neuropsychology: Current Status and Applications.* (Hemisphere, New York).

Rey A. (1941) L'examen psychologique dans le cas d'encephalopathie traumatique. *Archives de Psychologie,* **28,** 286–340.

Rey A. (1964) *L'examen clinique en psychologie.* (Presses Universitaires de France, Paris.)

Rorschach H. (1942) *Psychodiagnostics: A Diagnostic Test Based on Perception* (Grune & Stratton, New York).

Spreen O. and Benton A. L. (1969) *Sentence Repetition Test* (Neuropsychology Laboratory, University of Victoria, B.C.).

Taylor L. B. (date unknown) *Children's stories* (Montreal Neurological Institute, Montreal).

Thorndike R. L., Hagen E. P., and Sattler J. M. (1986) *The Stanford-Binet Intelligence Scale* (Fourth Ed.) (Riverside Publishing Company, Chicago.)

Wechsler D. (1945) A standardized memory scale for clinical use. *J. Psychol.* **19,** 87–95.

Wechsler D. (1967) *The Wechsler preschool and primary scale of intelligence.* (Psychological Corporation, New York.)

Wechsler D. (1974) *Weschsler Intelligence Scale for Children—Revised.* (Psychological Corporation, New York.)

Wechsler D. (1981) *Wechsler Adult Intelligence Scale—Revised.* (Psychological Corporation, New York).

Weiderholt J. L. and Bryant B. R. (1986) *The Gray Oral Reading Tests Revised* (Western Psychological Services, Los Angeles).

Wepman J. M. (1958) *Auditory Discrimination Test.* (University of Chicago, Chicago).

Witkin H. A., Oltman P. K., Raskin E., and Karp S. A. (1971) *Children's Embedded Figures Test—Manual* (Consulting Psychologists Press, Palo Alto, California).

Woodcock, R. W. (1973) *Woodcock Reading Mastery Tests* (American Guidance Service, Circle Pines, Michigan).

Woods M. L. and Moe A. J. (1985) *Analytical Reading Inventory* (Third Ed.) (Charles E. Merrill Publishing Company, Columbus, Ohio).

Index